# Reader Series

## in

## Library and Information Science

# Reader in
# Library Systems Analysis

Edited by

*John Lubans, Jr.*
*and*
*Edward A. Chapman*

 **Microcard Editions Books**
An Indian Head Company
**A Division of Information Handling Services**

1975

Published by Microcard Editions Books
P.O. Box 1154
Englewood, Colorado 80110

*Printed in the United States of America*

# Foreword

Unlike many other academic disciplines, librarianship has not yet begun to exploit the contributions of the several disciplines toward the study of its own issues. Yet the literature abounds with material germane to its concerns. Too frequently the task of identifying, correlating, and bringing together material from innumerable sources is burdensome, time consuming or simply impossible. For a field whose stock in trade is organizing knowledge, it is clear that the job of synthesizing the most essential contributions from the elusive sources in which they are contained is overdue. This then is the rationale for the series, *Readers in Library and Information Science.*

*The Readers in Library and Information Science* will include books concerned with various broad aspects of the field's interests. Each volume will be prepared by a recognized student of the topic covered, and the content will embrace material from the many different sources from the traditional literature of librarianship as well as from outside the field in which the most salient contributions have appeared. The objectives of the series will be to bring together in convenient form the key elements required for a current and comprehensive view of the subject matter. In this way it is hoped that the core of knowledge, essential as the intellectual basis for study and understanding, will be drawn into focus and thereby contribute to the furtherance of professional education and professional practice in the field.

Paul Wasserman
Series Editor

# Contents

## INTRODUCTION

## I
## HISTORICAL ASPECTS OF LIBRARY SYSTEMS ANALYSIS

## II
## GENERAL VIEWS OF LIBRARY SYSTEMS ANALYSIS

## III
## IDEAS, CONCEPTS AND PROCEDURES WITHIN SYSTEMS ANALYSIS

### Decision-Making Aids

### Management by Exception

# Sampling Techniques

# Questionnaires and Interviews

# Costs and Costing

# Work Measurement and Work Simplification

# Measurement of User Requirements

# Flowcharting

# Library Research and Systems Analysis

# IV
# APPLYING SYSTEMS ANALYSIS CONCEPTS

## Evaluation of Existing Systems

## Design of and Decisions for New Systems

## Implementing the New System

## Following-up on the Installation of a System

## Example of Cost Analysis in Systems Design

## Case Studies of Systems Analysis Applied to Library Problems

# V
# ELECTRONIC DATA PROCESSING AND SYSTEMS ANALYSIS

## Design and Implementation

## The Use of Computer-based Data for Management Decisions

## Computers and Libraries

## VI

## Selected Bibliography of References in Library Systems Analysis

## VII

## List of Contributors

# Introduction

The concept of a *Reader in Library Systems Analysis* came up when we were writing *Library Systems Analysis Guidelines*. It was our thinking then that such a volume, representing selections from the best and appropriate international literature on systems work, could be of value to librarians.

The present volume is the result of our initial thinking in 1969/70 and is the distillation of the numerous items on systems analysis we reviewed for the 250 item bibliography for the *Guidelines* book and have looked at since. It is our intent that this reader should be of value to students and faculty in graduate schools of library science, to practicing librarians, and to systems librarians. It could be used as a resource readings reference and textbook in library schools and in library staff training programs.

Journal articles, chapters of books, pamphlets, government reports and unpublished research reports have been the sources for the *Reader*. A dozen graduate schools of library science were consulted in 1972 in regard to their curricular readings in courses involving systems analysis and/or data processing with the result that a number of the recommended readings listed by several library schools have been incorporated into the *Reader*.

The scope of this work has not been limited to only the library literature on the topic but also includes selections of the most appropriate management literature on systems study. Essentially, what has been selected is because of its applicability and pertinency to current library problems. Included are case studies that describe systems analysis at work.

We have tried to select and arrange materials for the *Reader* along the lines of an actual systems study. The *Reader* is designed to present each aspect of systems work as it would be encountered by a systems analyst. This arrangement is meant to aid the library student or the experienced library manager in beginning or administering a systems study.

The *Reader* is divided into five major sections:
> *Historical Aspects of Library Systems Analysis*
> *General Views of Library Systems Analysis*
> *Ideas, Concepts and Procedures Within Systems Analysis*
> *Applying Systems Analysis Concepts*
> *Electronic Data Processing and Systems Analysis*

The perennial problems such as backlogs and understaffing in libraries appear still to be plaguing us. The hoped for solutions to the daily library routines by computer have not materialized (except in rare instances) beyond an experimental base and then at greater cost than previously considered. Researchers* in education are proclaiming that the "Sputnik" inspired era of educational expansion is now over. Librarians and others in many institutions are being directed to retrench. Demands for "accountability" are heard and will have to be listened to. Legislators across the country have become skeptical of the educational role in the face of inflation and many yet-unsolved social problems. Simply stated, libraries will have to streamline (with few additional staff) their procedures even further than they are now. But money will not be available to make what still appears to be a quantum leap from the typewriter to the computer. Rather, we will have to work with the resources at hand.

In a way, the envisioned retrenching may be of benefit to libraries. For the past few years many libraries have been expanding with such rapidity that there has been little time to evaluate and improve on internal operations. Perhaps the imposed introspection of stablizing budgetary adjustment will result in our studying and maximizing library work routines. It is hoped that this book can be of value in the process of refining library systems.

A bibliography of selected additional readings in library systems analysis is at the back of the book. Also, additional readings can be gleaned from the bibliographies and reference notes following the articles in the *Reader.*

John Lubans, Jr.          Edward A. Chapman
Boulder, Colorado        Naples, Florida

---

*Glenny, Lyman A. *The Changing Milieu of Post-secondary Education—A Challenge to Planners.* Keynote address delivered to the National Higher Education Seminar. October 16, 1972, Washington, D.C. photocopy 24 pp.

# Reader in
# Library Systems Analysis

# I

# HISTORICAL ASPECTS OF LIBRARY SYSTEMS ANALYSIS

## INTRODUCTION

As with many ideas in librarianship, scientific management or systems analysis has had its early advocates in the profession. Most historical surveys in cost analysis, work standards, etc., refer to certain events taking place in 1876. For examples, Charles Cutter wrote on "the cost of cataloging" in the first volume of *Library Journal* that year.

This section is meant to illustrate that the concern of librarians to better understand and streamline library operations is not new. The historical perspective should indicate the ties that libraries always have had to quantitative management. As managerial pressures were felt, solutions had to be found and frequently the methods have been borrowed from the business world. However, the methods borrowed have not been applied thoroughly or consistently. Rather, the historical picture reveals sporadic application of systems study concepts. Numerous local solutions to problems have been introduced and documented but in general there is a lack of whole-hearted acceptance, profession-wide, of systems analysis procedures.

Through the influence of the computer, programmers and systems analysts now appear on many library staffs. While this is admirable, it might be more effective for libraries if these same positions could be dedicated to analysis and re-design of manual procedures rather than the adaptation of current systems to computer-based procedures.

Ralph Shaw was one of the earliest proponents and practitioners of scientific management techniques in library management. In his selection, written while Professor Shaw was the Librarian at the United States Department of Agriculture, he stresses certain techniques of scientific management. The one method he particularly advocates is that of *integration*. This technique strongly suggests the "total systems" approach in the analysis and design of operations. Another interesting, but admittedly undesirable in most circumstances, approach to attaining efficiency is *reduction in force*. According to this concept, by reducing the workforce to below where it cannot do the job at all and then adding additional workers as needed, maximum efficiency can be attained in certain circumstances.

Fremont Rider's selection is made up of the preliminary material to the author's fifty page discussion of a three year time and cost study at Wesleyan University. His report, published in 1934, stands alone in the early library literature in its exactness and value to the profession. As with many other of Rider's ideas, such as the "microcard," those represented here were well ahead of their time. Rider's motive was to arrive at unit costs of processing library materials in order to have better control of library costs to maximize library efficiency. He cites the concern that if librarians will not do cost analyses then non-librarian managers will and librarians will suffer the consequences. A survey (beginning in 1876) of library attempts at cost accounting in cataloging is presented and provides some insight into the human difficulties with quantification. Many of the feelings expressed then, that library effectiveness cannot be measured, are still prevalent even if they are somewhat diminished.

SCIENTIFIC MANAGEMENT

IN THE

LIBRARY

*Ralph R. Shaw*

SOURCE: From Ralph R. Shaw, "Scientific Management in the Library," *Wilson Library Bulletin* 21 (Jan., 1947) pp. 349-352, 357. Reprinted by permission from the January, 1947 issue of the *Wilson Library Bulletin*, Copyright © 1947 by the H. W. Wilson Company.

DURING recent years there have been two main trends in the field of management. One of these has led to ever increasing specialization, and the other to growing emphasis on coordination and integration of the specialties into a unity of function and purpose.

While these two may at first glance appear to be conflicting philosophies, they really are not. The evolution of the philosophy of management appears to be spiraling upward by rapid stages—steps so rapid that frequently two or more of them may be evolving visibly at the same time. When management had evolved into a full-time job, for the first time men were employed to operate organizations in which the most intimate details of all operations were not known to the manager. Along with this development came similar enlargement of all the functions involved in the business and ever increasing specialization of effort on every level of operation from machine work to supervisory or management jobs. This tendency to ever increasing specialization, which continues to this day, may be termed micro-management. Micro-management, like the amoeba, is seen minutely at high magnification and continues to subdivide and to flourish. However, no organization can function efficiently in a hundred or a thousand minute segments which have no interest in other parts of the work

and are not integrated. This development of specialization forced the development of the generalist for, as pointed out by Paul Appleby, "The problems of greatest . . . difficulty are extremely broad problems, their character is such that they cannot be solved by breaking them into small parts. . . . The nature of the problem must first be seen and understood." [1] The larger management field, which may be termed macro-management would be worthless or unnecessary without a micro-management field; and specialization without macro-management can lead only to confusion.

Librarians have always been interested and indeed involved in both of these trends. We have thought and written a great deal about specialization of effort—and we have developed numerous esoteric activities. We have talked and written about division of labor among clerical, subprofessional, and professional staffs. We have made studies of building design to reduce the number of steps from the workroom to the loan desk, we have, in a few cases, made time studies and even, in at least one case, a micromotion study. We have even had evidence of conflict between the micro-management or technicians approach and the macro-management, or generalists approach, in our extensive and at times acrimonious discussions by catalogers and chief librarians on the problems of cataloging. Such manifestations as this indicate a failure on the part of the generalist more commonly

---

* Address at a meeting of the American Library Association June 21, 1946.

† Librarian, United States Department of Agriculture, Washington, D.C.

[1] Appleby, Paul H. *Big Democracy*, New York, Knopf, 1945, P. VII.

than they do excessive specialization—narrowness if you will—on the part of the technician.

## Survey Evolution

One of the best examples of interplay of the generalist and specialist in the librarian's philosophy of management is illustrated by the evolution of the library survey. Our early surveys were quite commonly made by an outstanding library administrator, a generalist. He gathered few statistics, he made few charts or graphs, and did not make title by title comparison of the collection with standard lists. Rather, he lived with the institution, its staff, and its public; he saw and felt the institution and prescribed for its ills on the basis of his broad experience and his own judgment. The middle period of surveys saw great flowering of the technique of the survey—and at times it appeared that statistical tables were truly the surveyor's end. The modern survey combines an outstanding generalist working with a team of technicians. The technicians may provide a more solid foundation than might be obtained by just living with an institution, and they serve to strengthen the ability of the generalist to see and feel the institution and to render sound judgments.

These indications that librarians are not unaware of the main currents of management thought must not, however, lead to complacency. If you have followed the text carefully you will have noted that it says, in most cases, that we have "thought," "talked," or "written" about all these things; it does not state that we have universally done all of them in well rounded operating programs in all, or in any material proportion of our libraries. These management principles do apply to us and we are not unaware of them but we have far to go in mastering them so as to make our over-all management programs compare favorably with the best in other fields.

It is obviously impossible to provide a handbook of management in all its aspects in one paper. The examples which follow report a few management approaches and their results. Before going on to these examples it must be emphasized that both micro- and macro-management require factual data for success, and that over-all policy determined by the generalists on the basis of local conditions results in variations of both types of management to fit those conditions. Furthermore, while the specialist tends more towards a mechanistic philosophy of management, both he and the generalist will normally agree that people are at least as important as systems. The best systems of operation require good personnel, working under conditions which make it possible for them to enjoy their work and want to do good work. Still another general prerequisite is adequate supervision. As was shown by Herbert Hoover's study of waste in industry, the greatest cause of waste is poor supervision. Good supervision implies study of performance on the job, and it, as well as budgeting, work-scheduling, planning, and all other aspects of the management task are based on determination of what needs to be done, how much of it must be done, how it can best be done, and what constitutes a fair day's work for an average employee working at an average rate.

The methods most commonly used, for the technical determination of work methods and work loads are (a) time and motion study, (b) production of the best worker, (c) experience of the group, (d) conference, (e) reduction in force, (f) group standards, (g) qualiquants, and (h) integration. Examples of each of these and the results obtained are given for one library.

## Time and Motion Studies

Time and motion studies are a very common management approach. The legend of the stop watch and the second-splitting camera may deter many from using this method. However, most library routine processes can be subjected to time studies with an hourglass or even a calendar in place of a stop watch or finer measuring devices. The study of the unit of time and unit of work to be used for each type of operation is very important, but it cannot be covered here. Motion study can be done adequately for many operations by experienced observers without any gadgets.

Photostating at the United States Department of Agriculture Library is a good example of results obtained from time and motion study. Since the latter was described in the *Library Journal* [2] the details need not be reviewed here; however its results are worth reporting. We used to expect a good photostat operator to produce about 250 exposures per day. The new process developed from this study now gives almost 4,000 exposures per man per day (sixteen times as

[2] "Continuous Fotoprinting at the USDA Library," *Library Journal*, September 3, 1945.

much) and, in addition, saves from one-half to two-thirds of the material formerly used.

### Production of the Best Worker

A second fairly simple method for determining a fair day's work is determination of the production of the best worker. Of course the maximum output of the best worker cannot be used as a norm. One example of this method follows. At a circulation desk two staff members charged 65 and 85 books per day respectively. The work differed from that at most charging desks since the charging involved about as much work as the average interlibrary loan charge. Four outstanding workers were set to charging books in turn and they averaged from 400 to 450 charges per day. Conference with the staff members normally doing the work regularly resulted in agreement on 200 charges per day as a fair minimum and they both stepped up their output to that level. Since that time, improved supervision, personnel selection, and incentives (in terms of credit for outstanding work) have resulted in an output of more than 800 charges in a day by one charger.

### Experience

A third simple method for determining work methods and setting standards is the experience of the group. Records are kept on what each person in a group does, dividing the work into logical units so as to permit specialized work and to eliminate nonproductive work. This method, applied in a reference division, resulted in an increase of work from about 120 reference questions per staff member per month to something over 900.

### Conferences

The fourth method is the conference method. This method is feasible only where the work is highly variable and irregular—such as in compilation of bibliographies. It consists in calling the group together and determining methods and standards from the judgment of the group. Spot checks to guide such judgments are highly desirable.

### Reduction in Force

A fifth method, highly undesirable under normal circumstances is the purely mechanical device of reducing force until the remaining staff cannot handle the work, even by superhuman effort, and then adding enough man-hours to relieve the strain. This method is justifiable only in cases in which other methods cannot fairly be applied; chiefly where cooperation is lacking; when the time to be saved does not justify the time of a trained observer; or when the total job is so big and complex that it defies analysis, i.e., control of total federal personnel.

### Group Standards

A sixth device, very coarse but useful in some types of work is group standards. The American Library Association standard for circulation per staff member is a good example of a group standard. This technique was used intensively in the Army Service Forces.

### Qualiquants

A seventh approach, highly experimental at the present, may be termed qualiquants or qualitative quantitative standards. While all of the above methods imply a certain level of acceptable quality, qualiquants attempt to express levels of quality of work in quantitative terms. The knotty problems relating to qualiquants involve policy rather than techniques. The theory of qualiquants in cataloging, for example, would set an objective level of quality for the revisers work. If the catalogers can be trained to meet that level of quality they could be given the same pay as revisers and the revisers would not be required. In the library experimenting with this method, one reviser is required for each two catalogers. If the promise of qualiquants can be achieved each cataloger who qualified could be given a grade raise (approximately $600 a year) and at the same time the cost of cataloging at the present level of quality would be reduced by more than 20 per cent. This method is still in early stages of experimentation.

### Integration

The final approach to be discussed differs radically from the others in that it lies entirely in the field of macro-management. While over-all management sets the framework for micro-management studies, this method, which might be termed the integration method, consists entirely of elimination of unnecessary duplication of work, clarification of policy and procedural problems, and fitting the whole work problems of the institution into closely interrelated work steps. While there are many examples, the one selected illustrates the interplay of a large number of different activities in the library. A considerable amount of work is necessary in the library to collect and prepare volumes for binding and to prepare binding instructions. Four or five staff members were formerly assigned to this work and the number of volumes sent to the bindery rarely exceeded 4,000 to 5,000 volumes per year.

One of the fundamental principles of sound management is the assignment of each task to a particular person or group and the elimination of responsibility for or authority over that work from all other groups. As done originally, bindery work cut across lines of work in a number of other sections and selection of materials to be bound was chance rather than purposeful or automatic. Duplicate checking records were kept for volumes which had been sent to the bindery, duplicate records were maintained to cover binding characteristics, and records of bound volumes duplicated the shelf list. The bindery staff frequently went into the stacks to get publications even though the stack staff is responsible for getting and replacing of publications on the shelves. Bindery charges were separate from the regular charges which meant that time was wasted in every search for a book not on the shelves because if it was not on the shelves it was necessary to check an additional set of records on the off-chance that it might be covered by those records.

## Revised Procedure

The revised bindery procedure is as follows: Since the first group to learn that a volume is complete and ready for binding is the group which checks in periodicals, responsibility for preparing bindery request slips was placed upon that group. As soon as a volume is completed a call slip, differing only in color from our regular call slip, is prepared at the checking desk. This bindery call slip is sent to the circulation section as are all other call slips. The circulation section gathers the numbers required and charges them just as they do everything else that leaves the stacks. The collected volumes are sent to the bindery assistant who checks them to make sure that they are complete and that they are in proper order for binding, adds the dummy or rubbing, and the bindery instructions, and prepares the binding order. A carbon copy of the bindery order is sent to the circulation section which indicates on its original charge the date on which the volume was actually sent to the bindery, and then sends

the bindery order to the preparation section which is responsible for preparing books for use. When the books come back from the bindery they go to the preparation unit which checks them against the bindery orders, sees that the books are prepared for use, revises lettering, etc., and forwards them to the circulation section where the returning books are discharged and returned to the shelves. The bindery order slips are forwarded to the shelf-list unit for revision of the shelf list, and are finally returned to the bindery assistant as a record of completed work.

While some of the minor details are left out of the picture for the sake of clarity in presentation, this does substantially cover the routine as operated at present. You will note that each group obtains full authority over the part of the work for which it is responsible and that there is no duplication of records or of work any place along the line. No group does anything that it would not normally do. The net result of this procedure was that last year one bindery assistant with part-time help sent more than 9,000 volumes to the bindery as compared with an average of less than 1,000 volumes per assistant per year in the past. This same approach greatly increases production of catalogers by providing them with the information about the book which was uncovered by the order unit in its order search.

These cases are, of course, reported merely as a small sample of what has been achieved in one library through the use of a few of the micro- and macro-management devices. In each case macro-management determined the purposes, policies, service delays permissible, over-all budgetary and planning, and similar aspects, while micro-management worked out techniques for achieving the purposes of general management.

These examples cannot be applied blindly to other management situations, but the lesson they teach is that sound application of broad and narrow management tools do pay dividends. It should be reiterated that micro-management studies, regardless of the competence of the technician, are worthless unless they aid over-all management, and over-all management which operates by custom and inertia cannot provide a framework for management improvement.

The need for sound management practices in libraries has never been greater than it is now. Interest in scientific management is manifested by librarians all over the country. A recent issue of the *Library Journal*, which excerpted a statement on increased production at one library, brought in inquiries from librarians all over the country who wanted to know by return mail (and probably in a one-page letter) how to achieve the result noted. That indicates a real danger in our present situation: The danger that superficial application of one or two devices of scientific management, good in themselves, may be confused with a well rounded scientific management program. If that is done, the results desired will not be obtained and enthusiasm for improved management will wane faster than it arose. A well rounded management program can result in increased salaries together with lower unit costs to the public, but it cannot be achieved by flourishing a wand, nor can its principles be learned in three easy lessons. It requires serious application of both micro-management and macro-management in a continuous, integrated, whole program.

## Money or Management

There are only two ways to meet increasing service demands. One is to get more money. The other is to do all the work that can be done by applying the principles of scientific management to effect most efficient utilization of the money and manpower already at your disposal. This latter is the soundest of all budget justifications, and it is the least that may be offered to those who support your institution. It may require outside management consultants or more adequate training of librarians in the field of scientific management. Above all, to be successful, it requires the cooperation of the entire staff.

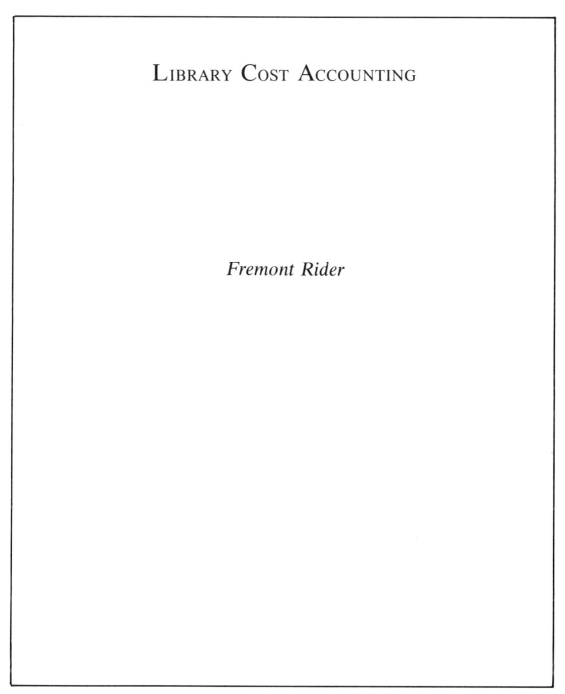

LIBRARY COST ACCOUNTING

*Fremont Rider*

MOST men and women hate "figures." Only a minority (Shall we say a slightly "psychopathic" minority?) really enjoy mathematical computation. Cost accounting is a matter of figures—of many of them. They are figures, furthermore, measuring *work*, and work is never a particularly blithesome subject. And, when a discussion of the measurement of work is in the hands of one not overskilled in its presentation, the net result is likely to be, to most readers, profoundly boring.

Yet there may be a few who, despite this altogether forbidding outlook, sufficiently feel the lure of the unknown, sufficiently delight in the thrill that comes when one looks for the first time upon great mountain ranges—upon mountains whose existence was known but whose exact location had never been charted nor exact altitude ever measured—to plow through the long barren wastes of their forelands, and to have patience to do this even though the wastes be only those of dry words upon white paper, and the hitherto unseen mountain ranges bear only some such unromantic appellation as "Fundamental Financial Facts"!

[1] Based on three years' experience at Wesleyan University, Middletown, Connecticut.

## I. IS COST ACCOUNTING APPLICABLE TO LIBRARY ADMINISTRATION?

"Cost accounting," as defined in that first aid of college reference rooms, the *Encyclopaedia Britannica*, is "a system of accounting designed to show the actual cost of each separate article produced or service rendered." And we have it, on the authority of its sister-encyclopedia, the *Americana*, that it "is the most difficult, and most technical, branch of accounting."

Why has business found cost accounting so essential a part of its administrative machinery? The answer is that with the growing complexity of business, the enormous increase in size of its component parts, and the constantly greater intensity of its competitions, there has developed, of necessity, a correlative intensification and refinement of all its accounting processes. A century ago, for example, the preparation of a general balance sheet once a year was considered quite adequate; now no important business considers it safe to postpone its trial balance more than a month, while all banks, and many thousands of other business firms, balance their account books daily. A century ago mercantile firms took inventory once a year; today almost all progressive organizations use some one of the methods of "continuous inventory," by which detailed statements of raw materials and finished merchandise on hand, and even of work in process of manufacture, may be obtained at, literally, a moment's notice.

And so it goes through every phase of accounting; for, without exact knowledge of all the factors entering into its operations, no business today can even survive. As a result business organizations spend hundreds of thousands of dollars, and employ hundreds of persons, in their cost-accounting departments alone, simply because the knowledge so gained is indispensable. Indeed, we have it on the word of one accounting authority, that "no part of modern industrial organization is of greater importance."[2]

Now this, merely by way of preamble, may sound a little

---

[2] D. S. Kimball, *Cost finding* (New York: Alexander Hamilton Institute, 1924), p. 1.

formidable. And so, since we shall endeavor to keep our material as simple as possible, it may clear the ground if, before we attempt to describe what cost accounting is, we try to eliminate some of the things that it is not, and to remove some common misapprehensions of its purpose.

The first of these misapprehensions is that, whatever its rôle may be in business, it is quite inapplicable to library administration. As one library reporting to the American Library Association Survey puts it: "The reason large commercial organizations find it necessary to keep exact accounting systems is that they are generally operated for profit and must have income as well as outlay."[3] But is it true that, because libraries are not operated at a profit, they therefore have no interest in reducing, where possible, their operating costs? As Paul N. Rice well says, speaking of one specific phase of this question of library costs: "The problem of cataloging costs must be attacked by catalogers themselves or it will be attacked by executives less able to judge fairly as to what should be modified or eliminated."[4]

Exactly what is it that cost accounting aims to accomplish? We are told that a satisfactory system of cost accounting assists the management to reduce working costs by pointing out waste and avoidable delay, to choose wisely between alternative methods of operation or production. It is the best possible safeguard against leakages. It stimulates work interest and competitive endeavor. Do these objectives have no appeal to the library executive?

There remains, however, an idea that library work is in some way so different from all other forms of organized human activity that even the basic principles of accounting do not apply to it. Nothing, of course, could be farther from the fact. All organizations which receive and expend money in order to manufacture products, or to distribute products, or to perform services are governed by identical fundamental accounting laws. The fact that some of these organizations may be profit-seeking

---

[3] American Library Association, *A survey of libraries in the United States*, IV, 137.

[4] "The cost of cataloging," *Library journal*, XXXII (1927), 239.

while others are intentionally philanthropic is quite unimportant; neither type can continue to exist unless it spends less than it receives, and both (theoretically at least) desire to keep such books of account as will accurately reflect the conduct of their operations. Libraries usually do very little manufacturing, but they are clearly engaged in both distribution and service; and they should, and usually do, desire to eliminate waste in the conduct of their operations.

It is true that, because a library's functions are not performed in hope of profit, the forms of its bookkeeping are usually not the same as those used by commercial enterprises; but variance in *forms* does not imply variance in fundamental *principles*. Just as every business has, so every library has, assets and liabilities, operating receipts and expenditures, labor costs and overhead costs, depreciation losses, and an interest on investment burden. It is quite true that it may choose to omit complete record of, or even to ignore entirely in its current books of account, some of these various factors. It may postpone consideration of them. It may confuse them or combine them. But the fact that they may be temporarily ignored, or postponed, or confused, or combined, does not mean that they are permanently escaped: they exist just the same. And, because they are fundamental, sooner or later, whether it wishes to do so or not, whether it knows that it is doing so or not, the library, like every other operating organization, has, in some way, to reckon with them. It is necessary to envisage this particular point unmistakably, for not otherwise can a clear understanding of the viewpoint and terminology of cost accounting be secured. And, to see it, a new orientation toward some of the conventional forms of our library bookkeeping may be prerequisite.

And I say "bookkeeping" rather than "accounting," because the distinction is important. As a rough-and-ready means of keeping track of expenditures, conventional library bookkeeping works, generally speaking, fairly well. But that does not deny that it is, from the standpoint of strict accounting, a sort of rule-of-thumb procedure, that it is incapable of analysis in any broad or deep sense, and incapable of analysis because it flagrantly violates most of the fundamentals of true accounting. Examine a typical library "statement." What does it show? On one side, "income" from "appropriations," "endowment funds," "gifts," "fines," etc.; on the other side, "expenditure," mainly for "salaries," "books," "binding," and "operating expenses." And if these two sides balance, we are satisfied. But no accountant, endeavoring to ascertain with analytical thoroughness the true financial inwardnesses of the running of a library, would be satisfied for a moment with such figures or with such bookeeping, because he would see at once that, from the standpoint of accounting, they are confused and defective.

In the first place, in all sound accounting there must be a basic differentiation between "operating accounts" and "capital accounts"; and, faced with the foregoing "statement," the accountant's first question would probably be: "Where's your balance sheet?" And, if you should reply to him that all your library's bookkeeping attempted to do was to present an "operating statement," and that you had no "balance sheet," he would probably pick up your "statement" and point out: (1) that, even as a statement of operations, it was defective because several items of expense, notably rent, depreciation, and interest on investment—and these the largest items of expense incurred by the library—did not appear on it at all; and (2) that, on the other hand, several items that were set down as "expenses"—notably "books," "periodicals," and "binding" (and possibly also the labor of cataloging, classification, etc.)— were really not operating expenses at all but capital investments.

But, in thus giving our imaginary accountant's comment on the conventional forms of library bookkeeping, in suggesting the need—for the purposes of executive analysis—of a "new orientation" toward library accounting, I would not for a moment be thought to be implying a condemnation of the use of this commonly found form of library bookkeeping. Our own library bookkeeping at Wesleyan is substantially of this conventional type. For what it is required to do, for the information we ask of it, it works pretty satisfactorily. All that is suggested here is that for those librarians who may be seriously interested in the technique of cost accounting the procedures followed by conventional library bookkeeping will be found inadequate. For any thoroughgoing, basic study of library finances, any study which attempts to see what lies behind and beneath the more or less fragmentary, incomplete, and superficial figures of our present library financial statements, it is necessary that we take a broader view, an accountant's view, of our accounting.

## II. SOME COMMON FURTHER MISCONCEPTIONS
### REGARDING COST ACCOUNTING

A second common misconception of library *cost accounting* confuses it with library *statistics*. Now all cost accounting results must, to be of practical usefulness, be comparable, i.e., they must be reduced to a common denominator; and the common denominator chosen is invariably dollars and cents. But, after *total* costs (in dollars and cents) have been obtained, these must, almost always, be analyzed into *unit* costs—i.e., in the case of library work, into cataloging costs *per volume cataloged*, into circulation costs *per volume circulated*, into bindery costs *per volume bound*, etc. It is perfectly obvious, therefore, that to secure unit costs it is necessary to provide the cost accountant with certain basic statistics of the *output* of the library in its various departments. In other words, although we cannot have practicable cost accounting if the library keeps no statistics (any more than we can have cost accounting if the library does no bookkeeping), cost accounting of itself is neither statistics nor bookkeeping; these are simply the tools which it uses, the primary data upon which it is dependent for the making of its own analyses.

Of the immense and invaluable work of statistical definition and correlation which has been carried on over many years by the American Library Association, it is quite unnecessary here to speak other than in praise. The bases of proper library statistics have already been fully defined by the Association's various statistical committees. For this reason such few further statistics as complete cost accounting would require, and their form, content, and definition, might properly fall into the province of the American Library Association Committee on Cost Accounting which is now functioning.[5]

A third misconception of cost accounting is that suggested by an earlier quotation—that it is such an esoterically complex and inextricably complicated subject that is quite beyond the range of the average librarian's mentality! I would rate librarians higher than that. But it should also be emphasized that in actual practice, in the average business, cost accounting is a matter not of abstruse theory but of very simple clerical routines. And in actual practice, in library work, the routines are more simple yet, because, compared with almost any sort of business, almost any library is far less complex in both its organization and its operation. When the library world has at hand its own body of confirmed cost data, its own standardized code of basic determinations, its own cost accountants experienced in library techniques and so able to step in upon request and instal simple but adequate library cost systems—all of which things the business world has today, but which the library world at present lacks—*then* library cost accounting will be a very simple matter. The real trouble is that we librarians are today where business men were fifty years ago, in the initial—i.e., in the theoretical and analytical—stage of cost accounting. We are more fortunate, however, not only that our determinations will be few and simple but also that we have all the prior cost experience of business to guide us in making them.

One other point—and an important one—library cost accounting will be almost entirely free of one difficulty that has always been one of the chief problems of the business cost accountant, viz., the ignorance, antipathy, and, occasionally, the actual dishonesty of business employees in the keeping of their

---

[5] At this point it is perhaps desirable to note that cost accounting, by its nature, can never more than approximate absolute accuracy. Bookkeeping books must balance to the cent; cost-accounting records, though in practice they must mesh into the books of account, can never arrive at exact balances. Nevertheless, the fact that cost accounting can never be more than 99 per cent or 99.9 per cent correct by no means denies its usefulness.

cost records. For library staff members are about as intelligent and loyal a class of employees as can well be imagined, and their enthusiastic and interested co-operation in any form of cost accounting undertaken may be taken for granted.[6]

A fourth misconception with regard to library cost accounting, and the one perhaps most commonly held, is that, if introduced, its procedures would be unwarrantably expensive. Five of the eight replies quoted in answer to the American Library Association Survey's questionnaire on cost accounting struck this note:

This library has never gone into exact cost accounting . . . . believing that it would be a process too expensive to justify the results. . . . . We do not consider the result obtained . . . . worth the time and expense required. . . . . Following out a cost accounting system in detail would be a large expense. . . . . A complete system by which every bill and every expense is analyzed and charged to the proper departments, branches, etc., has never been attempted, as the cost hardly seems worthwhile. . . . . The reason why we do not go any further in keeping the cost is that the expense entailed would not be justified.

To this so commonly held misconception of undue cost—for it is a misconception—there is obviously one irrefutable and definitive answer, namely, the determination, in a specific library, of the actual cost of its cost accounting! Being well aware that this very objection to library cost accounting had been raised, I was naturally curious to find out just what it would cost us to operate our own cost system at Wesleyan. (And it is hardly necessary to point out that this is exactly

[6] It should, perhaps, be mentioned at this point that such cost experience as we have procured at Wesleyan, and as is summarized in this paper, has been solely the work of the library. We keep our own books of account, which, must, of course, balance with the general books of account of the University. Our attempts at cost-keeping have, however, been entirely our own venture, for the accuracy of which the treasurer's office of the University naturally assumes no responsibility, and toward which its attitude has been one only of benevolent interest and the most cordial co-operation. Quite recently it began a cost determination of its own relative to the general administration of the college; and in the preparation of this article I was able to check a few of the library's prior determinations with the tentative ones recently made by the University's accountants. In the few cases where direct comparison was possible, we were gratified to find that there was close agreement between us, even though the University's cost setup was made from a different viewpoint and to secure quite different information.

and precisely what a cost system is for—to substitute, for assumptions and impressions, however sincerely arrived at, the actual facts of concrete experience.)

From our own cost records we *know* that it has cost us, over the last three years, an average of $102.66 a year to operate our cost system (i.e., approximately three-tenths of 1 per cent of our total pay-roll). And this $102.66 included, besides cost accounting, all work done in the preparation of statistics. As to whether the expenditure of this relatively insignificant amount has justified itself, it can only be said that, by means of otherwise unsuspected information supplied to us by our cost records, we have been able to effect economies in our administrative routines which have already saved us the cost of our cost accounting many times over.

It may be of interest to report in even more detail our experience in installing our cost system, pointing out the fact that we were obliged to develop it *ab initio*. So far as labor costs were concerned—and, as will be later emphasized, labor costs are, from a practical standpoint, nine-tenths of the story—its installation was hardly a formidable undertaking. The two simple forms (Forms A and C), which were all that seemed to us necessary, were first worked out. They were then explained to our staff at a staff meeting, which occupied, with all its questions and discussion, not over fifteen minutes. And ever since then our cost system has been carried on as an extremely minor current routine, with not even a ripple of difficulty. Perhaps our staff is unusually intelligent, but I am inclined to believe that the same, or a similar, system, *for labor costs only*, could be inaugurated in any library with no more difficulty than we had, and could be maintained in any library with proportionately no greater expense.[7]

[7] Our staff members enter records of their time, generally once, at the end of each day; they say that this takes them "a minute or two." Their sheets are collected monthly by my secretary, who is also our library bookkeeper, order clerk, and "statistician." She posts their totals and compiles our final monthly cost record. This posting process takes her about one day each month.

### III. WE ASK SIX QUESTIONS OF OUR COST ACCOUNTING

So far we have been telling what cost accounting is not. The converse question has now become rather insistent: What is it? Of what use is it? What specific questions—of actual practical value in the administration of a library—will it answer? The general reply to this last question is that it will give as much information as to costs as one is willing to pay for (in time and money).

But to be quite specific. Here at Wesleyan we asked our cost system to answer for us six questions—one for each of the six main subdivisions of the technical work of the library staff.[8] What does it cost us, we asked:

1. To "acquire" (order and purchase) a book?

2. To "accession" a book (i.e., to prepare it for the shelves through all its functional processes other than acquisition and cataloging)?

3. To catalog a book (taking, as is understood in all these questions, not a few arbitrarily selected books for a short period, but, over a year, *all* our books of all sorts and of all degrees of difficulty)?

4. To circulate a book?

5. To bind a book in our bindery?

6. To receive, check, and distribute an issue of a periodical?

Now these six questions are easy questions to ask—so easy that it might be thought incredible that, although one or two of them have, in the past, been answered in part, none of them has ever been fully answered, and most of them have never been answered at all.

But would answers to such broad questions, questions phrased in such general terms, have any practical value? Should not each question be further broken down, be made the subject of much more analytical scrutiny, so affording those comparative data which are always most informative? Of course it should be. But "first things first"! When no libra-

---

[8] We asked it a large number of additional questions regarding labor costs only, but on these six questions we asked for complete costs.

rian has the remotest idea what it costs him to "circulate" an average book, it is obviously our first duty to secure at least that simple basic data. Afterward, we can differentiate and analyze further to our full content—can determine the cost of circulating different classes of books (as, indeed, we have done to some extent), or of circulation to different classes of patrons, or of circulation at different times of the day, week, or year, or with various sorts of equipment, or what you will.

As a matter of fact, as already suggested, even in our own first very rough attempts at cost accounting, it did seem to us worth while to try in a few cases to secure some subdivisional data. In seeking our circulation costs, for example, we segregated reserve circulation from general circulation. In seeking our cataloging costs it seemed to us worth while to segregate filing costs, i.e., to determine just what, on the average, it cost us in labor to file a card in our constantly swelling card catalogs. And, perfectly obviously, we could, in any department we wished, and at a comparatively slight additional expense, have analyzed and refined all our costs much further. Possibly we shall later do so.

## IV. PREVIOUS WORK DONE IN LIBRARY COST ACCOUNTING

Before plunging into the details of our three years of cost-accounting experience at Wesleyan, it would seem to be pertinent to trace, very briefly, the history of library cost accounting.[9]

The first reference to library cost accounting in all our pro-

[9] And, in passing, it might be added that material for a history of library cost accounting is to be come by only by rather patient delving. In Cannon's *Bibliography of library economy* neither the term "cost accounting" nor, so far as I could find, any synonym of it appears at all, either in the Index or the classification. And there is the same entire lack of entry in the recent supplementary volume to Cannon. A few references to the cost of cataloging appear under the heading "Cataloging"; but, if there are references to cost accounting under any other phase of library work, I missed them. Similarly, in such a book as Drury's otherwise extremely comprehensive guide to the selection and acquisition of books for libraries, although the word "cost" appears in its Index, it is only in reference to the purchase cost of books, not to the cost to the library of carrying on those acquisition operations with which the book concerns itself.

fessional literature appears to have been in the first volume of the *Library journal*, where Mr. Charles Cutter, in reply to an inquiry, estimated "the cost of cataloging" at $0.50 per volume.[10] Mr. Cutter added: "This, it must be understood, is for books in various languages, of all ages and likely to bring up all the difficult questions. The ordinary cataloging of town libraries need not cost anything like this sum. The actual expense of cataloging one such library was $0.16 a volume."

In 1886, no less than ten years later, appeared Mr. Whitney's careful "Cost of catalogues,"[11] in which he estimated—but *estimated* still, you will note—the cost of cataloging (by which term it appears that he really meant [as did Mr. Cutter?] the entire cost of preparing a volume for the shelves, including accessioning and ordering) at $0.3575 per volume.

Although Mr. Whitney's paper aroused considerable discussion when it was presented, apparently no further interest in the subject developed, for another fifteen years passed before Dr. Steiner presented a paper on the cost of preparing library books for public use.[12] This paper showed some technical progress, for it at least pointed out some of the difficulties which would be encountered in the determination of library costs. Apparently, however, to the librarians of that day, the difficulties of acquiring cost information loomed higher than the advantages of having it, for two years later the American Library Association Committee on Library Administration reported that the attempt to arrive at the cost of cataloging had "had to be abandoned" as not "feasible."[13] Other scattered figures upon cataloging cost—and you will note that only cataloging costs are ever mentioned in all these references—which were quoted at various times during this period, ranged from $0.12. to $0.60 (i.e., in the money values of today, perhaps $0.20 to $1.25) per volume.

[10] *Library journal*, I, 219. To establish true comparisons of *value*, the relative value of money, as of the various dates of this résumé, must of course be kept in mind. In other words, Mr. Cutter's figure of $0.50 in 1876 might perhaps be the equivalent of $1.50 today. Catalogers in Mr. Cutter's day may have averaged $600 a year; today they may average $1,500 or more.

[11] *Ibid.*, X, 214.        [12] *Ibid.*, XXV, 32.        [13] *Ibid.*, XXVII, C/86.

In 1905, Dr. Bishop, then at Princeton, contributed a note-worthy paper, "Some considerations of the cost of cataloging,"[14] which, for the first time, offered the beginning of a firm basis for procedure. He suggested, for instance, that there were elements of cost other than labor involved, and clearly pointed out by way of introduction:

> The items which must be included in reckoning the actual cost of cataloging are numerous and diverse. Salaries vary, and must vary; heating and lighting present differing costs in different regions and buildings; thoroughness and extent of the actual work will differ with different systems; and so on through a long list of smaller and larger items. . . . . These elements of the actual cost of cataloging must of necessity vary with the individual libraries. . . . . When one library calculates its cost at 50¢ per title, another at 60, another at 20, another at 13, and so on, it is evident that until practical unanimity has been reached on the question of what must be reckoned as parts of the cataloging process and expense, it is well-nigh useless to compare figures.

Dr. Bishop's paper made no effort to secure any actual ascertainments of cost, although it did publish certain valuable cataloging statistics, and these latter not for a week or a month, but for fifteen months—long enough, that is, to establish really dependable averages. Even more important, he analyzed into their elements the statistical factors involved and showed clearly, as he implied in the foregoing quotation, the necessity of securing agreed definitions of terms and agreed delimitations of functional processes if comparison of results was to have any value.

Unfortunately, Dr. Bishop's clearly presented desiderata were largely ignored in what was to be the next landmark in library cost-accounting history, viz., the so-called A.L.A. "Cataloging Test" of 1913–16. Perhaps they had been forgotten, for it will be noted that another long period, no less than eight years, elapsed between Dr. Bishop's Princeton paper and this next recrudescence of professional interest in the subject. The A.L.A. "Test" was a co-operative effort, the co-ordinated answers to a questionnaire to which thirty-eight libraries replied. If it proved to be a failure—and it was generally felt at the time

[14] *Ibid.*, XXX, 10.

that it was a failure—this "Test" was one because of its disregard of what Dr. Bishop had previously pointed out to be essential. The experience recorded in the "Test" was in each case too limited to establish anything like a norm even for the library reporting, while, as between libraries, comparisons were valueless because there had been no prior agreement upon standards or terminology.

Each library had been asked to report the "cost" of "cataloging" "100 books" "selected at random." What were the results? Mr. Josephson thus summarized them, in a report presented at the Asbury Park Meeting in 1916:

(1) Three large libraries, each of which represents a type of its own, none of them easily compared with the other two. These libraries cataloged for the test a total of 302 books in 293 hours and 23 minutes at a total cost of $193.83, giving an average of 56 minutes in point of time and an average cost of $64\frac{1}{5}$ cents.

(2) Four university libraries which cataloged together 402 books in 139 hours and 16 minutes at a cost of $64.20, giving an average of $20\frac{4}{5}$ minutes in point of time and an average cost of 16 cents.

(3) Seven large public libraries with branch systems, reporting together 684 books cataloged in 399 hours at a total cost of $172.52, giving an average of 35 minutes in point of time and an average cost of $25\frac{1}{5}$ cents.

(4) Four smaller libraries, namely, three public libraries and one state library, reporting together 326 books cataloged in 73 hours and 31 minutes at a total cost of $36.14, giving an average of $13\frac{1}{2}$ minutes in point of time and an average cost of $10\frac{1}{10}$ cents.[15]

The foregoing record of "costs" becomes even more confusing when we learn that Yale reported a "cataloging cost" of $0.1733 a volume, and Columbia, the incredible one of $0.095![16]

Miss Prescott, commenting upon the "Test," said,

As a matter of fact, these varying results seem to me to point out very plainly that any cataloging test must necessarily fail to give us definite practical figures, because of the varying nature of libraries, and the multitudinous varying details which must be considered. The cost of cataloging depends very much on the available funds. We can make a very simple author card if

---

[15] *Ibid.*, XLI, 654.

[16] "The A. L. A. cataloging test at Yale and Columbia," *Library Journal*, XLII, 110. See also Hanson, "Cataloging test in the University of Chicago Library," *ibid.*, XL, 399.

we must; we can fill it with valuable bibliographical information and multiply subject cards if funds permit.

A cataloging test, therefore, which has as its aim to give the average cost of cataloging per title in any library, or in any number of libraries, cannot but fail in its mission. . . . .

And Mr. Josephson similarly commented:

Another factor that naturally influences the cost of the work is that of salaries. In this respect the five libraries stand as follows: No. X has a cataloging force of 24 persons, with an average salary of $906. No. XI has a force of 20, with an average salary of $581. No. XII had, in 1912, a force of 16 with an average salary of $985. The staff of this library has since been increased, but I have no report of any increase in salaries. No. XIV has a force of 12, with an average salary of $505. No. XV has a force of 19, with an average salary of $502. [Remember these are pre-war salaries.]

There are other factors that will influence the time consumed in cataloging and thereby the cost of the work: matters of organization, of local conditions, and the experience and alertness of the workers, the absence of which naturally result in waste of time.

It is rather obvious, as one analyzes the accounts of this "Test" and the comment upon it, that what it was seeking was really an empirical figure, a something that those in charge of the "Test" termed "the cost of cataloging." True cost accounting is entirely specific, entirely objective. What it seeks to ascertain is, not "the cost of cataloging" in general, but the definite cost of cataloging certain specific books in some one specific library under certain specific conditions. The books, the cataloging salaries, the working conditions, and the cataloging standards in that library may, or may not, be the same as those in some other library. The cost figures gained in the two libraries may or may not, therefore, be comparable. Unless all these factors are closely correlated, comparison is impossible, for wide variations in "cost" are certain, and any "average" of the two sets of costs is, so far as its practical usefulness goes, a mere statistical abstraction. But, if all the contributing factors in the two libraries reporting *are* closely alike, and there is still a wide variation in their individual costs, then administrative interest is properly aroused and cost comparisons may be, to both libraries, extremely helpful.

Of course, it is quite true that, having first secured accurate cost figures from *a great many libraries*, for *a large number of books*, over *long periods of time*, it is possible, by combining them, to secure an *average cataloging cost* for all libraries. But such an average as this represents a secondary and incidental cost figure, not a primary one; and, generally speaking, it would not be a very useful one. The important thing to remember is that real costs are not averages, that they can be determined for but one institution at a time, and that they mean nothing whatever to another institution except as the conditioning factors in both are known and due allowance for them is made.

Miss Mann, in her *Introduction to cataloging and the classification of books*,[17] comes to this same conclusion in these words:

Few definite results have come from the attempts to estimate the cost of cataloging a single book. . . . . Few libraries have yet grappled with the cost accounting problem; therefore, the cost of departmental work has not received separate scientific consideration. With the growing use of Library of Congress cards and the greater uniformity in methods resulting from them, it may soon be possible to arrive at more definite figures. When this is done, a combination survey might be made by librarians and expert cost accountants. About all that is known is the fact that the cost of cataloging increases with the size and character of the book collection. . . . . Catalogers cannot afford to overlook the economic side of classification and cataloging. It is an expensive business at every turn. Efficiency methods of management, leading to standardization where it is feasible, will reduce costs without interfering with the scholarly side of the work.

Although catalogers other than Miss Mann have been inclined more and more to feel that they "could not afford to overlook the economic side of . . . . cataloging," it was not until as recently as 1930 that there appeared anything approaching a consistent attempt at genuine cost analysis—even though still on the labor side alone—of the whole range of a library's functions. This was the extremely interesting and valuable pioneer venture of the library of the University of California. The report of this project, recorded in a paper read before the College and Reference Section of the American Library Association at Los Angeles in the same year, covered, it is true, only a six months'

[17] P. 345.

period; but six months is long enough to obtain an accurate cross-sectional result; and the technique followed, so far as one may gather from the report, was, from the standpoint of accounting, correct.

Miss Hand opened her report on this project by commenting: "In library practice, if there is one thing that seems to be less clear and more productive of misunderstanding than another, it is the question of costs." And she argued a little later most cogently:

Administrative officers and governing bodies have a right to know how, and how efficiently, libraries are spending the funds granted them. Librarians themselves can learn much from comparisons that really compare. . . . . The librarian who cannot produce the facts readily and in a form comprehensible to persons unfamiliar with the mechanics of library work may find himself responsible for a situation inimical to the institution under his care.

The latest—and in some respects a very useful—analysis of cataloging costs is that by Miss Buelow of the La Crosse, Wisconsin, Public Library.[18] She goes a step farther toward complete cost accounting by including, for the first time, a careful account of cataloging supplies as well as labor. She properly remarks that "the cost of cataloging depends on the individual book, and the routine of each library." But, for a medium-sized public library, using chiefly Library of Congress cards, not concerned with books of unusual difficulty, and, as in all these quoted cost figures, attempting no computation of overhead costs, Miss Buelow's article deserves careful reading.

[18] *Library journal*, LX, 657.

# II

# GENERAL VIEWS
# OF LIBRARY
# SYSTEMS ANALYSIS

## INTRODUCTION

This section consists of four papers treating systems analysis in an overall, concise view. Many of the concepts within systems analysis are mentioned and should give the reader some introduction to the intricacies of systems study.

Systems analysis is defined as "the logical analysis of the present system; the evaluation of the efficiency, economy, accuracy, productivity and timeliness of existing methods and procedures measured against the established goals of the library; and the design of new methods and procedures to improve the flow of information through the system."*

A number of important studies in library management techniques and systems analysis has been done in Great Britain. Frank Robinson and his colleagues provide an overall view in a few pages of most of the techniques employable in systems analysis. It is a condensation of many ideas and all are clearly and concisely stated. Among the techniques described are those of interviewing, input/output analysis,

---

*E. A. Chapman *et al.* *Library Systems Analysis Guidelines.* New York: Wiley, 1970. p. 19.

27

activity analysis (random samples of time/activity), decision tables and a section on the use of brainstorming for designing new systems. The style throughout this essay is commonsensical and at times humorous. For example, the authors warn against interfering too much with the work flow while making observations for time/cost studies: "There is a story of an observer who rang up at random intervals to ask what the subject was doing. The survey indicated that the subject spent 100% of his time answering the telephone."

Robert Burns essay is directed at the layman. General rules-of-thumb are provided for the many concepts within systems analysis indicating the intricacies and nuances of a systems study both for the analyst and the administrator. Also, a step-by-step discussion brings out the major phases within a systems study. The important bibliography suggests introductory and advanced or specialized reading material.

Thomas Minder writes of the responsibilities and requirements of the systems analyst. The scope of this staff position is a wide one, with responsibility in seven major areas from designing and implementing new library systems to monitoring and evaluating equipment for its potential value to the library system. In answer to the perennial question of whether or not a systems analyst should have formal training in librarianship, Minder states that lacking a person with a degree in both industrial engineering and librarianship "it is far better to have the formal training in the technical fields with experience in library problems, rather than the reverse (of a person with only informal training in systems analysis).

What is the role of management in the systems study, before, during and after? In Frank Robinson's second contribution to this section, management must play a leading role in systems study. One can recognize in this article reasons why, without intelligent participation by administration, many newly developed systems have been left spinning by themselves, rarely having the desired impact. The author strongly recommends that management involve itself in the systems study by sharing in: selecting the problem to be analyzed, determining the study objectives, planning and estimating (costing) the study, carefully studying the completed study and its recommendation and over-all, controlling the study.

THE TECHNIQUES

OF

SYSTEMS ANALYSIS

F. Robinson, C. R. Clough,
D. S. Hamilton
and
R. Winter

SOURCE: From F. Robinson *et al*, "Systems Analysis in Libraries," *Symplegades* number 2, (Series editors N.S.M. Cox and M. W. Grose) Newcastle-Upon-Tyne, Oriel Press, Ltd., 1969. Reprinted by permission of the publisher. Copyright © 1969 by Oriel Press Ltd.

### 6.1 Introduction

Anyone who is given the task of looking at a business system is forced to do certain things. Among other things he will talk to people concerned, take notes and look at the documents used. If an intelligent person is continuously engaged in such studies it is clear that he will develop methods to make these jobs easier. In fact many people have been studying systems for many years and a number of methods or techniques have come into general use and become highly developed.

Some of the main techniques are outlined in this section and a fuller list (associating them with the various stages of system development) is given in figure 3. It will be seen that the techniques are fairly evenly distributed over the data gathering to design phases.

| PROCESS | RELEVANT TECHNIQUES |
|---|---|
| Determine feasibility | Work flow analysis<br>Interviewing<br>Estimating<br>Communication (report writing) |
| Data gathering | Interviewing<br>Input/Output analysis<br>Work flow analysis<br>Work measurement<br>Documentation study<br>Activity analysis |
| Data analysis<br>(Procedure and performance analysis) | Flow charting<br>Use of codes in analysis<br>Mechanical aids in analysis<br>Critical examination<br>Decision tables |
| System design | Brainstorming<br>Flowcharting<br>Code design<br>Critical examination<br>Form design<br>Equipment selection<br>Timing calculations (clerical and computer)<br>System design |
| Specify system | Report writing<br>Estimating |
| Program specification | Decision tables<br>Flowcharting |
| System testing | Work measurement<br>Test data preparation<br>Test checking<br>Sampling |
| Implementation | Critical Path Scheduling<br>Bar charts<br>Estimating |

Fig.3    Techniques used in the System Development Process

The various techniques are described at a reasonable level of detail. It is hoped that it will be possible to practice them after reading the descriptions. It should be remembered however that the techniques are part of a longer process and will be of most effect when they are so used. Applying a particular technique to a problem may help ease it but may not alleviate the real cause which only deeper study will identify.

6.2 <u>Selection of Techniques</u>

Analysts have available a fairly wide range of techniques. Probably they will not all be needed for any particular investigations. The most common (almost ubiquitous) technique is interviewing, which exists in some form in nearly all investigations. Thereafter the selection of those most appropriate to the study is important: it can have a marked effect on the success or failure of the project and is a measure of the skill of the analyst.

In all cases the descriptions of the techniques which follow indicate the situation in which they may best be used. For some of the techniques, however, a summary of situations for which they may be suitable is given:-

1. Input/Output analysis. Multiple inputs with complex combinations for processing. Also useful when there is a complicated flow of documents between units.

2. Activity analysis. Large units with similar types of complex processing. Analysis will show where the significant effort is being applied. Also useful to test the utilization of facilities.

3. Flow Charting. Involved communications and document handling involving several groups. Also to record organizational links.

4. Decision tables are useful for definition of the rules, when choice of alternative courses of action depends on a number of determinable conditions.

In some cases the analyst will devise new 'custom-built' methods of solution of problems. These can be given the general heading of 'formalizing'; that is rearranging the data in some way that may point to some hitherto hidden factor. What books are ordered from one bookseller? - such a list may indicate a pattern which can be used to limit or extend the types which should be so ordered. Usually this requires the identification of a possible key feature and ranking the data in accordance with that feature to see if a pattern emerges.

One requirement for an analyst moving into an area is that he should make himself conversant with the background and methods used in that area. For instance an analyst moving for the first time into the study of libraries should spend some time studying how libraries work, the attitudes of librarians, the distinctions between different types of library and the various organizational structures concerned. This can be done by general reading or by discussions at a more general level that the interviews connected with the particular project. This essential preliminary does not merit the name of 'technique'.

## 6.3 Interviewing

Most of the knowledge that the analyst gathers about an existing system or requirements for any replacement is obtained from personal contacts. Many of these contacts will be informal, but for the systematic gathering of information from these sources formal interviews are essential. 'Formal' in this context merely indicates that the interview is pre-arranged, and that the analyst had the opportunity to prepare in advance in order to improve the changes of success. It has no implications on the conduct of the interview itself.

Interviews are obviously a personal matter; the styles of the successful ones will vary with the personalities involved. What is presented, therefore, is not a guide on how to interview, but rather a description of the planning and preparation, some points on matters relevant to the actual conduct of the interview and the things to be covered in following discussions. Some techniques which are adjuncts to interviewing (e.g. input/output analysis) are described in other sections.

The planning of a series of interviews is an important part of the planning of the analysis process because, as already mentioned, much of the information is to be found through interviews. This information is held at three levels; operator (clerk), supervisor and manager. In most cases the analyst will see the manager first in order to get a general view of the system. Thereafter he must plan the interviews so that his knowledge develops evenly. Too much discussion at the operator level, for instance, could tend to distort the picture unless it is qualified by other views. In planning there must, of course, be a time allowance for each interview, and this allowance will be less for managers than for operators. A rule of thumb on this is that an hour should be allowed for interviews at management level, a half-day for supervisors and a full day for clerks. This partly reflects the importance of the manager's time, and partly the need for more detailed discussions at the operator level.

Having prepared the interview plan, the analyst then makes his initial contacts which again will usually be done through the management of the department. Then individual interviews must be arranged. Again the level of the interviewee has a great bearing on this, and it may be necessary to replan because of the difficulty of making arrangements with senior staff. Nevertheless the main plan should be adhered to as far as possible in order to obtain a balanced view of the system. If full management support has been obtained the importance of access to senior staff will probably be accepted and this should ease the arrangements.

In the actual preparations, it is advisable to know the position in the organization of the person to be interviewed and to have some knowledge of that part of the system. This is so as to avoid the completely absurd question rather than to impress the other person. In most cases there will be a series of interviews; the first to establish the relationship, the second to obtain the necessary information, and thereafter follow-up meetings to clear detailed points. For the second, and subsequent, meetings, it is important to prepare a list of the points which have to be covered. A formal questionnaire may be possible, but the advantages of this could be outweighed by the bad effect it may have on the person interviewed. Try to avoid ringing up as soon as you have left with the phrase 'I forgot to ask'.

Conduct of the interview itself is also affected by the stage reached. The purpose of the first interview is, as already stated, to set the scene and to establish a relationship. This is probably most important at the operator/supervisor level. This is partly because the interviewee may be junior to the analyst, partly because he may be less able to express himself clearly and also because he may feel his position to be threatened by the study. At the first session the importance of the interviewee's experience and knowledge should be stressed. It is not difficult to achieve complete ignorance of the job which has to be discussed, but it is more difficult to display this in such a way that the person being interviewed does not assume total ignorance. Any attempt to hide ignorance by an overbearing attitude will fail completely, and it is better to evince a willingness to learn.

In subsequent interviews there will be a more purposeful approach as the details of the job are followed through and the gaps in the information given are filled. These gaps are inevitable and very rarely intentional. The man doing the job is very likely to have overlooked routine details and the value of methodical note-taking is that it enables these to be checked. Once again the input/output analysis technique is relevant. Particular attention should be paid to following discussions at the supervisor level as they are most likely to clarify the reasons for specific operations or decisions.

Some of the points to note in the actual conduct of the interviews are:-

1. Don't be overbearing.

2. Don't interfere with the running of the job, either by trying to continue the interview when something arises which has to be dealt with, or by using knowledge of the organization to override decisions or actions undertaken by the person being interviewed. The correct attitude at an interview when work requires to be done is to sit by and watch it being done. First because this will improve the relationship with the interviewee, and, secondly because it will add to the knowledge of the system.

3. Do not try to impress the person being interviewed with your own knowledge or techniques - don't try to 'blind him with science'.

On the 'do' side, one must remember that the person being interviewed has some knowledge which has been, and is, useful to the organization and, reasonably, has pride in that knowledge. Listen carefully to what the interviewee is saying and note carefully the points he is making.

The importance of this last point becomes apparent in the final stage of the process, which is writing up the interview. This should be done as soon as possible and should be concise rather than prosy. The notes are not the minutes of a meeting. It is far better to list the essential points describing any particularly important ones, and referring to documents used where applicable. If it is a complex situation then logic diagrams or decision tables may be used, but these should only be shown to the interviewee if he will not be misled, puzzled or annoyed by them. Above all do not say to him 'look how simple it is'. Bearing this in mind, however, it is often useful to discuss the notes before they are used in order to ensure accuracy.

There will inevitably be follow-up discussions, many of them completely informal, before the final picture is known. The next series of interviews will be on the design points of the new system. Here, a more restricted range of interview is usual. The exercise of getting comments on and acceptance of the new system will depend on relationships built up during the first interviews. If these are good, then it will augur well for the acceptance and operation of the new system.

### 6.4 Clerical Input/Output Analysis and Document Collection

The importance of careful note-taking during interviews has already been stressed. When interviewing at the detail level, much of the information will be descriptions of the way various routine operations are carried out. Many, if not all of these operations will have an input (form, visit, telephone call, etc.) some processing (add information, calculate, copy etc.) and finally an output. Clerical input/output analysis is designed to enable the recording of these operations to be dealt with on a common basis. (It is called 'clerical' to distinguish it from the input/output analysis of economies (coutries, firms, processes etc.) pioneered by Leontief.) In theory the technique could be used at any level but in practice the variety of inputs, processes and outputs make it unsuitable for senior staff. In some circumstances, discussed below, it will be necessary to include 'management' processing.

The technique uses a predesigned form for recording the elements of the analysis. An abbreviated example is shown as figure 4. Separate blocks are used for the different elements. A number of points should be made about the use of these blocks.

1. Input. All forms received by the unit being studied are recorded with the frequency of receipt. In addition any regular communications received by other means should be recorded.

2. Processing. Links processing operations with the inputs necessary for their performance and with the information already available on file (see 3 below). Processing covers even the simplest jobs such as taking two separately received forms, pinning them together and posting them on. No comment should be made, the form is strictly a record of things done.

3. File information. Note the records maintained for the purpose of processing the input. This will include such things as names and addresses (master files) and lists of orders (current working files). Information is drawn from and put into these files and the frequency of these movements should be recorded.

4. Output. Again the output may be documented or in other forms. The form should indicate the process generating each output.

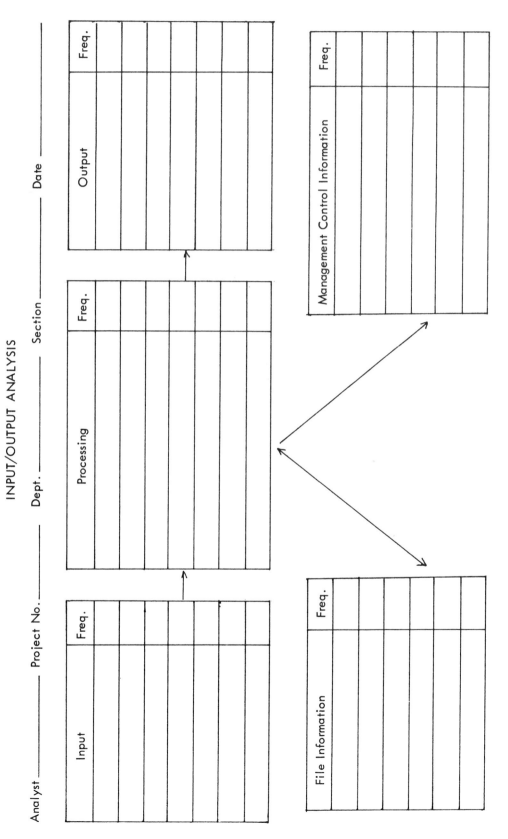

Fig.4    Clerical Input/Output Analysis Form

5. Management control information. Particular reports
which are prepared to enable management to control
the operation should be recorded here. These usually
cover such things as running totals of transactions
e.g. number of books issued. The information may well
have no other use and should therefore not be recorded
as either input or filed information.

This description of the forms is, to a large extent, a guide to their completion.
During the interview the form helps to concentrate attention on the flow of the
work and to indicate gaps in the information being given. There may be occasions when some work may be done which is irrelevant to the particular system being studied. Whether or not to record this information will depend on
the relationship of the project to the person being interviewed. If the project
covers part only of his work irrelevant information should be ignored. Any which
is wrongly omitted will be highlighted by other forms and can be followed through
subsequently. If, however, the whole or a major part of the work of the person
being studied is concerned with the project then there are effectively no
irrelevancies.

It is important to remember that the forms should not be used to comment on
the efficiency of performance or need for any of the work. Comments which are
made to the analyst or which occur to him at the time should be noted separately
and can be associated with the form. These points will be specifically discussed with the supervisor and management.

Following the interview the forms will be studied in conjunction with others in
the system. This examination will indicate gaps in the information and these
will be checked in follow-up discussions. Eventually the forms will be completed to give a full record of the operation of the existing system. Extraction
of data from all the forms will show the flow of data through the system and can
indicate the volumes involved. This may usefully be charted but in complex
systems the charts can be confusing. If charting is necessary in such cases
it is advisable to break down the system into intelligible sections.

### 6.4.1 Document Collection

It is neither possible nor advisable to clutter the input/output analysis
form with details of the content of each document used. A record of
the contents is, however, necessary and this can best be obtained by
collecting a copy of each document concerned. This may also be useful
later should it be necessary to redesign the forms. The forms should
include headings of pages in record books and lists of items transmitted
by letter or telephone. The forms should be referenced to the input/
output analysis sheet and, when they are used at several places, copies
will be obtained from each processing point. It is useful to indicate on
the form the sections completed at each point and a master copy might
well be retained for this purpose. It is necessary to watch for pertinence
in document collection as there is a danger that masses of paper will be
collected with a high degree of irrelevance. To avoid this in systems
where paper proliferates, deal with forms used for similar purposes on a
combined basis and ignore trivial differences between them.

There may be a temptation to display the forms showing which lead to which processes. In general, if the system is sufficiently simple for such a display to be comprehensible, the exercise is unnecessary. Usually, the only person who can follow such displays is the analyst who has prepared them. The work can be justified if it helps the analyst to understand the system, this being an example of 'formalizing' data for system analysis.

## 6.5 Activity Analysis

Activity analysis is a method of measurement which forms part of a more general set of statistical sampling techniques. There are many times in the analysis of an existing situation when these techniques may be of value. This is because it is sometimes impossible or unnecessary to examine in full detail all the activities, recordings or other events in a complex situation and a careful sample will reflect the composition of the whole. This paper is not the place to give a detailed description of sampling techniques. The specific form of activity analysis is however practicable without detailed knowledge.

The assumption on which the technique relies is the Law of Averages which, simply stated, is 'if samples of sufficient size are drawn at random from an unchanging mass of items/events (known as a population) successive samples will distribute themselves according to an approximately repeatable pattern which is a reflection of the "true" pattern of the whole population'.

The technique can be used to determine how a group of people is acting or working and for finding out the activities involved in their work. For example, it may be used to determine what proportion of a cataloguing section's time is spent in actually preparing catalogue entries or searching the existing catalogue. It may be applied to the commercial work or order preparation in order to find the critical activities which are causing delay. The technique may also be used to determine how the library facilities are being used; for example how many times a particular section of the reference library is used. When used properly, activity analysis will help to avoid intuitive conclusions based on unrecorded observation.

It is necessary first, to prepare some data sheets to record the activity and the 'readings'. These will be simple sheets with separate columns for each activity. Most or some of the activities may be prerecorded at the heads of the column but there must be provision for additional main activities to be entered. Grouping of minor activities is permissible but, in these cases there must be a clear definition so that recording is precise and quick. An example is shown as Fig.5.

The frequency and timing of the readings must then be decided. This is the point at which the Law of Averages is important. First the sample must be of sufficient size, it is clear; again without going into details of statistical technique, that confidence in the results will depend on the size of the sample. One should aim at a grouping of minor activities in association with a choice of sample size in such a way that at least ten recordings will be made in each activity group. This is obviously difficult to do in advance but it should be possible to specify a total number of recordings which will satisfy the minimum. The samples can be taken at any reasonable interval.

| Time | Book Reception | Book Issue | Shelving | Accessioning | Inquiries | Inter Library Loan | Overdue Books | Filing | Reader Registration | Bindery | Other |
|------|---|---|---|---|---|---|---|---|---|---|---|
| **MON.** | | | | | | | | | | | |
| 10.00 | | | | | | | | | | | |
| 10.30 | | | | | | | | | | | |
| 11.00 | | | | | | | | | | | |
| **WED.** | | | | | | | | | | | |
| 2.30 | | | | | | | | | | | |
| 3.00 | | | | | | | | | | | |
| 3.30 | | | | | | | | | | | |
| **SAT.** | | | | | | | | | | | |
| 10.00 | | | | | | | | | | | |
| 10.30 | | | | | | | | | | | |
| 11.00 | | | | | | | | | | | |
| 12.00 | | | | | | | | | | | |
| 12.30 | | | | | | | | | | | |

Fig.5    Activity Analysis –
Library Assistant

Next the samples should be randomly selected. This may be achieved by selecting times on some random basis, either in advance or by making observations on the occurrence of some external event.

Finally one must ensure that the population is unchanging. This requires for instance, that the period covered must include exceptional periods such as Saturday mornings or month end. It is, of course, possible to treat these as separate populations and, in some cases, this is the best plan. If, for instance, the activities at that time are completely different or if extra resources are needed to handle the exceptional load, the readings may distort those of the normal periods.

Having specified the times at which recording will be taken, the forms should be distributed to those taking part in the analysis. The activity being performed at each time interval is recorded. Additional forms may of course be used and it is reasonable to issue new forms periodically. This will ensure that the forms are kept in good condition. If possible, independent observers should be used to do the recording; this prevents interference with routine and improves the objectivity of the recording. (There is a story of an observer who rang up at random intervals to ask what the subject was doing. The survey indicated that the subject spent 100% of his time answering the telephone). At the end of the analysis the readings against each activity are totalled and the relative importance of the activity noted.

This type of analysis sheet may also be used for detailed time recording. In this case the columns will be divided into rows, these being either for different periods or different jobs. In each square thus formed is recorded the time spent on the activity during that day (or other period) and also against jobs. Used in this way the forms give a full record of all work carried out but are subject to the errors of apathy and of small items being missed. If this approach is used in analysis it is important that it should not be kept going for a prolonged period. One should also beware of the novelty factor. It is not a bad idea to disregard the first weeks results until the staff have resettled to their normal routine.

6.7 <u>Brain Storming</u>

At the point when the analyst knows the essential elements of the system he will start to look for ways in which it can be improved. It is important to remember that by this time he will have talked to many people about the existing system, looked hard at the documents used, studied the work flow and, almost certainly, have heard a number of comments on improvements and methods which should be made. There is a possibility therefore that his ideas for the replacement system might be influenced, by his studies. And there is also a tendency (more realistic with most analysts) to have developed ideas during the study which could be restrictive.

To guard against these dangers it is necessary to find some way in which completely unbiased ideas can be generated. A brain storming session is one way which this can be done. The name 'brain storm' gives the sense of unrestrained mental activity and this is virtually what such a session implies.

The important feature of such a session is that the ideas brought forward are not evaluated at the time by anyone. This allows for an unrestricted flow of ideas with rapid development of themes and equally quick changes of direction. After the session all the ideas are evaluated and the most practicable subjected to critical examination. The idea of brain storming can be usefully applied to procedures or to organization.

1. How many shall attend? Not less than four or more than seven are reasonable limits. A smaller group may tend to be stiff while a larger number will not generate group spirit. A useful guide is the numbers which would make a good drinking party.

2. Who should attend? In the selection of people a fairly wide range is advisable. Certainly the analyst doing the study should be there and preferably other analysts with no knowledge of the problem. Ideally someone who has worked in the area being covered should be included provided they do not have preconceived ideas. Such people are not difficult to find and their inclusion will help to convince the users of the good intention of the analysts. Finally someone with strong imagination, enthusiasm and interest can usefully be included. It may be necessary

to have a trial run at brain storming with the group on a different subject (say car parking) to ensure that everyone is happy about the technique and that the group works together.

3.  Who takes the notes? It is best if this is not one of the intended participants. A shorthand writer is useful. Don't think of using a tape recorder as it tends to freeze ideas (because people are wondering whether they are coherent enough) or loosen tongues too freely (because of a compulsion to speak while the reels revolve).

4.  Who should lead the discussion? Clearly the intention is that the discussion should be unrestricted and therefore a 'chairman' is not the answer. On the other hand there is usually a need for someone to guide the group away from geese and herrings (wild and red), albeit gently. He must also frustrate any attempts to evaluate the ideas coming forward. The task will usually devolve upon the analyst responsible for the project and this makes it important that he looks hard at his preconceptions.

In the session itself the analyst will describe the bones of the problem and will pose some possible solutions tending to make these primitive. Ideas should begin to flow; some wildly impracticable and some less so. One idea, while apparently ludicrous, may spark off some new approach which leads to a practical solution. But all are recorded irrespective of merit. The session should have an arbitrary time limit of about an hour but should be finished earlier if the process isn't working.

After the session the project analysts will go through the notes and select those worthy of further study. Obviously many of the ideas will be discarded immediately. Of the remainder (this could be 20-40% of the total) some will bear close examination and it is possible that these will include a radical, feasible and economic solution to the problem.

The unrestricted generation of ideas does not rely upon a formal session. As with other things formal sessions are useful in getting oneself used to the process; thereafter it can be practised by an individual as he approaches a new problem. However, in order to get the widest possible scope for ideas the session can be the best method.

## 6.8  Critical Examination

Objectivity is important in most walks of life but to the analyst it is essential. In fact the analyst must be rather more than objective; he must look at things critically. When working on a problem he must look at both the existing procedures and his proposals in this way. Looking at one's own proposals critically is difficult but is obviously essential.

The successful analyst tends to be naturally critical. He must develop this skill to the point at which it is applied automatically and methodically to any problem, the question of method being fundamental. In order to achieve the best results the critical examination should be based on the following primary questions:-

What? Why? How? Where? When? and Who?

The analyst must pose these questions on the principal features of the problem and consider all alternatives fully.

A formalized approach to critical examination can help to develop the automatic and methodical ability which the analyst needs. Various forms have been designed and these are useful in that they direct attention to the questions and alternatives. An example of one type of form is shown in figure 10. The discipline of completing such forms is invaluable and is recommended to all as an aid to developing the critical faculty.

When completing the forms there will be times when some of the questions seem inappropriate. It is important to examine this conclusion carefully for in practice the apparently inappropriate may be very relevant if one adopts a new approach to the problem. For instance in examining periodical binding the answer to 'when should it be done?' may seem irrelevant. Obviously one does the operation when the set is complete. However long binding delays may thus remove from the shelves issues still in relatively high demand. Therefore the question 'When?' may be a matter of delaying the binding rather than of bringing it forward. If the approach is altered in this way relevancy may be increased. Nevertheless some particular questions may not be relevant in some cases and, provided there is no attempt to use this fact to evade the effort of examination, these points can be ignored. This is the first stage in moving away from the formalized approach.

The second stage is to select from the possible questions those which are relevant without using the form at all. This requires rather more self-discipline in order to give the necessary concentration to each aspect and as this self-discipline develops so the need for reference to the list of possible questions decreases.

Eventually the critical faculties are developed to the automatic stage so that all manner of things are subjected to critical examination. This may sometimes prove embarrassing domestically. The occasional difficulties caused in this way however are adequately compensated in the improved approach to systems problems which all people meet whether or not they are analysts.

A final word on the need for concentration; when studying the existing or proposed alternative methods a satisfactory answer will only be achieved if the temptation to evade the difficult points, to hurry the whole operation or to jump to conclusions, is avoided. Critical examination must be thorough to be valuable and the result of a badly completed examination is almost certain to be misleading and harmful.

6.9 Form Design

When information has to be regularly recorded or communicated it is usual to use a standard layout. This may be the heading in a record book, an order form, or a form for date stamping stuck into a book. The ease of completion of these

## CRITICAL EXAMINATION

| PROJECT | PROBLEM | CRITICAL EXAMINATION | SHEET |
|---|---|---|---|
| What is done? | Why is it done? | What else could be done? | What should be done? |
| How is it done? | Why that way? | How else could it be done? | How should it be done? |
| Where is it done? | Why there? | Where else could it be done? | Where should it be done? |
| When is it done? | Why then? | When else could it be done? | When should it be done? |
| Who does it? | Why that person? | Who else could do it? | Who should do it? |

Fig.10   Basic Critical Examination Sheet

forms or layouts will have a marked effect on the efficiency of the system. Good forms must be easily completed and readily interpreted. The cost of a form is that of its use, not of its design. Clearly new systems are likely to introduce new forms, or to revise existing ones. The principles of form design are therefore important to the analyst.

The study of the clerical procedures in which the form is used will have been done in the analysis of the main system. This will have covered the need for the form and the information it contains. The possibility of eliminating or combining forms should have been considered and also the best place for their preparation and interpretation or use. Form design starts with a specification of the information it should use and the procedures governing its use.

As a first step it is important to eliminate unnecessary information from the form. A form is used to transfer data and each unit of data must be identified. If any of the data is irrelevant or if any of the identification is superfluous it should be eliminated.

Then consider the best layout for the form. This should minimize movement of hand and eye by the people inserting or extracting the information contained in it. When the document is completed in several stages at different steps in the procedure the design should, if possible, keep the various related entries together. The entry spaces should be arranged in the natural sequence of left to right and top to bottom. If additions or calculations are to be made the layout should facilitate these operations.

The entry spaces may be defined by boxes or columns. The size of the entry space is important. Too deep a space in proportion to length will encourage big writing which may overflow the space and will in any case be difficult to read. Spaces where a specified number of characters are to be entered should be carefully spaced to encourage clear entries. Boxes for variable length descriptions (such as book titles) must give sufficient space for the largest possible entry without construction. Recurring items should be pre-printed and where choices are to be expressed boxes should be printed in which a $\sqrt{}$ or 'X' can be placed. Clear identification of the entry spaces is important but try to avoid the waste of space which can occur when the heading of a column dictates its width. Don't separate the description from the entry by either irrelevant pre-printing or by wide spaces.

Similarly identification of the form itself is necessary, particularly in places where a number of different forms are used. This can be done by either distinctive titling or by different colours. Try to eliminate forms by combining similar ones; it is of course essential to retain clarity in the forms or the advantage of limiting the number of forms will be lost by the increased time necessary to complete them.

The final appearance of the form should be balanced and simple. This is helped by aligning entry spaces and by avoiding large blocks of printing. In a form with a large number of feint horizontal lines heavy printing should be used at intervals to ease line identification.

The choice of type face also affects the overall appearance and while size and boldness may be varied it is desirable to use only one style.

| PLEASE RENEW: | | | |
|---|---|---|---|
| Letter & Number on Date Card | Author | Title | Last date stamped on Card |

Books may only be renewed <u>once</u> by post or telephone

Date _____

Name _____

Address _____

_____

| PLEASE RENEW | | | | |
|---|---|---|---|---|
| Date Card | | | Author | Title |
| Letter | No. | Last Date | | |

Name _____

Date _____

Address _____

_____

Fig.11    Reservation Card and Possible Alternative

Finally, the presentation of the forms must be considered. This covers the number of copies, method of making copies (carbon paper, N.C.R., photocopying etc.), whether the form should be padded or loose and the quality of the paper used.

Figures 11 and 12 show forms currently in use in libraries. Figure 11 shows a reservation card and a redesigned alternative. The main weakness of the form is the small space provided for the title and, to a lesser extent, the author's name, when contrasted with the space for the data. This is an example of column width being dictated by the heading. The alternative rearranges the column to give more space for 'Title' and 'Author'. The note on the foot of the card has been omitted – it is suggested that this information should be inside the book; it will be of little help here to anyone wishing to renew by telephone.

Figure 12 shows two forms currently in use in the same library. Study of the information contained makes it clear that they could be combined. In the case of the Reservation use the recall date could be written in the 'For Library Use' line. Even if the two forms are not combined the better spacing of the lower form should be adopted for the other. This layout illustrates the benefits to be gained by using different size type.

It is interesting to note the different spaces provided in the three forms for addresses. This illustrates a common feature of forms provided for public use with the worst examples being the coupons provided by many advertisers in newspapers. People completing them are presumably in training for a 'writing the bible on a stamp' contest.

## 6.10  Code Design

Because codes are fundamental in making systems comprehensible and usable their design is important in systems work. This is not, of course, a new feature; in library work the various methods of classification have a long history. This history points to many of the significant facets of code design. Because of the impossibility of readily comprehending the nature of a book from its title, codes were devised to group together works dealing with the same subject. This shows the need for codes to identify and segregate. These codes were developed to the stage where they incorporated positional significance. Significance in codes is valuable if possible. Book collections tended to last and grow for a long time; this means that if constant change was to be avoided the codes had to be permanent and flexible. These two features are fundamental in most codes. The classification systems which are in most common use are remembered by the people using them: this too is useful.

This is not the place to discuss whether the various codes used in classifying books meet these criteria. The value of the codes is indicated by their widespread use. As important as these are the other codes used in a library. Accession numbers, readers' registration numbers and order numbers are important from the operating point of view and good design will be rewarded. Good design is equally important for these.

It is instructive to look at codes which have been developed for general use to identify the facets already mentioned. First look at these codes and see how many you can identify.

```
┌──────────────────────────────────────────────────────────────┐
│                      RESERVATION SLIP                          │
│                                                                │
│  Author       _____ │
│                                                                │
│  Title        _____ │
│                                                                │
│  Applicant    _____ │
│                                                                │
│  Address      _____    Date    _____ │
│  ════════════════════════════════════════════════════════════ │
│                                                                │
│  FOR        Class No. _____      Book No. _____ │
│  LIBRARY                                                       │
│  USE        Recalled  _____                           │
│  ONLY                                                          │
│             Recall on  _____                          │
│                                                                │
└──────────────────────────────────────────────────────────────┘
```

```
┌──────────────────────────────────────────────────────────────┐
│                                                                │
│  Author(block capitals)         │  Borrower (block capitals)   │
│                                 │                              │
│  ───────────────────────────────┼──────────────────────────── │
│  TITLE (and volume no.)         │  Class No.                   │
│                                 │  ─────────────────────────── │
│                                 │  Book No.                    │
│  ───────────────────────────────┼──────────────────────────── │
│  Signature  _____   │  Date    _____  │
│                                 │                              │
│  Address    _____ │
│  ──────────────────────────────────────────────────────────── │
│  (Library use only)                                            │
│  ──────────────────────────────────────────────────────────── │
│    The borrower is responsible for the book until the voucher is cancelled │
└──────────────────────────────────────────────────────────────┘
```

Fig.12    Two Forms in use in the same library

LPT 294C
AB 28 43 72 B
FGH 7155 755
65326

The first is obviously a car registration number. It satisfies the identification criteria and may be thought also to have some significance but this is only completely true in the case of the first group of letters. The final letter is often taken to indicate the year of manufacture, but this may not be so. The code used for early registrations has obviously failed on permanence because there have been several changes to the structure as the number of registrations outgrew the code. Car numbers are readily memorable. This is because of two features; the first is the mixture of letters with numbers and the second if the use of the letter/number split to break down the code into fields.

The second number may not be quite so generally recognizable. It is a National Insurance Number. In practice, the people who use these codes - and many members of the public - find them distinctive and memorable. There is really very little significance in the code. Its memorability rests again on letters/numbers and the use of fields. Here however the fields are also achieved by spaces between the numbers and people tend to remember numbers if they are led to think of them in small groups.

The third code is a National Health Number (formerly National Registration). This has the same features as the National Insurance Number but would appear at first sight to be slightly more memorable. This is probably a false assumption based on the inclusion of the same double number (55) in two places. Otherwise the length of the second field would probably make the code difficult to remember. The double numbers feature is a useful point to remember when constructing codes in which certain numbers having particular importance can be given double or other memorable number combinations.

As it stands the fourth number is meaningless. Add to it a word (say, Newcastle) and it becomes, fairly clearly, a telephone number. The fact that so many people do remember telephone numbers is interesting. In many cases this is a conscious exercise achieved by splitting the number into fields or groups having some rhythm or other pattern. It should be borne in mind when designing codes that if people wish to remember them they will find some way of doing so. This is less likely to be the case if many new codes are included in the system being designed. In that case each code will appear in the same way as the unadorned telephone number, i.e. meaningless without further identification. This further identification immediately reduces the value of the code as a means of abbreviation.

This brings in a further feature of code design. Many codes get added significance or even basic identification through combination with other codes. An example of this could well be the telephone number where a dialling code can give added significance to the extension code. More practical examples are in document coding when the meaning of one code may be entirely dependent on another code which indicates the type of document being handled.

The features of codes may be summarized thus:

| Essential | Desirable |
|-----------|-----------|
| Abbreviation | Significance |
| Identification | Memorability |
| Standardization | Distinctiveness |
| | Permanence |

In designing codes it is important to keep in mind the person who will be using the codes in a work situation. Bad design will mean incorrect coding with the problems that this will cause. Usually some codes are more vital than others and greater care must obviously be taken in the design of these.

The following are some points on the design criteria:-

1. Is significance required? - if so make sure that the rules for significance are clearly laid down and are understood by the people adding new codes to the list otherwise the code may be destroyed.

2. How important is the sequence of the entries being coded? If, for instance, the code is for reader identification it may be desirable to maintain the list of readers in alphabetic order, and to construct the code to assist in this. This requires careful planning and gaps must be left for later additions to the list. It is extremely difficult in these circumstances to avoid change to the code.

3. If permanence is required the code structure must allow for considerable increases in the number of items included. This affects both the length of the code and the spacing of entries (see (2) above).

4. Memorability will be aided if the code is distinctive, broken into fields and contains letters (particularly if the code is more than, say, five places long).

5. In order to check the accuracy of coding check digits (based on Modulus 11 or variants) should be used. Where this is impracticable and accuracy is vital then consideration should be given (in the system) to duplication of the number in an alternative form such as words or phonetic abbreviations.

## 6.11  Decision Tables

A decision table is a tabular representation of various possible conditions affecting a decision and the action to be taken when they are fulfilled. It is a valuable analysis tool, and can be used for program documentation.

Each tables will have five distinct parts:

1. A heading in which the decision is lucidly stated.
2. The various conditions which affect this decision.
3. The combination of these conditions.
4. The various actions to be taken.
5. The actions to be taken against any combination of conditions.

| PERIOD OF LOAN | | | | | | | | | | | | | | | |
|---|---|---|---|---|---|---|---|---|---|---|---|---|---|---|---|
| > 14 Days < 22 | N | N | N | N | N | N | N | N | Y | Y | Y | Y | Y | Y | Y |
| > 21 Days < 29 | N | N | N | Y | Y | Y | Y | N | N | N | N | Y | Y | Y | Y |
| > 28 Days | N | Y | Y | N | N | Y | Y | N | N | Y | Y | N | N | Y | Y |
| EXTENSION GRANTED | N | N | Y | N | Y | N | Y | N | Y | N | Y | N | Y | N | Y |
| SEND RECALL | | | | | | | | ✓ | | | | | | | |
| SEND REMINDER | | | | ✓ | | | | | | | | | | | |
| SEND ACCOUNT | | ✓ | | | | | | | | | | | | | |
| NO ACTION | ✓ | | ✓ | | ✓ | | | | ✓ | | | | | | |

Fig.13   Decision Table

At this point an example will be useful. Supposing that it is necessary to examine the status of books on loan, assuming the following rules: A book which has been on loan for more than 14 days will be the subject of a recall notice; one that has been out for 21 days will cause a reminder to be sent; and if out for more than 28 days an account to be submitted to the borrower; unless in each case an extension of the loan has been granted.

Figure 13 shows the result of combining these rules and actions into a decision table.

It will be noted that the table allows all the combinations of rules to be explored; indeed if it is remembered that as rules can have $2^n$ combinations then by applying the symmetrical approach all will be covered.

Each combination can be considered. Some will be invalid/impossible and these may be rejected. The appropriate action can be entered against the rest. All possibilities will have been covered.

This represents a great improvement over haphazard trial and error methods. It must be emphasized that the decision table depends upon the decision being precisely defined, and entry of all the conditions which affect that condition.

A warning must be given. If a defined decision is found to be dependent upon a large number of conditions then the resulting table may be so cumbersome that it confuses rather than clarifies. A practical maximum of five conditions may be established as a guide. Larger tables than this may be of some use for analysis, but will be of little use as program documentation.

If the reader wishes to develop his knowledge of this technique there is much literature available, but beware the fanatic. Some writers have elevated the technique to being the complete answer to both analysis and program documentation. This is too wide a claim, for while decision tables offer great help in analysis, and in their simpler forms can be used to instruct the programmer in certain applications they cannot, at least in their current state of development, be held to have wider significance than this.

# A
# GENERALIZED METHODOLOGY
# FOR
# LIBRARY SYSTEMS ANALYSIS

*Robert W. Burns, Jr.*

SOURCE: From Robert W. Burns, Jr., "A Generalized Methodology for Library Systems Analysis," *College & Research Libraries* 32 (July, 1971) pp. 295–303. Reprinted by permission of the author.

THIS ARTICLE IS DIRECTED toward the novice in systems work. Its purpose is to generalize at a very elementary level a methodology or approach which can be used in conducting a systems study. Systems work is discussed here as a point of view; a logical, coherent, from the top down, preface to decision-making and resource allocation which utilizes a very powerful body of sophisticated techniques. The approach and techniques reviewed in this paper, however, will be those on the most elementary level. No attempt will be made to discuss the techniques of queueing, inventory management, linear programming, simulation, marginal analysis, game theory, statistical inference, or any of the other highly sophisticated techniques available to the operations research/systems analysis (OR/SA) analyst. When the systems approach is clearly understood and properly used, it becomes a potent weapon in the arsenal of the administrator. Rather than a review of the tools themselves, a delineation of this systems methodology and point of view will be considered in this article. The methodology discussed here embraces a number of standard techniques used by the systems engineer, time and motion analyst, operations researcher, and occasionally, even the librarian. Examples of these techniques are scattered through the professional literature of librarianship/information science, management, industrial engineering, and operations research/systems analysis. Some of the more important references describing OR/SA in the library have been included in the bibliography which accompanies this article. Unfortunately, many of the most basic concepts of these twin fields remain poorly understood and as a result are seldom applied by the library profession. Two glossaries of terms have been added to the bibliography for the benefit of the user who wishes additional help in understanding the terminology of OR/SA.

In the past six to eight years, only a few publications of merit have appeared in the literature showing how and under what conditions a systems study can be conducted in a library environment. Some of these were prepared by librarians, but many of the best have been written in a highly technical jargon by individuals whose credentials are in fields other than library science. Indeed, one of the most significant developments has been the number of articles written about the library/information science field by individuals whose backgrounds are in other disciplines but who, nevertheless, have successfully used the library as a laboratory, and in doing so have given the library profession some of its most substantial contributions.

A true systems study should be able to document for the administrative officer the goals of the administrative unit being studied and the resources available to the unit, as well as suggest alternative methods for achieving these goals within a given set of constraints. All of this must be accomplished in such a fashion that the administrator is permitted to select the proper alternatives by manipulating resources to reach his preselected goals. Fundamentally, this is a process of balancing goals with resources based on the facts gathered by the analyst. Facts needed by the administrator include such items as unit costs, unit times, costs of materials and equipment, opportunity costs, configuration and availability of equipment, movement of staff and material, and staffing patterns. It is the job of the analyst/designer to ferret out these facts and present them to the administrator with a full display of available options.

A systems study must examine both the economic efficiency of the unit being studied as well as its operational efficiency, always being careful to study each *in vivo*. Economic efficiency can be judged in either of two ways: the ability of the system to produce or process the same number of units for less cost; or the ability to produce more units for the same cost. The savings achieved by library automation seem largely to accrue from the second advantage. Operational efficiency is a much more subtle concept and, indeed, involves many of the intangible values with which all librarians doing systems work are constantly confronted. One measure of operational efficiency derives from user satisfaction and can be determined by the questionnaire/interview method.

The achievement of maximum efficiency within a system is an extremely subtle process requiring the fine tuning and sensitive ear which one expects of a skilled violinist. It is in no sense of the word the obvious undertaking that some managers believe it to be. An efficient system is one which has reached a correct balance between the resources and the system's achievement of its goals, or performance. However, there is a distinction between efficiency and the measures of efficiency. It is quite common for the novice in systems work to confuse the ways of measuring efficiency within a subsystem such as decreased costs, increased production, etc., with the efficiency goals of the total system. True efficiency can only be discussed validly in the context of a total system's operation.

## What Is a System?

In discussing systems work, the first problem is to develop an unambiguous definition for the word system. Although it is used often and widely, the implications of this concept are seldom fully understood. As Nadler points out, there are almost as many definitions as there are people writing about the field.[1] The Random House dictionary stresses the concept of a system as "an assemblage or combination of things or parts forming a complex or unitary whole. . . ." The U.S. General Accounting Office, in its systems glossary, expands this to point out that "systems analysis may be viewed as the search for and evaluation of alternatives which are relevant to defined objectives, based on judgment and, wherever possible, on quantitative methods, with the objective of presenting such evaluations to decision makers for their consideration. . . ."[2] Bellomy refers to a system as "an assemblage of interdependent things and ideas necessary to achieve a set of related objectives . . . characterized by inputs which are processed to produce the outputs required to achieve specified objectives. . . ."[3] After examining these definitions, several ideas begin to emerge which are common to any systems effort, no matter what it may be called. The ideas of interrelated parts bound into a coherent whole possessing a common goal or objective are central to the systems concept. It is on these basic attributes that we shall build our methodology for a systems study.

In this article attention will be focused on the four steps or phases of a systems study which we shall call the systems survey, the systems analysis phase, the systems design phase, and the implementation/evaluation phase. This somewhat arbitrary division should not be taken to infer that these are discrete operations with a systems design proceeding only when the systems analysis effort has been completed. This would be a highly idealized solution since in actual practice the pressures to get on with the job will usually force the telescoping of these efforts. When this is done with care and in a recursive fashion, the chances of success are usually good. Each of these phases should be viewed as complementary to the others and, although they are similar and related, each must be performed in a sequential and discrete fashion, preferably in tandem. Some overlap is permitted, but the analysis phase always begins before the design phase, and the design phase always begins before the implementation phase.

### General Characteristics of a Systems Effort

Before discussing each phase in detail, several generalizations should be made about the entire systems effort. These will help the reader develop an understanding of the type of problem to which we are addressing ourselves; they are as follows:

(1) Attention to detail lies at the very heart of the systems effort and thorough precise work demands an intense preoccupation with every detail, no matter how small. Indeed, the entire systems effort hangs on the ability of the analyst/designer to unearth and articulate *all* the minutiae of a procedure. It would be difficult to overemphasize the importance of this aspect of systems work, for the most minute detail can jeopardize the success of an entire operation. This becomes even more critical when the systems effort involves machine planning, for machines, unlike people, will not tolerate ambiguity. This will suggest to the perceptive reader that it is wise, indeed essential, to plan several alternatives for each proposal, since the smallest miscalculation could force the scrapping of an entire proposal and change the direction of all work done up to that point.

(2) Every system is a subsystem of some larger system and each system is itself composed of a number of component subsystems. Therefore, all systems exist in both a micro and macro hierarchy depending on the perspective of the analyst. Knowing this, the analyst must constantly guard against suboptimization, i.e., the design of a component subsystem such that it operates in an optimum fashion to the detriment of the system as a whole.

(3) Systems work is a much more subtle process than simply fact gathering. It involves a thorough understanding not only of who, what, when, where, why, and how, but of the relationships which exist between the system under review and all of the other systems with which it interfaces, as well as the component subsystems which make up the system being studied.

(4) There is no single definitive measure for the effectiveness of a system—only circumstantial optimums, each of which must be weighed against all other possible options available to the manager.

(5) Systems are generally designed for the normal operation (quantitatively, the mean or median), and only rarely will the goals of the system permit design for the exceptional conditions.

(6) All systems work is by nature re-

cursive with each successive repetition performed either at the same or at a different level.

(7) Continuous feedback and monitoring are essential components of the systems effort.[4] One of the major difficulties in optimizing present manual library systems has been the lack of adequate provision for valid feedback.

(8) By definition, all systems must exist within an environment. The environmental factors are those which affect or relate to the system under discussion but which are not a part of that system. The analyst cannot fully describe the system without also delineating its environment.

(9) There is a danger in any systems work that it will attempt to quantify that which cannot be quantified—the intangible factors. Overquantification can become a very serious problem and often leads to a credibility gap in the entire systems effort.

(10) Documentation is as much an essential part of the systems effort as analysis, and to ignore or discount this aspect of systems work is to invite disaster.

(11) There is never any final phase to a systems effort, only iteration.

Library systems work provides us with excellent illustrations for each of the above axioms. For example, the failure to write down and describe all the steps in a systems effort *as they take place,* has forced many SA projects to start again whenever a change of personnel takes place. Or, how many librarians have unknowingly insisted upon a system which will handle all exception routines and then wondered why the system took so long to develop or refused even to work at all? Library systems are difficult to analyze, not because of their size, but because they are often unstructured, lack adequate provision for feedback and monitoring, and are always so interrelated and interdependent that the best descriptions of them are of dynam-

ic systems which have "evolved" over a long period of time through a trial and error process. Developing models for this type of a system, especially mathematical models, is a particularly difficult undertaking and can lead to very misleading conclusions unless the model builder understands the proclivity of mathematical modeling for oversimplification.

In fact, most of the dilemmas which plague all systems work also exist in the library systems effort. As with any systems work of magnitude, the analyst finds himself on the horns of a dilemma at the very beginning of his study. Machol has pointed out that the problems of designing a large system are often of such magnitude as to make the problem indigestible and even unsolvable if attacked all at once.[5] Yet the analyst cannot arbitrarily divide the problem to study it piecemeal without running the risk of losing the continuity of the whole. Where then does a realistic approach exist between these two extremes? A partial answer lies in the perspective of the analyst, in his ability to maintain a continuing balance between the unity of the whole and the detail of the part.

## STEPS IN THE SYSTEMS STUDY

The systems effort begins with a problem defined by the analyst as a system existing in an environment of other systems and bound by certain constraints. The first step is to isolate the system under review so that it can be described in an unambiguous fashion. This is the systems survey stage and marks the beginning of a series of successive partitionings which take place until the system has been divided into the smallest logical component still capable of being identified with the system being studied. This process of system dissection is analogous to the molecular theory of chemistry which defines a molecule as the smallest particle of matter still exhibiting all the characteristics of the larger mass (system) from which it came. After dividing the system into its molecular components, the analyst then proceeds to delineate the alternatives he has created by rearranging these component parts in whatever fashion the resources and goals of the system will allow, always being careful to work within the constraints which the system's environment dictates. The analyst then proceeds to evaluate these alternative solutions in the light of the stated goals or objectives and selects from them a preferred course of action which he recommends to the decision-maker. Thus, evaluation/implementation becomes the last sequential step of the systems effort and is followed by whatever iterations are deemed necessary by the decision-maker to reach the goals of his agency.

## SYSTEMS WORK FROM THE ADMINISTRATOR'S VIEWPOINT

At this point it might be appropriate to shift perspective and discuss systems work from the administrator's point of view; that is, in terms of the agency's goals, choices, resources, and inputs/outputs. Each administrator has at his disposal four categories of resources: staff, space, funds, and time. (To this some would add a fifth resource—information.) The mix a manager adopts to meet the goals of his administrative unit has depended in large measure upon his own judgment which up to now has been, at least in part, intuitive. In the course of getting the job done or reaching a goal these resources will of necessity be consumed to a greater or lesser degree. It is the responsibility of the manager to balance continuously the availability and consumption of these resources with his goals in order to assure that the goals are reached in the most efficient fashion possible.

Furthermore, a large portion of systems work consists of no more than asking questions about all those assumptions and operating norms which up to now have been accepted as obvious, axiomatic, or based on historical prerogative, and in so doing to pare away the obfuscation which tends to grow up around a deep-seated procedure.

But how does all this apply in a library environment where the goal is that nebulous entity "service"? In order to answer this question realistically, one must first decide what constitutes the library's service goal. The author has chosen to adopt the definition of the library's goal that Mackenzie has used: "to assist in the identification, provision and use of the document or piece of information which would best help the user in his study, teaching or research, at the optimal combination of cost and elapsed time. . . ."[6] Efficiency, when used in this context, becomes either answering more of the "needs" of a reader while holding costs and elapsed time constant, or meeting the same needs while cutting down costs and elapsed time. However, neither explanation of efficiency is entirely satisfactory when used in this fashion because the process described here is one using only quantification as the valid criterion for evaluating its success. This is not to imply that there are no areas in library systems analysis which can be evaluated in a quantitative sense—there most definitely are. It is merely to emphasize for the systems person that he cannot quantify all aspects of a library system. Indeed, insofar as any systems study attempts to use quantitative methods where they are not appropriate, the study will fail and, unfortunately, the reason will not always be clear to all concerned. What the analyst cannot do is quantify the intangible benefits from a course of action, and it is here that the administrator will need to depend most heavily upon his own experience and intuition for guidance. What follows is a generalized methodology for the systems approach to problem-solving.

### First Phase: The Systems Survey

In the first phase of the systems study, the analyst conducts what is called the systems survey, during which he relates the system under review to other systems in which it is embedded—to its environment if you will—by determining what is germane to the problem being studied. Once these boundaries have been established, the analyst begins to lay out the problem in very general terms, specifying the goals and functions of the system under review. This involves familiarization and departmental orientation of the analyst, preparation of such tools as a list of the files maintained, their contents, and the organization of each; a list of the forms being used with examples of each; and a description of their movement, and associated activities; a review of all procedural manuals and job descriptions; and finally a documented statement of the system's goals.[7] When used in this context, a goal can be thought of as either a direction or an objective or a combination of both. It can be a point to be reached or a line of march to be followed in moving toward this point. But each goal must also be defined in terms of the expected performance of the system. In fact, any discussion of goals which does not include a statement of the performance expected from the system is so innocuous as to be irrelevant and makes the entire discussion meaningless. Statements of performance coupled with goals have the added advantage of helping to prevent a dichotomy from developing between the real and stated goals.

### Second Phase: Analyzing the System

The analyst is now ready to begin the second phase of his study, preparation of a block diagram or system schematic, which outlines in a very general way the tasks performed by the system and the relationships which exist between the subsystems.[8] This is the first level of definition and is, of necessity, very gross. For a library circulation system these boxes might be charging, discharging, searching, shelving, etc. Each box is then further subdivided into its appropriate tasks down to the procedural level, showing the movement of people and materials through all subsystems. This is accomplished by using flow process charts first and then by using flow decision charts.

Construction of the block diagram and the flow charts are the first concrete expressions of an analysis effort which up to now has been primarily a data gathering and intellectual exercise. Flow process charts enable the user to visualize at once the movement of a person and, for example, the distances traveled in checking out a book. The chart will also point out for the user how many times a book is "inspected" as it moves through a given routine. The flow decision chart, on the other hand, uses a different set of symbols and shows at what points decisions are made and how these decisions affect the flow of materials/people. In his charting, the analyst works at a very specific level where he is concerned with discrete entities capable of quantification in terms of how long, how many, how much, and how often. Indeed, his next task is to begin the quantification of these steps by carefully tabulating the number of times a given symbol was used on the flow process chart and the time necessary to move through these steps. Parallel with this effort, the analyst should be identifying activities and compiling these into a document known as a standardized activities list. It is also customary to document the levels of personnel performing these tasks.

Thus far, the analyst has dissected the system—in this case, a library loan desk—through the activities (charging, discharging, etc.) and procedures (how a card is returned to a book in the discharging activity) levels with all the components enunciated at each level. As he does this he also begins the timing of these component subsystems at the procedural or task level. At the same time, the analyst should begin the process of deriving costs by determining what are the real wages (direct + indirect/productive time on job) paid to staff in order that he may translate unit times into unit costs. When this exercise has been finished, the analyst can measure quantitatively the available alternatives, at least in terms of costs, and offer these to the decision-maker for review.

There still remains the difficult problem of evaluating intangibles—those factors which cannot be quantified, such as convenience, availability, prestige, etc.—and if the cost studies have been close, intangibles become crucial to the decision-making process. Intangibles will add support to a program only when definable costs can actually be used to demonstrate a more efficient operation. In other words, the intangible factors can only be used to buttress an argument and never as the sole reason for modification of a system, experience and intuition aside. The point is that more subtle techniques of quantification must be used before funds can be invested in any change which intuitively appears to yield better results.

### Third Phase: Design of the System

The next phase, systems design, usually follows when the analysis efforts have been completed and carefully digested. In theory, these steps should be discrete. In actual practice, however, they seldom are, for the design efforts will often overlay the analysis studies. Usually design consists of a modification of the existing system—a rearrangement of the components in the old system—but with possible additions or deletions modifying any or all inputs of the resources discussed earlier, and always within the context of the systems goals.

### Fourth Phase: Implementation and Evaluation of the New System

The final phase begins with the implementation of the prototype system and its test/evaluation. This is often the most expensive single phase and its success depends on all earlier phases being in a state of completion. Up to this point the entire process has been a recursive one of dividing, measuring, charting relationships, defining, then repeating the whole process of quantifying the characteristics of the component systems, charting relationships again, and repeating the cycle. Because of economic constraints, however, the implementation and prototype phase cannot always be repeated easily. Therefore, it behooves the analyst to work with meticulous care once this phase of the systems effort has been entered. Another point which should be brought to the reader's attention here is that first-time processing costs, procedures, etc., are normally atypical and cannot be judged to remain constant throughout the life of a system. These are not the nonrecurring costs normally associated with the activation of a system, but those unit costs and unit times which would normally be expected to remain constant throughout the life of the system. The first complete operating cycle is never typical, no matter how carefully the planning and design work was done. There is always the problem of the unforeseen, and no analyst, no matter how good, is ever able to plan for all contingencies.

## CONCLUSION

Hopefully, the reader now has a better understanding of the intricacies and nuances inherent in systems work. It is obvious that such work is a prerequisite to library automation, but it does not necessarily follow that automation will automatically succeed the systems efforts. Indeed, the study can easily indicate that library automation is not appropriate given the existing resources of time, money, staff, or space. In essence then, systems work is a method—part science, part art—whereby one determines the correct balance between constraints and the resources necessary to realize predetermined goals, and leads to the establishment of realistic priorities based upon a thorough understanding of the total system being studied and its relationship to all other systems having a common interface.

## REFERENCES

1. Gerald Nadler, *Work Design* (Homewood, Ill.: Richard D. Irwin, 1963), p.87ff.
2. U.S. General Accounting Office, *Planning—Programming—Budgeting and Systems Analysis Glossary* (Washington: U.S. G.A.O., 1968) p.40.
3. Fred L. Bellomy, "Management Planning for Library Systems Development," *Journal of Library Automation* 2:187–217 (Dec. 1969).
4. "Feedback," *The Systemation Letter* 166: 4p. (1965).
5. Robert E. Machol, ed., *System Engineering Handbook* (New York: McGraw-Hill, 1965), p.1–5.
6. A. Graham Mackenzie, "Systems Analysis of a University Library," *Program* 2:7–14 (April 1968).
7. Ibid.
8. Robert Hayes, "Library Systems Analysis," in *Data Processing in Public and University Libraries,* ed. by John Harvey (Washington: Spartan Books, 1966), p.5–20; and C. D. Gull, "Logical Flow Charts and Other New Techniques for the Administration of Libraries and Information Centers," *Library Resources & Technical Services* 12:47–66 (Winter 1968).

## BIBLIOGRAPHY

*I. Basic and Introductory Material of Special Value to Librarians*

1. Herner, Saul. "Systems Design, Evaluation, and Costing," *Special Libraries* 58:576–81 (Oct. 1967).
2. Bellomy, Fred L. "Management Planning for Library Systems Development," *Journal of Library Automation* 2:187–217 (Dec. 1969).
3. Dougherty, Richard M., and Heinritz, Fred J. *Scientific Management of Library Operations.* New York: The Scarecrow Press, 1966. 258p.
4. Hayes, Robert. "Library Systems Analysis," in John Harvey, ed., Data Processing in *Public and University Libraries,* p.5–20. Washington: Spartan Books, 1966.
5. Becker, Joseph. "System Analysis—Prelude to Library Data Processing," *ALA Bulletin* 59:293–96 (April 1965).
6. Leimkuhler, Ferdinand F. *Mathematical Models for Library Systems Analysis.* School of Industrial Engineering, Purdue University, Sept. 1967. PB 176 113.
7. MacKenzie, A. Graham. "Systems Analysis of a University Library," *Program* 2:7–14 (April 1968).
8. Burkhalter, Barton R. *Case Studies in Systems Analysis in a University Library.* Metuchen: The Scarecrow Press, 1968. 186p.
9. MacKenzie. ibid.
10. Kilgour, Frederick G. "Systems Concepts and Libraries," *CRL* 28:167–70 (May 1967).

*II. Advanced Texts*

11. Morse, Philip M. *Library Effectiveness: A Systems Approach.* Cambridge: The M.I.T. Press, 1968. 207p.
12. Raffel, J. A., and Shishko, Robert. *Systematic Analysis of University Libraries: An Application of Cost-Benefit Analysis to the M.I.T. Libraries.* Cambridge: The M.I.T. Press, 1969. 107p.
13. Machol, Robert E., ed. *System Engineering Handbook.* New York: McGraw-Hill, 1965. p.1–5.
14. Nadler, Gerald. *Work Design.* Homewood, Ill.: Richard D. Irwin, 1963. p.87ff.

*III. Specialized Articles Covering Techniques*

15. Poage, Scott T. "Work Sampling in Library Administration," *Library Quarterly* 30:213–18 (July 1969).
16. Fazar, Willard. "Program Planning and

Budgeting Theory," *Special Libraries* 60: 423–33 (Sept. 1969).

17. Kozumplik, William A. "Time and Motion Study of Library Operations," *Special Libraries* 58:585–88 (Oct. 1967).

18. Gull, C. D. "Logical Flow Charts and Other New Techniques for the Administration of Libraries and Information Centers," *Library Resources & Technical Services* 12: 47–66 (Winter 1968).

19. Aslib Research Department. "The Analysis of Library Processes," *Journal of Documentation* 26:30–45 (March 1970).

*IV. General Material*

20. U.S. General Accounting Office. *Planning—Programming—Budgeting and Systems Analysis Glossary.* Washington: U.S. G.A.O. , 1968). p.40.

21. Spencer, Donald D. *The Computer Pro-grammer's Dictionary and Handbook.* Waltham, Mass.: Blaisdell, 1968. p.41.

22. Rivett, Patrick. *An Introduction to Operations Research.* New York: Basic Books, 1968. 206p.

23. Churchman, C. West. *The Systems Approach.* New York: The Delacorte Press, 1968. 243p.

24. "Feedback," *The Systemation Letter* 166: 4p. (1965).

25. "Analysis . . . The Second Essential Step," *Systemation: A Semi-Monthly Letter on System Trends and Techniques* 12:4p. (1 Oct. 1958).

26. Heyel, Carl, ed. *The Encyclopedia of Management.* New York: Reinhold Publishing Co., 1963. p.613. Taken from Pocock, John W. "Operations Research; Challenge to Management," Special Report no. 13. New York: American Management Association, 1956.

# LIBRARY SYSTEMS ANALYST— A JOB DESCRIPTION

*Thomas Minder*

SOURCE: From Thomas Minder, "Library Systems Analyst—A Job Description" *College & Research Libraries* 27 (1966) pp. 271-276. Reprinted by permission of the author.

Systems analysis has become an important part of librarianship. Courses are being offered in schools. Librarians want to add analysts to their staff, and the term frequently appears in library research and development literature.

There can be little doubt concerning the reasons for interest. The continuing pressure to introduce automation, especially electronic data processing, into the profession has caused librarians to look to the computer field for techniques. At the same time, there is an increasing awareness of the weaknesses in the traditional methods of library evaluation, design, and operation. Finally, operating costs within libraries are increasing to such an extent that libraries are being forced to look for improved methods of cost analysis for justification of budget requests.

Despite the common use of the phrase "systems analysis," very little has been written in library literature concerning it, its nature, the qualifications of the analyst, and the relationship of these two to librarianship. The following is intended partially to satisfy a need for a better understanding of systems analysis within the library profession. It is also a suggested working paper for those involved in establishing professional standards. Although the presentation is in the form of personnel qualifications, it is done within the context of the nature of systems analysis and the relationship to librarianship.

## JOB DESCRIPTION

A library systems analyst can be described as a staff person with the responsibility of applying the principles of scientific management to the library environment.

The restriction of an analyst to a staff position is significant for four reasons. The analyst has no direct relationship to the routine operations of the library. He stands apart and, it is hoped, observes these operations with an unbiased eye. He also makes these observations from at least two points of view. At the minutiae level, he may be expected to do time and cost studies of an operation in the smallest details. At the same time, he is expected to relate the minutia and their synthesized sets into a single unit. Such breadth of interest requires independence from operations activities.

This need for breadth implies a need to cross over organizational lines. This carries with it the need for authority or backing from a higher level. For example, a study of the catalog department cannot be complete without relating the department to its inputs, outputs, and place of department within the whole library organization. Such a study cannot be accomplished without freedom to study in depth the departments and managerial operations external to the catalog department.

This need for authority to conduct broad studies does not imply that the analyst should always operate at the highest managerial level—rather, at some particular level higher than the operation being studied. For example, a particular study of descriptive cataloging techniques may have relevance only to the internal activities of the catalog department. During such a study, authority may only be necessary at the departmental level.

Limitation of the analyst to a staff appointment formally excludes him from policy-making responsibilities. This responsibility belongs with management, including operating supervisors.

This lack of authority and responsibility may limit the short-range effectiveness of the analyst because he may not be able to overcome the lethargy, hostility, and weaknesses of operating personnel. At the same time, such lack of authority will help surface these weaknesses and serve as a warning of others more serious to be expected when plans become operational.

This possible hostility to systems analysis techniques and the implied need for personnel education rather than legislative action cannot be underestimated. Systems analysis has not been a recognized part of library training until quite recently. Supervisors are likely to continue to prefer the subjective judgment and *ad hoc* decisions about which they have considerable knowledge.

Scientific management is a term used to describe a whole new field of applying mathematical and scientific techniques to aspects of management traditionally considered to be creative in nature. It does not replace the decision-making functions of management. Rather, it provides management with better data in a syn-thesized form so that better decisions can be made. Scientific management also attempts to separate the truly creative decision-making operations from those that can be handled automatically or reduced to the clerical levels through the use of new tools and techniques.

Management science is also used to assist the analyst in reducing complex systems to the essentials, building new systems around these essentials, and then efficiently communicating these studies to library management.

Embodied within these general statements of scientific management is the implied use of probability and statistics, dynamic programing, time studies, flow diagraming, human engineering, and a host of other analytical tools. It would be misleading to define the field as only operations research or industrial engineering as suggested by these topics; however, these professions are the most prominent proponents of managerial science methodology.

Full justification for the use of management science within the library cannot be explored in this article. By way of partial justification, it might be stated that the library has most of the elements common to the disciplines where management science has been useful. Perhaps the only really major difference lies in motivation. For example, business is profit oriented, whereas libraries are service oriented. The library systems analyst exploits the tools developed in business applications and applies them to his own environment.

During the 1950's, considerable attention was given to the study of information retrieval. Research during this period tended toward theoretical studies of the statistical or mathematical nature of information and related topics. A hope was to develop ways to use the computer in the handling of concepts rather than the routine manipulation of alphanumeric data. More recently, the pendulum has swung toward the less glamorous, more traditional areas of technical processing, personnel records, and simple bibliographic compilations.

Systems analysis has often been assumed to enter the library as a tool to be used in the development of computerized systems, especially in the more mundane areas suggested above. This view is far too narrow.

Actually, systems analysis is a tool to be used in all departments of the library regardless of the computerization potentiality. It can be as effective in the analysis and design of a broad selection program as it is in the development of an efficiently run computerized technical processing department. For example, a real measure of the library's effectiveness is its ability to supply non-ambiguous responses to users' information needs within a time and cost limitation competitively set by other information media. The computer is considered in the analyst's study only insofar as it might contribute to the over-all solution.

The phrase "library environment" is meant to imply that any part of the library or its interfaces with sources, users, and parent institution are legitimate areas of study. The proper placing of book return boxes on campus or a study of the overlapping between two reference tools are no less of interest than the design of a completely computerized technical processing department.

## PREREQUISITES

The prerequisites for the position of library systems analyst is approached with some degree of trepidation. It is rather hard to state categorically what makes a good analyst. Perhaps the only thing that can be said with any certitude is that he must be a born skeptic about the status quo, a dreamer about the future, and a realist in the implementation of these dreams. This is hardly adequate for a recruiting brochure! Despite the implications that an analyst can only be evaluated in terms of his temperament,

there are some general guidelines that can be helpful. The following is a statement that seems to include the major qualifications that he should be able to demonstrate.

The curriculum in modern industrial engineering appears to supply most of the technical background for a library systems analyst. It emphasizes applied statistics and probability, work analysis, management organization, the design of abstract systems, automation, and systems evaluation techniques. Most of these topics are used as tools by the analyst. They require a certain amount of formal classroom study and development. For this reason, formal classroom exposure to these subjects is desirable.

One need not look deeply into librarianship before he sees a large data base, many repetitive operations, and highly systematized set of operating rules. It is in such systems that computers are potentially useful tools. We are only now beginning to exploit this potential, however, and the library profession is not fully aware of the possibilities. The systems analyst should be adept in the use of the computer because of its usefulness as a library tool, not as an end in itself.

This suggests a requirement of some training in electronic data processing. This should include at least a good understanding of the basics of computer design, construction, and operation. It should also include flow charting and the ability to program one of the common computers in one of the common languages. Depth in programing and hardware understanding is not necessary, since the analyst is more interested in knowing applications than he is in actual programing efficiency and computer design. It should be emphasized, however, that some programing experience is invaluable. The effect of actually programing a computer is somewhat like swimming. You can read about it, but you will never really learn it until you jump into the water and swim. The actual depth of training necessary in programing is an open question. Probably this training should stop short of real proficiency so that the analyst does not lose his perspective. To him, an efficient program is an efficient tool, not an end in itself.

Electronic data processing training is valuable for other reasons. It develops the student's ability to formulate problems and think in a formal deductive way. It also teaches him to communicate by using a formalized, well-defined language.

The prerequisite that will meet with the least acceptance concerns the amount of required training in librarianship. Industrial engineers and operation research specialists can be expected to demonstrate the close similarity between librarianship and business or military problems. Thus, little training is required. On the other hand, librarians frequently complain that outsiders do not understand the uniqueness of the library. Thus, a degree in librarianship is required.

Both are right and both are wrong. To be sure, the library has a classical management and operating structure; however, its standards of economy and service have no parallel in other fields. What company would accept a million-item inventory with an average use expectancy of once every five years? The tendency of outsiders has also been to criticize the library profession for its ignorance of the information problem and lack of creativity in the search for solutions. On the other hand, the record of success by outsiders has left a lot to be desired. There are numerous examples of their unused automated systems, re-inventions of the "book" (though now automated), and systems that violate the most basic axioms of information handling and service.

Perhaps the most outstanding recent example of this last item was demon-strated at the Airlie House Conference on Library Automation.[1]

The nonlibrarians repeatedly stated that the design of an automated library system must be done as a unit. That is, all the inputs, outputs, and internal operations for the whole library must be determined before implementation begins. Yet such an approach is not possible in the library because the system as a whole is open, not closed, and the specifications are never static. A library system must be open-ended and designed to adjust easily to change while in an operating environment. It is the uncontrollable and unpredictable outside factors, such as changes in the organization of knowledge by scientific advancement, that negate the applicability of the closed-unit systems design axiom.

At the same conference, both sides displayed a deep dependency on the other. In effect, one side said—tell us what to automate and we will build the machine to do it. The other's reply was—tell us what the machine will do, and we will define the jobs to be done. They left without even resolving the impasse. Ideally, the systems analyst should be a catalytic agent between the two.

It is obvious from this discussion that an education in librarianship and in the technical tools is essential. Recognizing the fact that it is unrealistic to expect degrees in both, however, it is far better to have the formal training in the technical fields with experience in library problems, rather than the reverse. Insight into library problems can be gained through experience and observation. The technical tools must be acquired in formal training.

## FUNCTIONS

The job description and prerequisites suggest that the library systems analyst has responsibilities beyond the analysis of existing operations and the introduc-

[1]*Libraries and Automation, Proceedings of the Conference on Libraries and Automation held at Airlie Foundation, Warrenton, Virginia, May 26-30, 1963.* (Washington, D.C.: Library of Congress, 1964).

tions of automation into the field. Consistent with the analysts' philosophy of broad vision and unity of operation, the following list of functions is suggested.

1. Analyze existing library policies, procedures, equipment, literature content, and human interfaces systematically in qualitative and quantitative terms.
2. Design and implement new and/or improved library systems in terms of the library's physical limitations, funds, personnel, available equipment, and available techniques.
3. Coordinate these analyses and designs with the library's management and professional objectives. Act as technical advisor at the various levels of management in the evaluations.
4. Design, implement, and operate management data systems that will provide library management, operating personnel, and designers with data to assist them in library control and evaluation.
5. Monitor and evaluate equipment, procedures, and new systems of potential value to the library.
6. Conduct technical liaison between the library and outside services such as the institution's computer facilities and equipment manufacturers.
7. Train operating personnel and library management in the characteristics and operations of the library's newly installed systems, new equipment, on the market, and new techniques being developed.

The analysis function is broader than that normally considered in the computer field. It encroaches on an area generally considered to belong under management and library operations. For example, the analyst has techniques available for the partial analysis of the subject content of a collection. He can also provide improved statistical data which the librarian can use as a powerful tool in establishing his selection policy. In virtually every area of library operation, the librarian is in need of better analyses of his library to assist with his decision-making activities. The analyst can assemble the significant data in a meaningful way.

Systems design is the synthesis of elements from the analysis function, with additional useful outside elements to form a meaningful system. Usually systems design is more than an engineering or mathematical coordination of elements. In complex systems such as those found in the library, the best solution also involves qualitative factors and subjective judgment.

Coordination with the library's policies and objectives is in many ways a part of systems design. Separate recognition is made here to emphasize the unity of technical design and management policies within the whole system. Although the two complement one another in systems design, they are separate and distinct. This separate recognition also re-emphasizes the staff rather than policy-making role of the analyst.

A management data program is understood to mean a separate subsystem designed to: (1) accept statistics and other data relating to the operations of the library; (2) correlate, synthesize, technically evaluate the data; (3) feed the results back into the operating system for its automatic modification or send the reports to management so that it can evaluate and modify library operations. This is an extension of the traditional library statistics but raised to a level where the data become an active agent in the control of the library. Management data is essential if the library is to exist as a dynamic system.

The need for a management data program (over and above traditional statistics) is frequently overlooked by both systems designers and management in the enthusiasm to get the "new system" operational. Sometimes the reasons for omitting management data go deeper.

The profession has practically no experience or background on management data systems design. Furthermore, traditional library statistics contribute very little. When contractors have over-all systems responsibility, they have little incentive to add costs that provide little evidence of a return. Also, management data could very well bring out embarrassing weaknesses in design.

The importance of evaluation data in library systems today cannot be overemphasized. First of all, every new system today is a prototype even if it is operational. We simply do not have adequate data gained from experience in order to design permanent systems, especially when automation is heavily used. Secondly, much of the data needed for making design decisions can only be obtained from observing an operating system. Yet we cannot get the data until the design is complete. In time, this first reason for management data will diminish in importance. The second reason will have a continuing role in the library.

Many byproducts of the nation's large research and development expenditures have relevance to problems in information handling. Frequently these advances are cloaked in the robes of the physical sciences, engineering, and mathematics. The analyst should not only keep himself aware of possible contributions from these other disciplines, but he should also translate these relevances into meaningful library terms. This article is such an example. It is an attempt to take some characteristics of industrial engineering, operations research, and computer technology, relate them to librarianship, and synthesize them into the profession. The analyst should do this continuously for the institution with which he is attached.

The analyst's technical librarianship bilingual capability places him in a unique position to act as technical liaison for the library. This duty comes to him almost by default.

Just as coordination between the new system and library policy is a part of design, so also the training of operating personnel is part of implementation. Beyond this, the analyst has a continuing responsibility to keep the operating staff and management up to date on new products and techniques being developed. This responsibility is especially significant at the present time, since most librarians have had no experience in technology and systems methods. Library schools are just beginning to integrate these newer tools into the profession. The systems analyst can serve a continuing education function to older staff members and at the same time feed back data to the schools for course and curriculum improvement.

## CONCLUSION

This job description has been formulated independent of any particular institution. Special institutional requirements would most certainly impose local variations. For example, a university or large public library system with a staff to assist the analyst would probably emphasize managerial ability and ability in the art of seeing both library and technique points of view. On the other hand, the analyst for a small college might be a jack-of-all-trades. No matter what emphasis is placed on individual needs, however, it should be remembered that systems analysis is never merely part of what has been described in this article; systems analysis rather encompasses it all. Also, regardless of local variations, the systems analyst must always have one quality. He must be a catalyst between librarianship and technology.

Finally, it is probable that until the library profession solves the problem of training librarians in depth in both librarianship and technology, it will be necessary to seek analysts outside the profession. Perhaps this is as it should be. ■■

# SYSTEMS ANALYSIS
# IN LIBRARIES:
# THE ROLE
# OF MANAGEMENT

*F. Robinson*

SOURCE: From *Interface; Library Automation with Special Reference to Computing Activity*, edited by C. K. Balmforth and N. S. M. Cox, "Systems Analysis in Libraries: The Role of Management" by F. Robinson, Cambridge, MIT Press 1971. pp. 101-111. Reprinted by permission of the publisher. Copyright © 1971 by Oriel Press Ltd.

There seems to be a strong view that systems analysis is a subject that can be discussed with a special relationship to particular interests. This is not true. One can properly discuss systems design in a particular field; a number of papers of this type are presented elsewhere in this volume. These are the papers on British *MARC* (Coward), and on the state of progress of computer applications (Cox). It is also possible to discuss problems or situations met during particular studies and these can be concerned with library systems. The value of knowing 'how I climbed Everest' is however limited unless one intends to adopt the same objective and the same route. It may be useful, when planning to climb K3 to know the climbing techniques used to overcome the conditions met on Everest and to be able to identify similar conditions elsewhere. Knowing the location of the ice fields on Everest will be less useful. In the same way descriptions of library studies, the situations in which they were made and the solutions applied are often extremely interesting. They are of limited value in planning studies in other libraries and are not descriptions of systems analysis.

This paper presents systems analysis in that way. The role of the systems analyst will not vary significantly when studying warehousing methods or library routines. It is his job to study an existing system in such a way as to be able to design the most efficient replacement system. The conditions may differ—the objectives of the unit being studied could be tremendously different—but the problems to be them recognized and techniques used to overcome might well be the same in both cases. This should not be taken to indicate indifference. The initial study of any new project will include a study of the problem area. This gives a general impression of library organization, traditions and practices, which serves as a background to the detailed study the analyst carries out and, as this progresses, so his knowledge increases. The knowledge gained by analysts in particular areas of operation explains why many become specialists in the study of those areas. However, many analysts move from problem to problem successfully and they are able to do this because the essential qualities for analysis are patient investigation and careful design.

The analyst, therefore, has two basically simple things to do: first to find out what happens now, and then to work out how to do it better. This is difficult when the situation being studied is complex and techniques have been developed to assist in both the analysis and the design of systems. The techniques range from the apparently obvious (e.g. interviewing) to the apparently obscure (e.g. decision tables). He also has knowledge of ways in which things can be done to help him in designing the replacement system. Nowadays this includes knowledge of computers and methods of using them, but if this is the analyst's only expertise he will almost certainly produce stiff, unworkable, clerical systems.

There is a temptation to discuss techniques, and computer system design only. It is at least as important to consider the impact of system analysis on libraries and particularly on the library management.

To do this is really to take a close look at the role of the 'user' department in the development of a new system. It is sometimes thought that the role is a passive one, but this is clearly not so. Library management must play a part in the following aspects of a project:

    1. Selecting the application;
    2. Determining the objectives;
    3. Project estimating and planning;
    4. Studying the proposals made and specifications developed;
and 5. Controlling the project.

Each of these warrants separate treatment.

*Selecting the application:* In considering library operations the number of possible project areas is rather more limited than is the case in an industrial or commercial organization. Nevertheless, it is probable that at any given time there will be a number of operation areas that merit study, ranging from the redesign of commercial operations to research studies in bibliography. Assuming that resources are limited there are various factors which must be taken into account in selecting applications, and these are detailed below. More important, there are two approaches to the selection process.

The first approach is to make the selection only when resources become available to work on a project. At that time the potential projects are studied and the most suitable (based probably on the factors discussed below) chosen. This assumes that the choice is made by the management. It is possible for this prerogative to be lost either with or without the consent of management. If this has happened without consent (e.g. by an approach from a particular library section to the analyst staff) it calls for some control of the process. If management has abdicated its responsibility by allowing others (possibly analysts) to select applications for development, it is unlikely that the choice will be made with all the interests of the library in mind. Even if the choice is made by management, this method of selection has major drawbacks since it affords little time to study all the aspects of all the potential projects, and in particular because it is difficult to see the future pattern of development.

The second approach is based on the need to study this pattern and must involve management. It requires all potential applications to be studied as a special, but preferably continuous, exercise. This produces a long-term plan for the development of the library system. This plan should be continuously amended as conditions change and as projects are completed. Therefore when resources become available they can be allocated to the most appropriate project in the plan. A long-term plan is probably the only way in which an integrated library system can be achieved.

No matter which approach is used (and there may be libraries where the first is appropriate) the factors which govern the choice of application are still important. These factors will vary principally with the objectives of the library, and include:

1. *Cost savings—the basic economic considerations:*
   Projects are chosen on the savings they make in the cost of running the library;
2. *Services improvement:*
   The benefits of improvement to the service provided to library users are intangible and the choice between alternatives in this area will be difficult. It can be based on the number of users of the various services but this may exclude improvements which could lead to heavier usage of a particular section and also projects in important minor activities. The assessments of benefits will take notice of the numbers of users of the various services, any

potential for increasing the use following the improvements, and the needs of minor, important activities. In this field the work of the Durham University PEBUL (*Project for the Evaluation of the Benefits of University Libraries*) team must be mentioned (Hawgood and Morley, 1969).

3. *Long-term research:*
   As a matter of policy it may be decided that research offers considerable potential advantages and that some resources should be devoted to it. This is even more intangible than improvements to service. The alternative is of course to allow other people to do the work or to obtain external finance. Money will not necessarily provide all the resources needed; it will not provide, for instance, more library management time;

and 4. *External circumstances:*
   Changes in external circumstances (e.g. building changes) can make it important to replace certain systems which would not otherwise warrant attention.

Balancing these factors against the resources available either now or in the future is the only satisfactory way either to develop a long-term plan or select a project for immediate develoment. Anaysis is a necessary part in the process of proposing solutions and weighing alternatives. Systems analysts can usefully be employed to help library management in this problem.

*Determining the objectives:* Clear definition of objectives is an essential preliminary to the development of a project and is also important in the setting of a long-term plan. Without this clear definition it is unlikely that either the system or the plan will be satisfactory. This is obvious—the best system will not be appreciated if it is doing the wrong things. It is possible to exaggerate this for at a routine level the objectives may be so clear that the trap is avoided, but when one thinks of cataloguing, user services and other fundamental activities, the possibility of providing good answers to the wrong questions is fairly high.

A good analyst will try to avoid this happening. He can however be thwarted by the members of the library staff with whom he discusses the problem. The analyst will be looking for the network affected by the system under study and will consider for what purposes the information which it generates is being used. He will also be looking for external factors influencing the network. This cannot however be done by the analyst in isolation. The library management must be clearly involved, and the staff made aware of the importance of the job.

The main hurdle for library management is an over-confiden

assumption that the objectives are obvious. This may be so in the routine operations, as has already been discussed, but in the more abstruse areas of the library it is probable that the effects of tradition, academic variations and changes in staff procedures will have obscured the original objectives without identifying any new ones.

In library catalogues, for instance, the effects of changes in standards and practice over the years will make the definition of the objectives of the cataloguing service difficult. Many of the changes will be associated with changes in staff and some will have been made because of changing conditions. Whatever the causes, the effect of change without planning is to obscure the purposes of the work.

The task of defining objectives has the advantage that it allows management to consider whether the current objectives should be retained. It may be easier to define existing objectives but this carries the risk that the needs of some small but important areas may be overlooked.

Whether one is defining current or setting future objectives, it is essential to:

1. look at the objectives of the total organization;
2. determine how the particular unit contributes to the total;
3. define the internal objectives of the particular unit;

and 4. consider any effects outside either the unit or the total organization.

For example, a library may be attached to a research organization and have its main functions determined by that fact. It will have some internal objectives principally in the maintenance of records for future use and could well have some external links with other research libraries. The purpose and nature of all these will need to be closely investigated. At present the objectives of many libraries are ill-defined, particularly in academic institutions. Because of this, and because the definitions are fundamental in systems development, library management must be prepared to play a leading part in the study.

*Project estimating and planning*: Estimates must be prepared for the cost of developing and operating the replacement system. Operating cost estimates should be considered with the proposals to which they apply and may be an important factor in the final acceptance of a proposal. Estimates of the costs of developing the system should be considered at the same time and library management should insist on these being realistic, since failure to produce good estimates of development costs can be a major cause of dissatisfaction. In computer systems many of the estimates will be difficult to challenge. In those parts of the project linked more closely with the library however the estimates should be studied carefully to ensure that they are realistic. This is particularly relevant in such things as the conversion of existing records and catalogues into computer-usable form.

Costs are easier to assess, but rather less important in non-computer systems. This applies also to project planning; in non-computer systems planning is directed mainly to a smooth change-over from the old to the new. One aspect which will require planning is the change to records and files which may be necessary in a new non-computer system. In computer systems the complex process of development and the more prolonged period over which it takes place make planning essential.

Library management must be involved in this planning. The planning target is of course the implementation date. The major stages preceding this are:

   Systems Analysis and Design,
   Programming and Testing
and Trials and Parallel Running.

Some elements in the planning and performance of each phase will require effort by the library.

In the analysis and design stage it is important for the eventual success of the project that the existing system is properly studied and management should ensure that the planning makes adequate provision for this. In particular the study will need the active participation of at least some member of the management and the plan cannot be achieved if this time is not made available. In this stage plans will probably be prepared for the conversion of the major files. This is vital to the success of the overall plan and management will be in a good position to help in both the planning and performance of the work.

If project objectives have not already been decided, analyst time must be allowed for them to be finalized before the design of the replacement starts.

Planning of programming and testing is obviously a matter for the computer specialists. The library management may be able to assist by ensuring that plans for the preparation of test data are realistic. During this stage the selection and training of staff for their new or revised duties will also be proceeding. Once again it is important to make staff available and this requires forethought. It is usually made more difficult by the need for work to be done on converting the system files and additional staffing should be provided if possible.

The planning of trials and implementation is a complex matter and many decisions will be taken as the stage develops. As the work depends on the needs of the new system it cannot be planned until the design stage is complete. Again there is a strong need for additional staff to be available because of the possible need for parallel running and for the checking of results.

It should be remembered that planning is an essential preliminary to estimating and that the accuracy of both will be improved as work on the project proceeds. This indicates that firm implementation dates should not be fixed before the project starts, and that management should retain the right to cancel the work if costs are escalating.

A major responsibility of management, which must be considered during planning, is for the well-being of the staff concerned in the system changes. This is affected in several ways. The staff may feel that their future is affected by the system, and that their skills which have been valuable for years may no longer be required. In many cases this will not be so and management must make this clear to avoid unrest and lack of interest in the scheme. In those cases where staff redundancy is inevitable this fact must be made clear and steps taken to minimize the personal disturbance this causes. Lack of interest will adversely affect any timetables made and this factor can be easily avoided by a little forethought.

The need for staff involvement in the project goes deeper than this however. It is important to remember that much of the knowledge of the existing system and of the needs it fulfils is held by the staff operating it. It is the analyst's job to find out about the existing system and this means close contact with the staff. Management can help by making it clear that the knowledge and opinions of the people will be respected and taken into account when the new system is designed. Management must also help in assessing what training is necessary to fit people for their changed duties. This, in particular, needs careful planning.

*Studying proposals and specifications:* As the project develops the analysts will submit to the library management various proposals for and specifications of the new system. The proper consideration of these is the responsibility of library management. Failure to do so will mean that the management cannot be sure that the proposals meet their requirements and this could lead to an unsatisfactory system. Proposals will probably be submitted at the end of the feasibility study and when the new system has been designed.

The first of these documents is a system proposal or feasibility report, following a survey of the operations area to determine the practicability of the application. The system proposal contains a summary of the existing procedures, a statement of the information seen to be required and a proposal for the replacement system with notes on any alternatives considered. The document is a proposal for further work to be done. It cannot be a complete design based on full analysis as detailed work cannot have been done. Management should look at the proposal in this light. In respect of the existing system the document will show whether the analyst has discovered the facts and this should be verified either by the management or by the library staff. Errors of detail are not so important as misinterpretation of the general structure. Failure to appreciate the basis of the existing system makes any proposals meaningless.

Library management must ensure that the information requirements are identified correctly. The information requirements are to some extent an expression of the system's objectives, insofar as these can be stated as information stored or reported. In projects such as book ordering systems they will include the routine reports. Such a project will tend to have three levels of information requirements. At the lowest level these will include the orders and the lists for checking receipts. The intermediate level might have summaries of books ordered on a particular day, possibly by university departments or library sections, so that the ordering supervision can check that allocations are correct.

At the highest level the system might summarize current order levels, include budget checks, and numbers of books on order by class with cumulative figures of holdings. The interpretation of the requirements at the highest level is the most difficult. This difficulty is also apparent when considering systems dealing with library services such as cataloguing. It is up to the library management to ensure that the analyst's interpretation of these requirements is valid.

Finally, the proposed system must be examined to ensure that it satisfies these requirements and appears likely to be satisfactory in operation. It is often difficult to comment on the operation of the computer element of the proposal (this difficulty is discussed more fully below) and this often prevents users from commenting on the non-computer operations. As the latter most affect the library it is important that these parts of the proposal are studied for simplicity, effectiveness and practicability.

Acceptance of the system proposal (and its associated cost estimates) is authorization for full system analysis and design to be carried out. At this point the outline design has been accepted. This will be completed and possibly amended in the light of the full analysis. If the scheme is significantly amended a revised proposal should be submitted. Normally the amendments are slight and eventually the system will be fully designed and a complete specification submitted to the library management.

The detailed system specification of a major project is a considerable document. It will contain layouts of all the reports, notes on procedure in the library, recommendations for organization and full estimates. Clearly by this time a lot of work has been done and heavy costs incurred. While rejection is a serious step this fact should not deter the library management from giving the specification a very close examination. This is the first time that the full system has been available for study and a detailed specification deserves the study. In practice senior library staff are unlikely to be concerned with checking detail but should delegate this to the supervisors of the functions concerned. Library management should ensure that the principles agreed during the preparation and discussion of the system proposed have been implemented in the specification. The supervisors must ensure that the functions currently carried out manually are adequately covered in the specification. Satisfactory answers to omissions must be obtained from the analysts; these should have been discussed during the design period.

Finally, the library management must decide whether or not the system specified is acceptable. Discussion is inevitable and valuable but eventually this must come to an end with a decision to proceed or cancel. This decision can only be taken by the management.

There is one final chance for the management of the library to examine the project formally before it is implemented. This comes when the trials of the new system have been completed and some form of parallel or pilot running has been carried out. These final tests of the system will inevitably bring to light minor faults. There may be a fair number of these errors; the causes may be programming, design or inadequate specification of requirements by the library staff.

Such errors are unlikely to invalidate the basic principles. If they do management has to decide whether to write off the whole project or to re-start at the beginning 'sadder and wiser men'. This is an extremely serious decision and the need for it reflects badly on the organization and control of the project. In practice a more serious risk is that the number of errors leads the management to believe that the principles are wrong. This belief is likely to be fed by those who tend to exaggerate the effect of the errors. It is important that management take a balanced view of this, weighing the probable optimism of the project team against the possible pessimism of the people who are to operate the new system.

In most cases the worst that need be decided is whether the operational date of the project should be deferred. If the problems are likely to cause delay of more than one to two months, then suspension of parallel running may be necessary. This might also be necessary if the nature of the errors is making the operation impossible. If suspension is not considered necessary the extension of the trials will probably be beneficial in the long run by getting the staff fully conversant with the new system.

These are the three major check points during the development of a project. After some months of operational running the system must be checked against the specification. This audit of the system will be reported to the management and should include suggestion for corrections and amendments.

*Controlling the project:* It is usual for the leader of a project team to be a systems analyst. He should, if possible, be supported by a member of the library staff seconded full time to the project. The team must report and be responsible to senior management. It has been the usual practice for this higher contact level to be the management of the computer department. This has the advantage in that the technical aspects of the job are more readily appreciated. Nevertheless one of the most important factors in a computer development is the involvement of library management. If there is a feeling in the library that the development is something which 'they' are doing, involvement is unlikely to occur.

There are two alternative ways of tackling this problem. The first is to have a steering committee consisting of management members of both departments to which the team reports progress. If this is to be the only control the meetings will need to be fairly frequent. It is better to have the team report to the chairman of the committee whose responsibility it is to call meetings if necessary but who is authorized to manage the project in normal circumstances. The chairman of the committee will normally be the manager of the computer systems department but there is no reason why he should not be a library manager.

This leads to the second method of tackling the problem; namely to make the team report to the management of the library with functional responsibility to the computer department. This will clearly lead to active involvement in the project. It has however the disadvantage that library management may get too deeply drawn into the technical computer aspects, and at the implementation date, be too involved in overall project matters to concentrate on the situation in the library itself.

The steering committee approach has much to offer. Properly used it will ensure that both sides are aware of the views and problems of the other and can build up a relationship between the management to match the close links that the project team should be establishing at the operational level.

*Summary*: This paper has discussed some particular aspects of system development which affect library management. There is one outstanding moral: the development of a new system for a major library operation is going to affect library management. The impact will be greater if management do not interest themselves from the very beginning of the job. Ideally this interest should start with the production of a plan for future library system development but it must not be delayed beyond the time of the system survey. Early involvement will not guarantee a trouble-free development but will help to achieve this and will also do much to ensure that the operational system is what the library wanted.

# III

# IDEAS, CONCEPTS
# AND PROCEDURES
# WITHIN
# SYSTEMS ANALYSIS

**Decision-Making Aids**
**Management by Exception**
**Sampling Techniques**
**Questionnaires and Interviews**
**Costs and Costing**
**Work Measurement and**
      **Work Simplification**
**Measurement of User Requirements**
**Flowcharting**
**Library Research and**
      **Systems Analysis**

## INTRODUCTION

This section is provided to illustrate many of the procedures employable in a systems study. The aspects of systems analysis included are:

**Decision-making aids:** The logic of systematic analysis and its value in the decision-making process as contrasted to "intuitive" arrival at decisions.

**Management by exception:** An interesting principle under which management is notified only about what it *needs* to know.

**Sampling techniques:** The time-saving method of studying a selected few items instead of everything in systems analysis when dealing with large numbers of observations.

**Questionnaries and interviews:** The finding out of what is happening with either people or things in a systems study.

**Costs and costing:** The important principle of cost-control. What does a certain procedure cost from a man/material viewpoint and what, if any, are the benefits of the procedure.

**Work measurement and work simplification:** The techniques of measuring job performance and of refining it objectively.

**Measurement of user requirement:** User needs are an important part of any library system and need to be known for effective improvement of library services.

**Flowcharting:** The useful method to depict what happens in a system and what could be happening in a refined or newly designed system.

**Library research and systems analysis:** Throughout a systems study, research opportunities frequently occur. The information discovered in the systems analysis of certain procedures can be of significant value in better understanding how libraries serve their users.

A system study may involve most, if not all, of the above concepts. Also, apart from a formal systems study, certain of the procedures can be of value in day-to-day library problem-solving.

Errett W. McDiarmid's article is selected from the January, 1954 issue of *Library Trends*. The entire issue, dedicated to "Scientific Management in Libraries" and edited by Ralph Shaw, is a good starting point for the beginner in systems analysis. Errett McDiarmid indicates the value of the scientific method in administrative decision-making. To the author, this application has particular value in that whatever aspect of administration it is used, it can serve to clarify and quantify the most difficult and subjective problems. However, it is not seen as a cure-all or a formula by which answers can be found magically but rather as a valuable aid in analysis so that better decisions can be made.

A. G. Mackenzie reports on the preliminary and on-going work in mathematical modelling at the University of Lancaster. It was decided at the outset of their operational research that a mathematical model of what happens in a University library would be the best way to analyze its effectiveness. Once the library is represented in mathematical terms, it is thought that further tests of projected changes in a system can be evaluated by "plugging" different items into the model and seeing that over-all effect. An important concept throughout this selection is the improvement of library service for the user through a better awareness, by systems analysis, of the user's needs.

The "management by exception" principle is frequently applied by librarians in their everyday responsibilities. As with many philosophies in management, some people do come by it naturally. For those of us who don't, the literature of management can be helpful. Lester Bittel suggests that managing by exception contains six elements: measurement, projection, selection, observation (control), comparison and finally, decision-making. The principle has meaning particularly in the revision of old systems or the design of new. It is at this time that parameters should be built into, for example, a reporting system that will alert a library department head of his staff's best and worst efforts.

Dougherty and Heinritz's selection from their well-known text *Scientific Management of Library Operations* addresses itself to the time-saving device of sampling and some of its guiding formulas. Through sampling the authors show that studies involving large groups of files, etc., can be done expeditiously and with reasonable accuracy. Explanations are given for the use of random digit tables, random permutations, sample sizes and either/or sampling. Sampling has direct application in system study since rarely is there time to observe the complete universe of operations in any situation.

After one decides the number of people to be interviewed or the number of observations to be made the next task is that of data collection. Maurice B. Line's selection from his book *Library Surveys* delves into the various techniques of collecting information such as the on-site survey and unobtrusive observation, the diary method, etc. The author treats in some depth the construction of questionnaires, particularly stressing problems of question-wording, and the scheduling of interviews, either by mail or in person. With quotes such as, "Reader watching can be as rewarding an occupation as bird watching" his essay makes for not only enlightening but also entertaining reading.

The answer to the question of what it may cost to employ a worker is provided in the concise selection by Burkhalter. The concepts in his "memo" are those applied in the studies using systems analysis done at the University of Michigan and published as *Case Studies in Systems Analysis in a University Library.* What the memo brings out is that there are considerable costs in an employee's hourly wage beyond the stated rate.

The one group that has taken Ellsworth Mason's satirical essay most to heart is that of computer-oriented librarians. And yet, this selection strongly implies the need for tough-minded cost-awareness regardless of the object, be it computers or pressure-sensitive labels. The original article had the following abstract:

> In which are Exposed the delicious Delusions of those will-o-the-wisps; the Echoes in computerization of Phrenology, Haruspication, and other discredited Ancient sciences; and the moral and Mental decline of our Profession.
> "If it costs you twenty-five percent more, will you stop it?"
> "No,"
> "Why not?"
> "Because we believe that sooner or later all libraries will automate."
> -From a real-life, absurd conversation.

John Kountz provides insight into the inner workings of cost analysis. He describes the steps of costing from identifying and quantifying the process to the on-going review procedure. Of particular value is the section on what each step or element in an operation requires in the way of manpower, supplies, equipment, supervision and space and their effect on costs. An irreverence at times, and annotated bibliography is offered under the caption: "Zens and Kelving: An Open-Minded Bibliography on Cost Techniques for Libraries."

The interesting topic of time study is described in another Dougherty and Heinritz selection. The authors illustrate some of the tools, guidelines and formulas that can be employed in time studies. The approach is seen as valuable because of its ability to establish time standards for library tasks, and since time is money, the cost of such operations can be ascertained. Also, employee standards (e.g. average book shelving rates) can be arrived at through time study. The authors suggest that the pursuit of highly precise (industrial) standards in library time studies is not necessarily desirable but that general measurements should produce adequate and applicable standards.

The "survey of requirements" as described by Chapman *et al.,* is seen as the first part of the analysis phase in a systems study. During it, the various demands placed on a system and manifested as files and/or reports are listed and explained from the total systems viewpoint. Through the use of the *Worksheet for the Survey of Requirements* the survey should provide knowledge about such aspects as the "information definitely required from each person at each level in order for him to fulfill his functions within the system" and also to discover any unnecessary (redundant, etc.) requirements. A major feature of this approach to work study is that it is largely self-administered by the employee for summation by the systems analyst.

The item from the Association for Systems Management by Wallace presents a practical ready-to-apply discussion of work simplification. It is defined as the process of "systematically eliminating unnecessary work and streamlining that remaining so that it will move faster and better." The author sees the process as involving four steps: selection of area to be simplified; the breaking of the current procedure into its components; charting the tasks on a *procedural analysis worksheet*; and, then eliminating tasks and streamlining the procedure wherever possible. The *procedural analysis worksheet* (flow-process chart) is fully defined and used in an example.

Work measurement for libraries is evaluated in the article by Woodruff and its adoption is advocated. The advantages of work measurement are stressed and these can be seen as particularly applicable today in the increasing climate of library accountability. Elaine Woodruff explains the usefulness of work measurement mainly in terms that this approach can help quantify and as a result control that which has largely been left up to intuition or chance. An illustration is the keeping of statistics on work flow, based on standards, that can be useful in predicting when any slack will appear in an operation. When the slack occurs, plans can have already been made to deploy the unit effected to other tasks. The development of a work measurement program is described in detail.

The figures in Voos' study using micro-motion techniques are generally regarded as reliable examples of some standard times in library clerical activities. Among the author's conclusions are that: standard times for common clerical routines can be predicted in many circumstances and, once available as in industry, standard time data will be of value to management in arriving at costs and work standards. Specific examples of possible improvements attributed to time study are listed. These include the statement that "typing can be done more effectively by trained personnel who spend full time on" it than by fragmenting the responsibility among a group of workers with different tasks.

Carole Bare's paper reports on a user study at a large special library made by the author. Although Dr. Bare's report centers on this particular user requirements survey, the rationale and procedures set forth "can be used if tailored to the special needs of each library." Sampling, of course, would be resorted to in the case of large populations served by a library. The purposes of a user study are presented and the sequential steps outlined. Consideration is given the following related factors: who should conduct the study, identification of all library activities and procedures impinging on user services, enlistment of governing administrative support and importantly, the enlistment of library staff interest and support. The major study methods—the questionnaire, the diary and the interview—are described and evaluated with examples of each given. Concurrent use of the three methods is recommended, each applying to a different level of information and specificity. The importance of reporting findings to the library's users with all speed is emphasized as is a repeating of the study from time to time to keep communications open between the library and its users. Computer tabulation of data is of considerable help in reporting findings promptly to users.

C. D. Gull's paper is mainly concerned with the techniques of flow charting library operations integrated with the older techniques of library administration such as staff manuals, job descriptions and so forth. This integration is illustrated by Figures 1-4. Professor Gull's directions for making block diagrams and logical flow charts are those developed by him for library school classes. How to prepare these graphic representations of systems analysis, evaluation and design, is set forth in step by step detail including definition and purpose of each. The use of the various charting symbols also is detailed with emphasis on standardization and consistency in their use. The pitfalls in constructing flow charts and how to avoid them are enunciated continously throughout this paper. It is pointed out that flow charting is an uniquely important tool for analyzing, evaluating and designing operations for effectiveness and efficiency. This is true whether such operations are to be manually performed or computer-based. An under-scoring treatment of this subject is found in Chapman *et al, Library Systems Analysis Guidelines*, Chapter 6. (Wiley, 1970).

Cammack and Mann report on a pioneering and currently significant instance of library research using machine-readable data specifically gathered to yield timely information about the use and non-use of the library; to test the library's circulation and service systems design for effectiveness and efficiency; and to determine how closely collection development was geared to user requirements—indeed a "total systems" approach through circulation data. The data used can be found in most circulation records but in their study EDP is used to extract the required information. To do the same thing by manual methods would be too time consuming and costly in terms of timeliness of the information and penalities on current operational requirements. Although their analysis is confined to an university student environment, the questions asked and answered seem adaptable to the needs of any library: more efficient and economical scheduling of service personnel; identification of the most demanded portions of the library collection from time to time as user demands change, identification of the classes of readers and their proportional distribution, and other information contributing to the increased effectiveness of services.

SCIENTIFIC METHOD

AND

LIBRARY ADMINISTRATION

*Errett W. McDiarmid*

SOURCE: From Errett W. McDiarmid, "Scientific Method and Library Administration," *Library Trends* 2 (Jan., 1954) pp. 361–367. Reprinted by permission of the author and the publisher Copyright © 1954 by the Graduate School of Library Science, University of Illinois at Urbana—Champaign.

SOME PEOPLE VIEW the terms scientific method and library administration as incompatible. According to their view scientific method is something that may be used in a vague area of business or manufacturing or industry, but has no place where humans are concerned or where values are important and results are gauged in terms of the conveyance of ideas rather than of items. The assembly line can be studied scientifically and its cost accurately appraised, but one must not use measuring devices on things as intangible as information received or understanding gained. There are those who do not agree, and as with many differences of opinion, both sides of this argument have points in their favor. Only after one defines his terms and clarifies the issues involved is it possible to look objectively at the questions—does scientific method have any place in library administration, and if so, what place?

Library administration, broadly defined, includes all the things that go on from the time a group of citizens establishes a library to serve its needs, to the moment that library does something which helps a citizen. The objectives of the institution, and its methods, its facilities, and its personnel—all are involved. To paraphrase a well-known definition, library administration is as much concerned with men and materials as it is with their use in fulfilling accepted purposes.

Implicit is another point not always recognized. Library administration is charged with accomplishing a job, but more than this it is expected to perform its task as well and at as low cost as it can. Nor are these aims antithetical; for a good job can be done economically, even if in ways that differ in terms of expense. Obviously good administration chooses the cheaper manner, or, if it selects a more costly way, satisfies itself that the latter actually is better.

This, in effect, poses library administration's major problem: given certain things to be done, what are the ways in which they can be done best, and what will be the cost of each? And rarely is the answer crystal clear. Even for a very simple or limited function, there are diverse degrees of excellence to be achieved and numerous variations in expense. One does not find that the most excellent way always is the cheapest, nor that the cheapest is always the poorest. One way gives certain advantages and involves certain outlays, while another entails other merits and costs. How balance these facts, and how decide what is best?

There is a considerable body of folklore regarding administrative "hunches." Prominent in this is the belief that a chief's feelings generally determine his conclusions. With it goes the picture of the busy administrator who listens to a staff member, then stares out the window for a moment and barks out a decision. The philosophy seems to be: "One has to have a feeling for this sort of thing, a hunch, and once the hunch comes settle the matter and go on to the next." Naturally, many questions come along in a busy administrator's day that have to be decided quickly—where someone has to weigh alternatives and reach a judgment. But administration by hunch soon becomes administration by guess, and society has yet to discover many people who can guess correctly even a fair proportion of the time.

Actually, of course, the popular stereotype has more to it than appears above. For even the administrator who claims he is acting on his hunches is doing much more than that—he is using, perhaps without being aware of it, a body of experience, facts, and information built up over years. He may not consciously isolate, tabulate, and total up all that he brings to bear on a given problem, but his decision is inevitably influenced by it. In one degree or another he is applying scientific method to administration. That method is nothing more or less than the collecting, evaluating, and applying of facts. That it is done in the cerebrum rather than on a Friden makes no difference.

But this too is an oversimplification. Though scientific method does involve the assembling and evaluating of data, it implies much more than recalling those facts that happen to come to mind when a certain problem is under discussion. Such a procedure would be fragmentary, and subject to the chance of memory or influenced by the recency with which something had been reported or discussed. One could never be sure that all the pertinent evidence had come to light, or was used in reaching a decision.

The real purpose of scientific method, thus, is to see that the best and most complete factual information is discovered, made available, and brought to bear upon a given problem. To whatever extent this process is complete and objective, scientific method is being employed in library administration. Accordingly we now can answer the first question posed above—does scientific method have any place in library administration?—with an unqualified yes.

The next question then is, what place can scientific method occupy in library administration? Where can it be used, and what will it achieve? Though as implied above scientific method should not be thought of as one thing in one area and another elsewhere, for purposes of clarity it may be desirable to separate two of its aspects: (1) scientific method as employed in appraising objectives and programs; and (2) scientific method as a tool for superior administration. Later papers in this issue will develop the latter theme extensively; the first is taken up in the following paragraphs.

How can the scientific method help the administrator decide matters of broad policy, of program, or even of objectives? First, the scientific method as a way cf studying problems helps the administrator appraise even qualitative and subjective questions more carefully and accurately than otherwise would be possible. For instance, the question arises whether a library should undertake a new line of activity. How can the scientific method help in such a situation? It can aid at least by outlining clearly and carefully the steps to be taken in arriving at a decision.

These are:

1. Defining the problem.
2. Identifying and stating the assumptions.
3. Breaking the problem down into its component facets.
4. Assembling all the pertinent facts available.
5. Collecting and analyzing facts not already in hand, but needed.
6. Evaluating and appraising the facts and their relation to the problem.
7. Constructing a hypothesis regarding the best solution.
8. Testing the hypothesis in the light of various aspects of the problem and the pertinent facts.
9. Final analyzing, and the reaching of a conclusion.

Perhaps a simpler way of discussing this subject would be to say that the scientific method means the application of logical reasoning to a problem. But unless one remembers clearly what goes into the process and takes great care to see that no step is omitted, he is likely to fall back closer to administration by hunch.

It should be clear that at every step in the above process, judgment and intelligence are required. The scientific method is not like a chain reaction, where one step leads automatically to the next until an end result is achieved. The person employing it must decide how to define his problem clearly and completely, and a considerable leeway of judgment is possible. This judgment, however, is much more likely to be sound and correct if one consciously sets out to approach the problem systematically. Thus even where there is a necessary dearth of factual information and of quantitative data, the method helps.

Second, as implied above, the scientific method aids the administrator by refining and clarifying the problem. Very few questions exist upon which no factual information is available or could not easily be obtained. Even in the most theoretical problem, where broad objectives and purposes are the major consideration, there are likely to be some data that would illumine some aspect of it, however minute. If this can be done within the limits of time and money available, and the facts necessary can be collected, obviously the decision is simplified just that much. To cite one specific example, if we know how much it costs to circulate phonograph records, or how many people have phonographs on which to play them, we have advanced one step toward learning whether or not to inaugurate a program involving such records. This does not mean that collection of a few facts will decide an issue, but it does mean that they can help to make a decision more sound.

Third, scientific method will contribute to rendering a decision more widely accepted, or, if it is to be made by someone other than the person charged with studying the problem, will greatly increase the likelihood of the right decision being reached. Using the example cited above, suppose the librarian decides that it would be desirable to circulate phonograph records, but he must have endorsement by the library board. Of course the board will weigh carefully the librarian's statement of purposes and objectives and his appraisal of the good to be accomplished, but it will enter into a decision with more wisdom and certainly with more enthusiasm if it has even a few facts to guide it. For instance, if there can be a demonstration that the cost will be so much, and that this is well within the library's resources, the chances will be increased that the board will approve.

Finally, the use of the scientific method can assist the administrator in strengthening morale. If it is known that a given matter has been studied objectively, and that pertinent facts have been collected and analyzed, a decision is apt to be regarded as sounder than otherwise would be the case. Reaching a conclusion by the scientific method may not be necessary to satisfy those who agree with it, nor will such procedure always convince those opposed to the decision, human nature being what it is. But it may elicit and justify recognition and acceptance. One can accept a judgment even contrary to one's interest if a modicum of facts points in that direction.

How can the scientific method serve as a tool for better administration? As suggested above, this subject is developed in other papers in the present issue and need not be dealt with extensively here. Workload analysis, standardization, time and motion studies, all of which are treated in other papers, illustrate the application of scientific method to administrative problems. The point to be made is that in any aspect of administration there are facts to be obtained, analyzed, and evaluated, and that when they are utilized, better decisions can be reached than could be anticipated otherwise. However, lest the foregoing imply that application of the scientific method, or the use of scientific management, offers an easy solution to all problems in libraries, several cautions may well be noted.

First, it cannot be overemphasized that scientific method is an aid to administration and not a substitute for it. No matter how detailed the facts, or how complete the analysis of a given matter, someone has to take the final action regarding it. Indeed, unless there is some happy utopia where all data are easily available, someone has to make a decision as to which facts to collect. For rarely is there lack of question as to the relative importance of the materials needed. Often one has to admit that, although it would be nice to have certain information, to obtain it would be unduly difficult and expensive. Use of the scientific method should help an administrator to reach a wise decision, but it will not relieve him of the necessity of making a decision.

Second, use of the scientific method can be overdone, or, perhaps more accurately, improper use of the scientific method may handicap an administrator. What is referred to here primarily is the tendency to seek facts for the sake of collecting facts, rather than for their relevance to a given problem. It is very easy, particularly when one does not have to do the job oneself, to assemble data indiscriminately. This is not wise, however, unless careful judgment enters into the decision regarding the relevance of the data before they are assembled. Perhaps the most common criticism of the questionnaires developed by graduate students for their thesis problems, or indeed by administrators when they wish to find out how others are doing things, is that the pertinence of the information requested has not been seriously considered. Why collect it? What will one have if it is obtained? Of what use will the facts be in solution of a problem?

It should be clear that caution in deciding what facts to collect does not imply that one must know in advance what they will reveal. It means simply that one must be sure that they will be relevant, and will aid in resolving a problem.

In the judgment of some, librarians have erred on the side of neglecting factual information. Admitting that librarianship is a profession where there are many intangibles, and for which objective data are not always easily available, one still can argue that this does not condone neglect of facts. Even if information is hard to obtain, it still should be sought where it can be secured and where it will be relevant.

A library's policies and programs are close to its *raison d'etre*. Routines, methods, devices, equipment—these are used to enable it to do something. They are fundamental to the work undertaken, and the better they are the better the results. Furthermore, because they are tangible, they seem to be best subject to application of the scientific method, or of the principles of scientific management. In a sense this is so, since we can measure more adequately the cost of cataloging, or the expense of photocopying as compared with other reproductive procedures, than we can measure the aesthetic satisfaction derived from reading a poem. But administration entails more than deciding the most economical way of carrying out a given operation well. It must regularly weigh the values derived from reading a poem, against the values derived from reading a novel, or a biography, or a magazine of current history.

Scientific method or scientific management will not give one the answers. Even if one had endless arrays of facts regarding the relative cost of circulating poetry and novels, regarding changes of attitude before and after reading, regarding the interests of readers, and regarding community needs, someone still would have to weigh and appraise the data. But if relevant facts can be obtained on even a part of a given question, if the problem can be analyzed carefully and critically, and if logical reasoning can be substituted for emotional impulse, scientific method can become a valuable aid to administration.

# SYSTEMS ANALYSIS
## OF A
# UNIVERSITY LIBRARY

*A. Graham Mackenzie*

SOURCE: From A.Graham Mackenzie, "Systems Analysis of a University Library," *Program* 2 (April, 1968) pp. 7–14. Reprinted by permission of publisher. Copyright © 1968 by Aslib.

In presenting some account of the work being done at Lancaster on the systems analysis of a university library, I wish to make it quite clear at the beginning what I do not mean. We are not concerned with "scientific management" as defined in the recent book by Dougherty and Heinritz(1) - time and motion study, work measurement or work simplification, the re-design of forms and stationery, and all the other paraphernalia of low-level industrial or commercial management. These admittedly have their place, but only a minor one: it profits a library little if its procedures are all perfect, but all directed to the wrong ends.

We are not at this stage concerned with information retrieval on the grand scale for two reasons; first, we do not believe that it can ever be an economic possibility for individual large general libraries to do this - apart from any-thing else, the problem of input is immense; secondly, we do not believe that it would fulfil all of our needs: it is possible to set up a simplified model of a library as a store of discrete bits of information, each bit being identifiable by one or several index-entries or "addresses". The user is assumed to approach the store knowing what he wants, and the problem considered is how to deliver to him what he wants quickly, cheaply and with certainty. We might conceive as a future possibility the establishment of some vast central store of information, computer controlled, accessible by coded instructions passing over telephone lines from any part of the country or of the world, and delivering information by some teleprinter or television technique. In a limited field, such a study has been done; the Library of Congress report(2) recommends an expenditure of some $70 million for the complete transfer to computers of its catalogue and internal processes. This is excellent news for all concerned with information retrieval, "conventional" librarians and information officers alike, but to my mind it is putting the cart before the horse: it seems risky to spend sums of money of this order on work which does not go to the root of the problem of library use.

We agree that it may become economic to use the computer-controlled central store to deal with the accelerating increase in the mass of discrete items of information; but such a store would fulful only part (and perhaps only a small part) of the functions of a library. Some people - perhaps as many as half the total(3) - come to a library with a precise question:- "What was the value of British imports of bananas in 1962?" But many others come with a more general kind of question:- "Has anything been written about banana imports which may throw light on my theory of commodity trade?" They browse, they leaf quickly through a book to see if it may help them, they follow up footnote references from one book to another; they are interested in ideas or methods of presentation as well as gobbets of information. There are strong indications that successful library use at times depends on serendipity - or lucky discovery - rather than on painstaking directed search. It is not clear that computer-controlled information stores have any relevance to this more general type of library use; in spite of Project MAC and Project INTREX(4) you cannot readily browse in the memory of a computer. Much of the work on scientific documentation - indexing and retrieval, abstracting, the information needs of scientists, the improvement of scientific journals, fundamental work on the nature of communication through language, and on coding and translation seems to us to have in mind the simplified model of the library as an information store to which precise questions are addressed. But library policy should be based on what people actually need from libraries. Comparatively little work has been done on this question; there have been many attempts to find out what the library user actually uses, but what he uses and what he needs are not necessarily the same things(5)-(7). The consideration of actual needs may imply a complete rethinking of the technical methods which are appropriate and economic. On the face of it, indeed, the printed book which can be held in the hand is well adapted to the needs of those who seek answers to questions which are general and not precise. But it is not easy to transmit, without introducing delay or inconvenience, and the mass of printed knowledge is too great, and growing too fast, to make it possible to assemble fully adequate collections in (say) 60 university institutions in the United Kingdom. This is the dilemma: and it is not answered by dealing only with the problem of retrieving and transmitting precisely identified factual information.

   The methods of use-study commonly employed have only a limited relevance to the problem stated here: they may be able to state with some accuracy what happens at present, and analyse the scholar's (rather than"scientist's" because we must consider the humanist as well) own evaluation of what he believes he needs; but this is only a subjective judgement, and will usually be made without a full knowledge of either the current situation or of potential developments. In information work it has often been shown in the past that the supply of a facility very rapidly creates a previously unthought-of demand, and the field has become so specialised that few practising scholars are in a position to understand the present, let alone forecast the future.

For these reasons, the University of Lancaster has in mind a long-term programme for the investigation of the research worker in relation to his sources of information, and the interaction between them. This undoubtedly varies from subject to subject and probably even from person to person. If a specific question cannot be answered immediately, is there a penalty in wasted time whilst the answer is obtained from another library? This will depend on the researcher's methods and whether he works on more than one project or aspect of a project at the same time. How does the penalty vary with the delay? Is there a cumulative effect so that the researcher becomes semi-frustrated and uses the library less frequently than he might? If he does, then is the problem of duplication of research severe and does the quality suffer? Is the quality of research reduced when browsing is restricted? Is there a limit to the amount of browsing which a man will do? Could the browsing be done at intervals in a regional library instead of in the local library? All these questions must be answered in a quantitative form. Once some generalisations such as these have been established, it would then be possible to derive some optimisations for such variables as the size, function and location of individual libraries.

For example, should an individual library purchase obscure books and journals if they are unlikely to be used frequently or by more than one man? Should there be a potential minimum usage for single books or should all the books on a topic be considered together? The results of such a study might be used to design a hierarchy of libraries, for which a balance would be struck between wasting money on purchasing "unnecessary" books and causing frequent duplication between libraries on the one hand, and wasting researchers' time and causing duplication of research on the other hand.

Such an investigation, however, still lies in the future; it is undoubtedly the most important section of the whole problem, but initially the University has restricted itself to a more immediate object, namely the study of a library system as it is. Little or nothing is known, in objective mathematical terms, of how existing libraries operate, and what has been published(8)-(10) is based on experience in the U.S.A., where budgets are much more lavish and operating conditions and users' needs are very different. In addition most

of this work has been conceived as a necessary prelude to some form of partial
or total mechanisation of the domestic economy of a library, rather than as
a tool to aid management in decision-making - an example of this attitude
is contained in the excellent pioneer study of the University of Illinois by
Schultheiss(11), which analyses in extreme detail the technical processes
involved in acquiring, cataloguing and issuing books, but ignores almost
completely the purposes behind all these activities. This limited sense of
systems analysis does not satisfy us; it needs to be done if we are eventually
to mechanise our processes, but in a different and more restricted - even
although a more detailed - way.

What we mean by systems analysis is much more far-reaching in its effects -
in a few words, it is the use of scientific method to study the effects of
managerial decisions in large-scale, complex situations. A librarian, like
any other administrator, has to make many decisions in his routine work;
generally at present these are the resultant of many conscious or unconscious
forces - his previous training and experience, his innate prejudices, the internal politics of his organisation, even sometimes the state of his liver!
Now the ecological system formed by a library and its readers is complex,
delicately balanced and expensive; therefore decisions taken without full
knowledge of the facts, and of the probable consequences of any given line
of action, are likely to result not merely in trial and error, but in trial and
disaster.

We need considerable justification for experimenting in vivo with a
library system which has achieved a reasonable balance; the potential dangers
are great. On the other hand, there is no such thing as the perfect library.
How can we improve, then, without risking damage to the system? Obviously
it is possible to tinker with small sections of a system - to redesign a form here,
change a loan period or a rate of fine there - but the effects on some other
segment of the library may be quite unexpected. Modern techniques of
operational research are at our disposal, including the one essential tool of
the mathematical model, which bears the same relationship to the object of
study - in our case the library - as does an engineer's force diagram to the
girders of a bridge he is designing.

How can we set about an operational research or systems analysis of a
library? The normal procedure for such an investigation, let us say in a
large industrial concern, is fourfold: first, to define the objectives of the
company (perhaps maximum long-term profit, within certain constraints
such as the law of the land); second, to flow-chart all the processes which

form the company, in whatever detail is required; third, to construct a mathematical model, using appropriate equations to describe each process or operation; and lastly, to use this model to optimise the company's operations so that the highest possible proportion of each of its stated objectives is attained.

Unfortunately, most O.R. studies are conducted in terms of profitability, where it is relatively simple to determine the objectives; but an academic library is a service organisation, and cannot look at its operations in this light. It is, of course, fairly easy to arrive at a first approximation of a library's purpose - the tentative definition which we have adopted is "to assist in the identification, provision and use of the document or piece of information which would best help a user in his study, teaching or research, at the optimal combination of cost and elapsed time"; but unfortunately this raises many more questions than it solves. For example, bibliographical training of students will result in increased library use, and thus in higher operating costs in book purchases or inter-library loans; and yet this training itself is also expensive in terms of salaries. If the sole criterion were financial, obviously we ought not to try to increase library use - but we should then be producing inferior students. Similarly, to invent an extreme example, we might interpret "at the optimal...elapsed time" in the definition of our objectives as meaning that we should maintain a helicopter and pilot to fly to the National Lending Library four times a day to collect loans for our scientists; the delay would thus be minimised, but at a cost which might prevent us from buying any books at all, and which in any case would far outweigh the value to the university of the readers' time which had been saved.

One solution to this problem which has been suggested is quite ingenious, and is in fact also being studied by a parallel investigation in the University of Durham: various aspects of a library's services (additional book purchase, improved inter-library loan service, improved reference service, SDI, free photocopying, extended opening hours etc.) are assigned prices which, although nominal, bear some relationship to the true costs. Members of the university are then asked to imagine that they have at their disposal a certain fixed amount of money and to declare how they would wish to spend this on varying combinations of the proposed benefits. Although this is a standard technique in marketing operations, we see certain objections to its use in a library situation; in particular, that those asked are unlikely to be able to visualise in any detail the results, good or bad, which would come from these additional services if they were provided.

It is becoming clear to us that there are serious difficulties in formulating an academic library's objectives in relation to its research readers, and hence in optimising its performance, unless we can find out more about the function and performance of the research worker himself, as outlined earlier in this paper. We are therefore trying to proceed to the second and third stages of the analysis, by flow-charting and by constructing a simplified model which can be made more sophisticated as our knowledge grows.

A few of the equations for the construction of this dynamically balanced model have already been formulated, and a start has been made on quantifying some of the variables which will have to be built into it. It is our expectation that when the model is complete, even in its simplified state, we shall be able to inject into it any number of differing managerial decisions and investigate how these will affect the system with the passage of time. (It would, of course, be equally possible to inject existing decisions to see whether they are likely in the long run to have the intended effect - the aim of this type of model is to simulate either backwards from the optimal end-state or forwards from any set of decisions, provided that the model itself and its associated computer programs are sufficiently sophisticated and initially represent all the required inputs and outputs). An example of the basic structure of one part of the model is given in the appendix, although since the project is still in its early stages this must be regarded as highly provisional and unsophisticated - for example we have not yet built in the space requirements for storage, staff and readers, or the financial implications of different grades of workers.

Simultaneously with the investigation of the technical processes of a theoretical model of a library we are looking at the actual provision of textbooks for students at Lancaster. We are perhaps in an unusually favourable position to do this, since there is a collection of over 2,000 volumes representing material which departments consider to be essential reading for undergraduates' lectures, essays and seminars. This collection is stored on closed-access shelving, and books are issued for periods of up to four hours or overnight; we can therefore easily analyse the use made of it (as a library which merely has a "reference" collection of this type of material cannot do), and, equally important, we can study with complete accuracy what we call the frustration factor - the proportion of enquirers who do not get the book they want at the first attempt. To our surprise, during two weeks just before an examination period, this proportion averaged only 11%, indicating that the system was working reasonably well. We have plans to correlate the use of this short-loan collection with a record of lecturers' recommendations to students, and hope eventually to be able to predict with some accuracy the number of copies of a title which would guarantee a given availability rate in any particular set of circumstances; and it should also be possible to extend this theory to the provisions of all books required by undergraduates, although data collection is much more difficult in an open-access collection. This type of experiment demonstrates that quantification which is originally undertaken for the sake of model-building has an immediate relevance to the routine operation of a real library - our subjective impression had been that the average frustration factor was about twice as great as the figures demonstrate.

This exercise in data collection also suggests a possible index of a library's performance, and hence a quantifiable set of objectives, which might be difficult to specify in any other way. If we assume a constraint imposed on the system by the maximum financial support available to the library, and specify levels of availability of printed materials for different classes of users, it should be reasonably easy to predict from the model what the results are likely to be in terms of the system as a whole, and by managerial decision to adjust frustration factors to optimal levels. We do not believe that this is the complete answer to the problem of defining objectives, but it is certainly an improvement on the hit-and-miss method of reaching decisions which is currently followed.

A further valuable result of model-building is that it will inevitably draw attention to areas where further research is needed, and thus aid us in formulating more precisely the major investigation into the total dynamic ecological balance between information and its users; we realise that we are only at the beginning of a project which may take many years, and that it may not be possible to achieve the results which I have optimistically foreshadowed in this paper. Nevertheless, we believe that the exercise is well worth while; the mere process of analysing in detail what we do from day to day must bring a deeper understanding of our eventual objectives, and this in itself will help to smooth the channels through which information flows.

Acknowledgement: The work of model building is being financed by a grant from the Office for Scientific and Technical Information; I wish to record my gratitude to that body, and to Ian Woodburn and M.K. Buckland (Research Fellow and Assistant Librarian (Research) respectively), who have contributed many useful ideas to this paper.

## REFERENCES

1. Dougherty, R.M. and Heinritz, F.J. Scientific management of library operations. New York, Scarecrow Press. 1966.

2. Automation and the Library of Congress: a survey sponsored by the Council on Library Resources. Library of Congress, 1963.

3. C.W. Hanson. Research on user's needs: where is it getting us? Aslib Proceedings, 16, ii, 1964, 64-78.

4. Overhage, C.F.J. and Harman, R.J. (eds.). INTREX: report of a planning conference on information transfer experiments. Cambridge, M.I.T. Press. 1965.

5. Information methods of research workers in the social sciences: proceedings of a conference ... edited by J.M. Harvey. Library Association, 1961.

6. Science, government and information: Report of the President's Science Advisory Committee, U.S.G.P.O., 1963

7. Menzel, H. The information needs of current scientific research. Library quarterly, 34, i, 1964, 4-9.

8. Becker, J. System Analysis - prelude to library data processing; ALA Bulletin, April 1965, 293-296.

9. Leimkuhler, F.F. Operations Research in the Purdue Libraries in Automation in the Library, when where and how, Purdue University, 1965, 82-89.

10. Leimkuhler, F.F. Systems analysis in university libraries. College and Research Libraries, 27, 1, 1966, 13-18.

11. Schultheiss, L.A., Culberston, D.S. and Heiliger, A.M. Advanced data processing in the university library. New York, Scarecrow Press. 1962.

## APPENDIX

*Simple model of some consecutive technical processes with a built-in optional decision rule*

The system is represented as a number of stages through which material passes for processing. We assume that the time scale is divided up into intervals of time (e.g. hour or week) referred to as slots during which the flows of work and labour are constant.

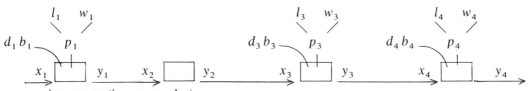

$x$ = input rate (items per slot)
$l$ = labour allocated (manhours per slot)
$w$ = processing rate (items per manhour)
$p$ = processing capacity (items per slot)
$b$ = backlog at end of slot
$d$ = delay to next item assuming serial processing (slots)
$y$ = output rate (items per slot)

Consider the system in time slot $(t)$.
Decision rule allocates just enough labour to clear backlog in one time slot.

1. *Ordering*

If decision rule then $l_1(t) = \dfrac{b_1(t-1) + x_1(t)}{w_1(t)}$; otherwise $l_1(t)$ given.

$p_1(t) = l_1(t) w_1(t)$

$\left.\begin{array}{l} y_1(t) = p_1(t) \text{ if } b_1(t-1) + x_1(t) \geqslant p_1(t) \\ y_1(t) = b_1(t-1) + x_1(t) \text{ otherwise.} \end{array}\right\}$

[Because input rate might be less than processing capacity.]

$b_1(t) = b_1(t-1) + x_1(t) - y_1(t)$

$d_1(t) = b_1(t)/p_1(t)$

2. *Bookseller*

Assume distribution of supply times.
$f(i)$ = fraction of books supplied during the $i$th slot since order despatched.

$x_2(t) = y_1(t)$

$y_2(t) = \displaystyle\sum_{i=o}^{i=\infty} f(i) x_2(t-i)$

3. *Accessioning*

$x_{31}$ input rate of purchased material (items per slot)
$x_{32}$ input rate of donated material (items per slot)

$$x_{31}(t) = y_2(t)$$
$$x_3(t) = x_{31}(t) + x_{32}(t)$$

If decision rule then $l_3(t) = \dfrac{b_3(t-1) + x_3(t)}{w_3(t)}$;

otherwise $l_3(t)$ given.

$$p_3(t) = l_3(t)\, w_3(t)$$
$$\left. \begin{array}{l} y_3(t) = p_3(t) \quad \text{if} \quad b_3(t-1) + x_3(t) \geqslant p(t) \\ y_3(t) = b_3(t-1) + x_3(t) \text{ otherwise.} \end{array} \right\}$$
$$b_3(t) = b_3(t-1) + x_3(t) - y_3(t)$$
$$d_3(t) = b_3(t)/p_3(t)$$

## 4. Cataloguing

$$x_4(t) = y_3(t)$$

If decision rule then $l_4(t) = \dfrac{b_4(t-1) + x_4(t)}{w_4(t)}$

otherwise $l_4(t)$ given.

$$p_4(t) = l_4(t)\, w_4(t)$$
$$\left. \begin{array}{l} y_4(t) = p_4(t) \quad \text{if} \quad b_4(t-1) + x_4(t) \geqslant p_4(t) \\ y_4(t) = b_4(t-1) + x_4(t) \text{ otherwise.} \end{array} \right\}$$
$$b_4(t) = b_4(t-1) + x_4(t) - y_4(t)$$
$$d_4(t) = b_4(t)/p_4(t)$$

### Print-Out from Model Programme

#### WOODLAND UNIVERSITY LIBRARY

WEEK: 11

| | LABOUR | W/RATE | BACKLG | INFLOW | WK/CAP | OUTPUT | DELAY |
|---|---|---|---|---|---|---|---|
| ORDERING | 42 | 10 | 60 | 450 | 420 | 420 | 0 |
| BOOKSELLER | 0 | 0 | 0 | 0 | 0 | 348 | 0 |
| ACCESSIONS 1 | 35 | 10 | 0 | 348 | 350 | 348 | 0 |
| ACCESSIONS 2 | 10 | 50 | 1930 | 20 | 500 | 500 | 4 |
| BRIEFLISTING | 10 | 20 | 1000 | 450 | 200 | 200 | 5 |
| ALPHA STORE | 0 | 0 | 1900 | 200 | 0 | 0 | -1 |
| CATALOGUING | 80 | 5 | 298 | 398 | 400 | 400 | 1 |
| CARD PROCESS | 48 | 8 | 176 | 400 | 384 | 384 | 0 |
| CARD FILING | 20 | 100 | 0 | 1920 | 2000 | 1920 | 0 |
| CATALOGUE | 0 | 0 | 21120 | 1920 | 0 | 0 | -1 |
| BOOK PROCESS | 18 | 20 | 440 | 400 | 360 | 360 | 1 |
| DESP TO BIND | 10 | 12 | 0 | 119 | 120 | 119 | 0 |
| BINDERS | 0 | 0 | 0 | 356 | 0 | 333 | 0 |
| BIND RECEIPT | 8 | 12 | 0 | 0 | 96 | 45 | 0 |
| BOOK STOCK | 0 | 0 | 2986 | 286 | 0 | 0 | -1 |

WOODLAND UNIVERSITY LIBRARY

| | LABOUR | W/RATE | BACKLG | INFLOW | WK/CAP | OUTPUT | DELAY |
|---|---|---|---|---|---|---|---|
| ORDERING | 42 | 10 | 60 | 420 | 420 | 420 | 0 |
| BOOKSELLER | 0 | 0 | 0 | 0 | 0 | 355 | 0 |
| ACCESSIONS 1 | 35 | 10 | 5 | 355 | 350 | 390 | 0 |
| ACCESSIONS 2 | 10 | 50 | 1430 | 0 | 500 | 500 | 3 |
| BRIEFLISTING | 10 | 20 | 1250 | 450 | 200 | 200 | 6 |
| ALPHA STORE | 0 | 0 | 2100 | 200 | 0 | 0 | −1 |
| CATALOGUING | 80 | 5 | 298 | 400 | 400 | 400 | 1 |
| CARD PROCESS | 48 | 8 | 192 | 400 | 384 | 384 | 1 |
| CARD FILING | 20 | 100 | 0 | 1920 | 2000 | 1920 | 0 |
| CATALOGUE | 0 | 0 | 23040 | 1920 | 0 | 0 | −1 |
| BOOK PROCESS | 18 | 20 | 480 | 400 | 360 | 360 | 1 |
| DESP TO BIND | 10 | 12 | 0 | 119 | 120 | 119 | 0 |
| BINDERS | 0 | 0 | 0 | 0 | 0 | 428 | 0 |
| BIND RECEIPT | 8 | 12 | 332 | 428 | 96 | 96 | 3 |
| BOOK STOCK | 0 | 0 | 3323 | 337 | 0 | 0 | −1 |

WOODLAND UNIVERSITY LIBRARY

| | LABOUR | W/RATE | BACKLG | INFLOW | WK/CAP | OUTPUT | DELAY |
|---|---|---|---|---|---|---|---|
| ORDERING | 42 | 10 | 60 | 420 | 420 | 420 | 0 |
| BOOKSELLER | 0 | 0 | 0 | 0 | 0 | 355 | 0 |
| ACCESSIONS 1 | 35 | 10 | 11 | 355 | 350 | 350 | 0 |
| ACCESSIONS 2 | 10 | 50 | 950 | 20 | 500 | 500 | 2 |
| BRIEFLISTING | 10 | 20 | 1500 | 450 | 200 | 200 | 8 |
| ALPHA STORE | 0 | 0 | 2300 | 200 | 0 | 0 | −1 |
| CATALOGUING | 80 | 5 | 298 | 400 | 400 | 400 | 1 |
| CARD PROCESS | 48 | 8 | 208 | 400 | 384 | 384 | 1 |
| CARD FILING | 20 | 100 | 0 | 1920 | 2000 | 1920 | 0 |
| CATALOGUE | 0 | 0 | 24960 | 1920 | 0 | 0 | −1 |
| BOOK PROCESS | 18 | 20 | 520 | 400 | 360 | 360 | 1 |
| DESP TO BIND | 10 | 12 | 0 | 119 | 120 | 119 | 0 |
| BINDERS | 0 | 0 | 0 | 119 | 0 | 428 | 0 |
| BIND RECEIPT | 8 | 12 | 236 | 0 | 96 | 96 | 2 |
| BOOK STOCK | 0 | 0 | 3660 | 337 | 0 | 0 | −1 |

Management

By

Exception

*Lester R. Bittel*

The time is the present. Four managers have just taken action in order to solve a pressing business problem:

Charles Ryerson, purchasing agent for East Coast Electronics, rejects a shipment of cathode tubes because a systematic sampling of the lot showed that the percentage of off-specification tubes exceeded a predetermined allowable number.

Avery Prentice, chairman of Amalgamated Alloys, calls a meeting of the board when a monthly statement shows that the brass division has failed to meet its sales target for the third successive time.

Elmer Bond, plant superintendent of the Seattle Stamping Works, authorizes overtime in the plating department when his weekly "late report" pinpoints plating delays as the cause for late shipment of six orders.

Joseph Schaeffer, office manager for Lifetime Insurance Mutuals, has a heart-to-heart talk with his chief clerk after an attitude survey reveals growing employee unrest in the clerical section.

Charles Ryerson, Avery Prentice, Elmer Bond, and Joseph Schaeffer have been solving problems that differ in nature and degree. But all four executives have one thing in common. They are all practicing management by exception. Every day thousands of other business managers practice management by exception, too. Most of them do so, not so much by conscious intent as by instinct. Fortunately this instinct is soundly based upon their own experience.

Experience has taught most seasoned managers that much of what happens in business life is a repetition of what has occurred in the past. Consequently, an alert manager can project yesterday's occurrences into today and tomorrow. This way he can know ahead of time what is likely to happen. Knowing this, he can preplan what to take for granted and what to get excited about. Those *routine*, expectable occurrences permit him to set up a plan of action for dealing with them. Only the *exceptional* occurrences—and not even all of these—need bring on new and difficult problems.

For example, let's take another look at the case of Charles Ryerson, the purchasing agent. Ryerson was employing a statistical technique based upon a "normal," or typical, distribution of product characteristics throughout a "lot" of any size. Ryerson was able to predetermine, to a high degree of accuracy, how many tubes would be off-quality. When his sampling showed that the actual number had exceeded the expected—that it was, in fact, "exceptional"—Ryerson's decision was easy. He followed an established buying policy of rejecting an entire shipment when sampling showed an exceptional amount of off-quality goods.

How did Avery Prentice practice management by exception? His method was technically simpler than Ryerson's. With the aid of the division manager, he had previously set up monthly profit goals for brass products. Experience, however, had taught him that if the division failed to meet profit goals for more than two months running, something unpredictable was happening to the brass market. As company chairman, he knew it was his responsibility to deal with the unpredictable. While his action technique was simple, even unscientific, his problem to be solved was difficult. This is so because no amount of accumulated data has taken market prediction out of the realm of art and into science.

Elmer Bond, too, was using a simple, time-saving management method. Instead of examining the progress of each of the thousands of orders as they moved through the shop, he studied only those items that were not on schedule. A reporting system that flagged these exceptional items helped him to make sense out of what would otherwise be an unintelligible mountain of data. The action he took was also relatively simple. It was dictated by existing company policy. In the face of a tough competitive situation, late deliveries could not be tolerated. Bond knew he was expected to correct the delivery problem before he could take long-term measures to produce on-time deliveries without the extra cost of overtime.

Finally, Joseph Schaeffer was managing by exception, too, because he was comparing measurements of attitudes in his department with average measurements in his company and in his industry. The survey earmarked the chief clerk's employees as having a lower than average—or exceptional—attitude. In this instance, the measurement method isolated the exception, and existing knowledge of human behavior indicated an approach Schaeffer might follow.

### Management by Exception Defined

Management by exception, in its simplest form, is a system of identification and communication that signals the manager when his attention is needed; conversely, it remains silent when his attention is not required. The primary purpose of such a system is, of course, to simplify the management process itself—to permit a manager to find the problems that need his action and to avoid dealing with those that are better handled by his subordinates.

It is easier to tell you what the *ingredients* are in management by exception than to provide a simple definition. For instance, one might say that management by exception has six key elements. They are measurement, projection, selection, observation (or remeasurement), comparison, and decision making.

*Measurement* assigns values—often numerical—to past and present performances. Without measurement of some sort, it would be impossible to identify an exception.

*Projection* analyzes those measurements that are meaningful to business objectives and extends them into future expectations.

*Selection* pinpoints the criteria management will use to follow progress toward its objectives.

*Observation* is that phase of measurement that informs management of the current state of performance.

*Comparison* of actual performance with expected performance identifies the exceptions that require attention and reports the variances to management.

*Decision making* prescribes the action that must be taken in order to (1) bring performance back into control or to (2) adjust expectations to reflect changing conditions or to (3) exploit opportunity.

In many ways the concept of management by exception is inseparable from other management essentials. All planning, for instance, is based upon this concept. The philosophy of management by exception is also deeply rooted in the principles of the division of labor, delegation of responsibility and authority, and span of control. But it has as its taproot the concept of measurement, to which a later chapter will be devoted.

## Its Historical Background

You have only to look around you in business today to find dozens of modern management practices that directly utilize, support, or reflect the management by exception principle. To name just a few, there are:

Statistical control of product quality

Economic order quantities and order points for control of inventories and supplies

The Gantt chart for production control

Break-even points for determining operating levels

Trends in ratios of indirect to direct labor used in apportioning overhead

Attitude surveys for gaging employee morale

Dozens of financial ratios (such as net sales to net working capital) to measure the economic health of a business

There is no managerial function that cannot practice management by exception in some form or other. In fact, management by exception has become such an obvious—often unconscious—way of life that it is hard to realize that it is relatively new as a crystallized concept. There is little doubt that its application could be documented by a historical search of early cultures—such as the Roman's. The world's military organizations and many religious organizations have used it for centuries. But it was not until the end of the nineteenth century that management by exception was identified as a principle that could be applied to business situations in general.

In order to enlarge your own foundation in applying this principle, it is helpful to see it as some of management's more notable pioneers have seen it. Here is a sampling of their views.

## Frederick W. Taylor

Under [the exception principle] the manager should receive only condensed, summarized, and invariably comparative reports covering, however, all of the elements entering into the management, and even these summaries should all be carefully gone over by an assistant before they reach the manager, and have all of the exceptions to the past averages or standards pointed out, both the especially good and the especially bad exceptions, thus giving him in a few minutes a full view of progress which is being made, or the reverse, and leaving him free to consider the broader lines of policy and to study the character and fitness of the important men under him.[1]

## Henry Robinson Towne

The primary object [of records and paperwork] is the systematic recording of the operations of the different departments of the works, and the computation therefrom of such statistical information as is essential to the efficient management of the business, and especially to increased economy of production.[2]

## Henry Laurence Gantt

In order to be sure that they all got the assistance possible from the foreman he, too, received a definite premium for each machine under his charge that made its bonus, and in order that the poorer men might receive sufficient instruction from the foreman, it was made to his interest to give them special attention. That was accomplished in this way: While the foreman was given a definite amount for each machine that earned its bonus, he was given an additional fifty per cent if all the machines under his charge earned their bonus, thus making it to his interest to give special attention to the men most likely to fall behind.[3]

## Leon Pratt Alford

Experience is the knowledge of past attainment. It includes a knowledge of *what* has been done, and also *how* it has been done. It is inseparably associated with standards of performance, that is with the ideas of quantity and quality in relation to any particular method of doing something. . . . The great instrument of experience, which makes progress possible, is comparison. By systematic use of experience is meant the careful analysis of what is about to be attempted, and its reference to existing records and standards of performance.[4]

## Frank Bunker Gilbreth

The personal work of the executive should consist as much as possible of making decisions and as little as possible of making motions. General recognition of this fact has resulted in the common practice of assigning to the executive one or more secretaries, or clerks, to relieve him of certain parts of his work which involve mere motions and less important decisions than that part of the work retained by the executive. This procedure varies in degree according to the kind of work done by the executive and how well he realizes the possibilities of eliminating waste through the use of the "exception principle" in management.[5]

## Peter Drucker

Any serious attempt to make management scientific or professional is bound to lead to the attempt to eliminate those disturbing influences, the unpredictabilities of business life—its risks, its ups and downs, its wasteful competition, the irrational choices of the consumer—and, in the process, the economy's freedom and its ability to grow. It is not entirely accident that some of the early pioneers of Scientific Management ended up by demanding complete cartelization of the economy (Henry Gantt was the prime example); that the one direct outgrowth of American Scientific Management abroad, the German Rationalization movement of the twenties, attempted to make the world safe for professional management by cartelizing it; and that in our own country men who were steeped in scientific management played a big part in Technocracy and in the attempted nation-wide super-cartel of the National Recovery Act in the first year of Roosevelt's New Deal.[6]

This was management by exception as the pioneers of scientific management saw it. Its big advantage lay in making management more effective and in preventing management from "overmanaging." Today, however, management by exception has taken on a new dimension. In an era of phenomenal technical growth and harassing economic change, a management by exception system stimulates a manager to actively search for exceptions that allow him to apply his creative ability. Consequently, management by exception is not only a technique of control but also a method for seeking out opportunities.

The early pioneers—and critics—of scientific management and the exception principle have had their say. Because of them, or in spite of them, the principle has stood the test of time well. Today it is applied in broad scope to every conceivable management function in every kind of organization—in America or abroad. This is so because its rewards to management are so generous.

## What Management by Exception Can Do for Management

Taylor's analysis of the exception principle highlights its obvious advantages—executive timesaving and an objective view of progress made toward established goals. A more penetrating examination justifies another dozen more. For instance, the practice of management by exception:

• Saves personal time. You apply yourself to fewer problems—the ones that really count. It minimizes time-consuming work on trivia and details that others on your staff can handle.

• Concentrates executive effort. Instead of spreading managerial talent thinly across all sorts of problems, you place your effort selectively only where and when it is needed.

• Reduces distractions. The management by exception system only flags attention to critical areas. It remains silent on matters that are under control or that are delegated automatically to your subordinates.

• Facilitates broader managerial coverage. Concentrated, more effective management effort enables you to increase the scope of your activities and your span of control. This frees you to tackle promising projects that otherwise might be left undone.

• Lessens frequency of decision making. The system makes most of the minor decisions for you. You don't have to check every item under your supervision every day to see if it is okay or not. The system passes along to you only the few important problems for you to rule on or act on. But these decisions are, of course, the more difficult ones.

• Makes fuller use of knowledge of trends, history, and available business data. Management by exception, when applied systematically, forces an executive to review past history and to study related business data, because these are the foundations upon which standards are derived and from which exceptions are noted.

• More fully utilizes highly paid people on high-return work. Because delegation is planned, it more carefully relates and assigns the more complex problems to the more talented and highly paid people.

• Identifies crises and critical problems. As much as anything, management by exception helps you to avoid uninformed, impulsive pushing of the panic button. Yet a crisis is almost always recognized because seemingly unusual variations can be reliably and quickly compared with anticipated conditions.

• Provides qualitative and quantitative yardsticks for judging situations and people. Management by exception takes much of the prejudice out of performance appraisals by making individual and organizational goals and measurements tangible and specific.

• Enables inexperienced managers to handle new assignments with a minimum of related experience and training. A new man benefits from measurements and projections that have been established as standard for his function in the past. He can depend upon the system to alert him to problems rather than having to rely solely upon his own experience.

• Alerts management to opportunities as well as difficulties. Managing by exception helps to counterbalance purely negative, control-minded management. Attention is directed to "breakthrough" variations as well as to omissions or shortcomings.

• Encourages more comprehensive knowledge of all phases of business operations. You cannot practice management by exception systematically without continually gathering and updating real facts about your organization and its operation.

• Stimulates communications between different segments of an organization. With its focus on results, management by exception seeks to relate causes, regardless of their place in the organization, with overall organizational results. As such, it encourages exchange of measurements between functions as well as between a function and the cost or profit center to which it reports.

## How Companies Are Using Management by Exception

Hardly a successful company exists today that does not practice management by exception. As indicated earlier, this conclusion is more often inferred by observation than confirmed by positive evidence of an active system. Nevertheless, the evidence is considerable. For example, see how these topflight companies are making management by exception work for them:

*In Decision Making.* Bell & Howell Co. uses the practice of management by exception to make its policy of delegation work. Its policy is threefold: (1) Delegate to the lowest possible level. (2)

Step in only when problems warrant. (3) Tap optimum capabilities of everyone. Charles H. Percy, company chairman, says that under this system, executives have been able to remove up to 90 per cent of the routine work from their desks. Furthermore, he points out that it has additional value because it serves to "get management people *out* of the act" and to let others have an opportunity to participate to the extent of their capabilities.[7]

*In Marketing.* Smith-Corona Marchant, Inc., directs executive attention to exceptional marketing problems by classifying products annually according to their sales potential. In the first group are those products that are expected to hold their own in the ensuing year. These get only superficial attention from top management. In the second group are those products with the greatest immediate potential. Top management examines these intensively from many corporate viewpoints, including engineering, finance, and production. The intention here is to exploit an opportunity. The third group of products are those adversely affected by a dwindling market. These also come up for sharp executive consideration. The problem here is one of improvement and/or control.[8]

*In Corporate Finance.* Westinghouse Electric Corporation greatly reduces the time required by top management to review capital expenditure requests by establishing a table of amounts that may be approved by lesser management without higher approval. For instance, individual projects up to $50,000 go to the division manager, projects from $50,000 to $100,000 go to headquarters management, and those over $100,000 go to the board of directors. In one year, as an example, about 500 projects, representing in dollars 15 per cent of the total program, went to the division managers, while only about 150 projects, which represented 85 per cent of the dollars, required headquarters or board approval.[9]

*In Production.* St. Regis Paper Company expects its managers of production to delegate most day-to-day matters to subordinates. To quote William R. Adams, president, "Most of the prime duties of the production manager will necessitate his being freed of routine functions (but) there are routine matters . . . of which he will need to inform me. Usually they are the result of decisions arrived at long before. Reports of this kind serve to acquaint me with the progress made in his area. Sometimes they give me the option to exercise control, sometimes to inform managers in other areas." [10]

*In Organization Staffing.* International Business Machines Corporation after World War II expanded its engineering force by over 400 per cent. Most of the new people held M.A.'s and Ph.D.'s. However, an appraisal of executive resources brought to the attention of president Thomas J. Watson, Jr., the fact that practically none of the engineering and research management positions were staffed by people of similar educational background. In order to bring IBM management into phase with a rapidly changing technology, a policy was instituted to promote to top administrative posts those people who best understood the technology with which they were dealing. Consequently, IBM promoted large numbers of the advanced-degree men into positions of executive responsibility.[11]

*In Upgrading Performance.* Otis Elevator Company operates throughout the United States in ten districts. In setting standards of performance for each zone, the company holds up for example the best performance set by any of its components. It then makes comparisons on the basis of a composite picture made up of the selected best features to be found in the entire ten districts.[12]

*In Reducing Costs.* Rockwell Manufacturing Company recently performed a conclusive experiment documenting the value of measurement, standards, and management by exception. Two plants were tooled up at the same time to make the same product. At one plant methods were studied and controls prescribed by the industrial engineering department. In the second plant the industrial engineering department did not see the job until it was in operation. *The result:* the first plant's production averaged 50 per cent more as a result of this preplanning for management by exception for the executive group who took over the plant.[13]

## Where Its Limitations Lie

Peter Drucker's criticism of management by exception expressed a fear that management might interpret the exception as only a nuisance to be reduced through refinements in measurement, more careful planning, and stricter controls. Management by exception has other pitfalls, of course. Here are some that come to mind:
- It breeds "organization-man" thinking.
- It is often dependent upon unreliable data.
- It requires a comprehensive observing and reporting system.
- It tends to proliferate paperwork.
- It often assumes an unnatural stability in business affairs.

• In the absence of exceptions it can give management a false sense of security.

• It is silent about conditions predetermined not to be critical.

• Standards of comparison tend to become obsolete (such as the ratio of indirect to direct labor).

• Some critical business factors (such as human behavior) are difficult, if not impossible, to measure.

Finally, the process of management by exception cannot be a substitute for thinking—nor for decision making. Its big advantage lies in the fact that much of the time-consuming process of thinking and decision making can be done in *advance*. A progressive system of action can be prescribed beforehand, much as a troubleshooter's manual can give an instrument repairman step-by-step directions to isolate a fault. As a result, only radical exceptions, either good or bad, need be interpreted and acted upon under pressure of time.

The executive's problem, therefore, is threefold. First, he must convert his instinctive, unconscious approach to management by exception to a positive, systematic way of handling every kind of management problem—men, money, machines, and materials. Secondly, he must guard against the conformity and false sense of security that systems of any kind tend to nourish. Thirdly, freed from the demands of routine work, he must fill his time with creative effort directed toward improving his plans, organization, staff, and decisions.

## References

1. Frederick W. Taylor, "Shop Management," 1911, reprinted in *Scientific Management*, Harper & Row, Publishers, Incorporated, New York, 1947, p. 126.
2. Henry R. Towne, "The Engineer as an Economist," *Transactions of the American Society of Mechanical Engineers*, vol. 7, p. 428, 1886.
3. Henry L. Gantt, "A Bonus System of Rewarding Labor," *Transactions of the American Society of Mechanical Engineers*, vol. 23, p. 341, 1901.
4. L. P. Alford with Alexander H. Church, "The Principles of Management," *American Machinist*, May 30, 1912.
5. Frank B. Gilbreth, "Graphical Control on the Exception Principle for Executives," Paper 1573a, American Society of Mechanical Engineers, New York, December, 1916.
6. Peter F. Drucker, *The Practice of Management*, Harper & Row, Publishers, Incorporated, New York, 1954, p. 10.

7. R. R. Conarroe, *The Decision Makers*, The Bureau of Business Practice, New London, Conn., 1958, p. 16.

8. Gordon H. Smith, "Direction and Control of Expansion," in H. B. Maynard (ed.), *Top Management Handbook*, McGraw-Hill Book Company, New York, 1960, pp. 1026–1029.

9. Ross G. Walker and Russell B. Read, "Capital Investment Control," in Edward C. Bursk and Dan H. Fenn, Jr. (eds.), *Planning the Future Strategy of Your Business*, McGraw-Hill Book Company, New York, 1956, pp. 85–102.

10. William R. Adams, "Top Management and Production," in Maynard (ed.), *op. cit.*, pp 559–567.

11. Thomas J. Watson, Jr., "Promotion Innovation," in *ibid.*, pp. 514–515.

12. LeRoy A. Petersen, "Establishing Objectives," in *ibid.*, p. 186.

13. Willard F. Rockwell, Jr., "Managing a Highly Decentralized Organization," in *ibid.*, p. 930.

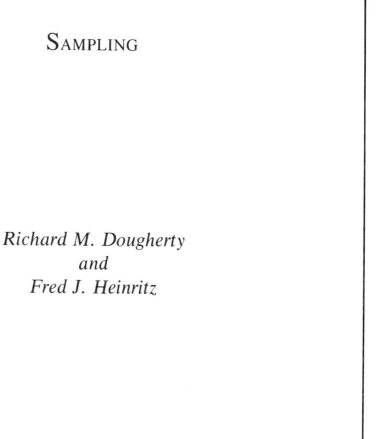

Sampling

*Richard M. Dougherty*
*and*
*Fred J. Heinritz*

SOURCE: Reprinted by permission of the publisher from *Scientific Management of Library Operations* by Richard M. Dougherty and Fred. J. Heinritz (Scarecrow Press, 1966). Copyright © 1966 by Richard M. Dougherty and Fred J. Heinritz.

In making a management study the magnitude of the total group (universe) about which we wish to make generalizations frequently makes the examination of every member of the group prohibitively costly and time-consuming. To examine the total circulation record for a year in a library circulating 218,000 volumes per year (the circulation of the main library alone in the medium size public library system whose circulation system is described in Chapters XII and XIII) would be a herculean task. Fortunately, the examination of every member of the total group (in this case a "member" being the record of an individual circulation) is usually unnecessary because it is possible to select a relatively small number of the individual members and generalize from them with reasonable assurance that they are representative of the entire group. The picking out and analyzing of the members is known as sampling. The members selected are a sample.

Let us suppose that there was an urgent need to know the number of cards currently in the 1152 drawer (all of the drawers contain cards) North Carolina Union Catalog.

Counting all the cards (assuming we counted carefully) would provide an accurate answer. However, the time required to do this would be enormous. This work could be reduced to a small fraction of that required to count all the cards by using the following device: Since it is easy to ascertain the approximate number of cards per inch, it is possible to substitute the quicker method of measuring inches of cards. This figure can then be converted into numbers of cards by multiplying it by the number of cards per inch. But even to measure the cards in 1152 drawers, not to mention having to sum their measurements, would require considerable labor. Therefore, instead of counting all of the cards a sample will be drawn.

If all the drawers in the catalog contained the same number of cards, the sample would need to consist of one drawer only. We would merely measure the inches of cards in the drawer and multiply this number by the number of cards per inch and by the total number of drawers.  Since it is inevitable that some drawers will contain more cards than others, the answer would be incorrect.  Therefore, several drawers, although still only a small fraction of the total must be included in the sample.  The number of inches of cards in the individual drawers of the sample will be determined; this sum will then be multiplied by the number of cards per inch and divided by the number of drawers in the sample. This arithmetic will yield an average number of cards per drawer. Finally, this figure will be multiplied by the number of drawers in the catalog to yield the total number of cards in the catalog, which is the information we need.  Let us decide arbitrarily to begin with a trial sample of twenty drawers.

### Random Selection of the Sample

The chief precaution to observe in picking a sample -- in this case, 20 out of 1152 drawers -- is to make certain that the members comprising it are chosen at random.  This means that it is necessary to choose the members in such a way that the probability of choosing any one is the same as the probability of choosing any other.  In practice the most efficient way to choose a random sample is to use a table of random numbers.  Before turning to a random number table, it will be necessary to assign numbers to the total group.  The North Carolina Union Catalog drawers happen to be numbered (1-1152).  If they were not, it would be easy to assign a number to each of them.  If the numbers must be assigned a quick sketch is a convenience.  This would enable quick location of specific drawers when observations are being made.  If we were dealing with a total group with no obvious geometric or other pattern -- such as a group of books not circulated in the past year or a group of users who had books overdue last January -- the names would be recorded in any arbitrary manner and then numbered consecutively.

Random Number Tables

A table of random numbers consists of digits arranged randomly in column and rows. . Tables vary in length and arrangement. The table in this chapter (See Figure 8-1) contains 5,000 digits. Rand Corporation has produced a table of 1,000,000 digits. There is a table by Kendall and Smith of 100,000 digits. There are other printed tables of 41,600 digits (Tippett) and 15,000 digits (Fisher and Yates). The Bureau of Transport Economics and Statistics of the Interstate Commerce Commission produced, in 1940, a table of 105,000 digits. It is not easily available, but one will find excerpts from it in other books.

For any one sampling problem the table employed should be long enough to provide an adequate sample without using all of the numbers. The 100,000 digit table by Kendall and Smith is available for only $1.50.

Where to Begin in the Random Number Table

The next order of business is to determine where and how to begin in a random number table. One starts initially at the beginning of the table and reads straight through in any given direction. When the end of the table is reached, the user returns to the beginning and starts over again. Since a single use should not carry one all the way through the table, the point where the reading ended for any particular use should be carefully marked and the next subsequent reading begun at that point. Some individuals find it most convenient to read random tables down the column, probably because they are used to seeing numbers in columns in other contents. Others prefer to read across to the right, the same way a book is read. Unless the table stipulates otherwise the user should select whatever direction is most convenient to him and then be consistent about it.

Reading the Random number table to Obtain the Sample

Returning to the North Carolina Union Catalog example, let us now use Figure 1 of this chapter to pick the 20 drawer sample. The highest numbered drawer is 1152. Since this is a four digit number, we must read the table digits in groups of four. There

*Random Sampling Numbers Produced by the Machine*

### 1st Thousand

| | | | | | | | |
|---|---|---|---|---|---|---|---|
| 23157 | 54859 | 01837 | 25993 | 76249 | 70886 | 95230 | 36744 |
| 05545 | 55043 | 10537 | 43508 | 90611 | 83744 | 10962 | 21343 |
| 14871 | 60350 | 32404 | 36223 | 50051 | 00322 | 11543 | 80834 |
| 38976 | 74951 | 94051 | 75853 | 78805 | 90194 | 32428 | 71695 |
| 97312 | 61718 | 99755 | 30870 | 94251 | 25841 | 54882 | 10513 |
| 11742 | 69381 | 44339 | 30872 | 32797 | 33118 | 22647 | 06850 |
| 43361 | 28859 | 11016 | 45623 | 93009 | 00499 | 43640 | 74036 |
| 93806 | 20478 | 38268 | 04491 | 55751 | 18932 | 58475 | 52571 |
| 49540 | 13181 | 08429 | 84187 | 69538 | 29661 | 77738 | 09527 |
| 36768 | 72633 | 37948 | 21569 | 41959 | 68670 | 45274 | 83880 |
| 07092 | 52392 | 24627 | 12067 | 06558 | 45344 | 67338 | 45320 |
| 43310 | 01081 | 44863 | 80307 | 52555 | 16148 | 89742 | 94647 |
| 61570 | 06360 | 06173 | 63775 | 63148 | 95123 | 35017 | 46993 |
| 31352 | 83799 | 10779 | 18941 | 31579 | 76448 | 62584 | 86919 |
| 57048 | 86526 | 27795 | 93692 | 90529 | 56546 | 35065 | 32254 |
| 09243 | 44200 | 68721 | 07137 | 30729 | 75756 | 09298 | 27650 |
| 97957 | 35018 | 40894 | 88329 | 52230 | 82521 | 22532 | 61587 |
| 93732 | 59570 | 43781 | 98885 | 56671 | 66826 | 95996 | 44569 |
| 72621 | 11225 | 00922 | 68264 | 35666 | 59434 | 71687 | 58167 |
| 61020 | 74418 | 45371 | 20794 | 95917 | 37866 | 99536 | 19378 |
| 97839 | 85474 | 33055 | 91718 | 45473 | 54144 | 22034 | 23000 |
| 89160 | 97192 | 22232 | 90637 | 35055 | 45489 | 88438 | 16361 |
| 25966 | 88220 | 62871 | 79265 | 02823 | 52862 | 84919 | 54883 |
| 81443 | 31719 | 05049 | 54806 | 74690 | 07567 | 65017 | 16543 |
| 11322 | 54931 | 42362 | 34386 | 08624 | 97687 | 46245 | 23245 |

### 2nd Thousand

| | | | | | | | |
|---|---|---|---|---|---|---|---|
| 64755 | 83885 | 84122 | 25920 | 17696 | 15655 | 95045 | 95947 |
| 10302 | 52289 | 77436 | 34430 | 38112 | 49067 | 07348 | 23328 |
| 71017 | 98495 | 51308 | 50374 | 66591 | 02887 | 53765 | 69149 |
| 60012 | 55605 | 88410 | 34879 | 79655 | 90169 | 78800 | 03666 |
| 37330 | 94656 | 49161 | 42802 | 48274 | 54755 | 44553 | 65090 |
| 47869 | 87001 | 31591 | 12273 | 60626 | 12822 | 34691 | 61212 |
| 38040 | 42737 | 64167 | 89578 | 39323 | 49324 | 88434 | 38706 |
| 73508 | 30908 | 83054 | 80078 | 86669 | 30295 | 56460 | 45336 |
| 32623 | 46474 | 84061 | 04324 | 20628 | 37319 | 32356 | 43969 |
| 97591 | 99549 | 36630 | 35106 | 62069 | 92975 | 95320 | 57734 |
| 74012 | 31955 | 59790 | 96982 | 66224 | 24015 | 96749 | 07589 |
| 56754 | 26457 | 13351 | 05014 | 90966 | 33674 | 69096 | 33488 |
| 49800 | 49908 | 54831 | 21998 | 08528 | 26372 | 92923 | 65026 |
| 43584 | 89647 | 24878 | 56670 | 00221 | 50193 | 99591 | 62377 |
| 16653 | 79664 | 60325 | 71301 | 35742 | 83636 | 73058 | 87229 |
| 48502 | 69055 | 65322 | 58748 | 31446 | 80237 | 31252 | 96367 |
| 96765 | 54692 | 36316 | 86230 | 48296 | 38352 | 23816 | 64094 |
| 38923 | 61550 | 80357 | 81784 | 23444 | 12463 | 33992 | 28128 |
| 77958 | 81694 | 25225 | 05587 | 51073 | 01070 | 60218 | 61961 |
| 17928 | 28065 | 25586 | 08771 | 02641 | 85064 | 65796 | 48170 |
| 94036 | 85978 | 02318 | 04499 | 41054 | 10531 | 87431 | 21596 |
| 47460 | 60479 | 56230 | 48417 | 14372 | 85167 | 27558 | 00368 |
| 47856 | 56088 | 51992 | 82439 | 40644 | 17170 | 13463 | 18288 |
| 57616 | 34653 | 92298 | 62018 | 10375 | 76515 | 62986 | 90756 |
| 08300 | 92704 | 66752 | 66610 | 57188 | 79107 | 54222 | 22013 |

Fig. 8-1 A Table of 5,000 Random Digits

Reproduced from M. G. Kendall and B. B. Smith. "Randomness and Random Sampling Numbers," Journal of the Royal Statistical Society, 101 (1938), p. 147-66. By permission of the Royal Statistical Society and the authors.

*3rd Thousand*

| | | | | | | | |
|---|---|---|---|---|---|---|---|
| 89221 | 02362 | 65787 | 74733 | 51272 | 30213 | 92441 | 39651 |
| 04005 | 99818 | 63918 | 29032 | 94012 | 42363 | 01261 | 10650 |
| 98546 | 38066 | 50856 | 75045 | 40645 | 22841 | 53254 | 44125 |
| 41719 | 84401 | 59226 | 01314 | 54581 | 40398 | 49988 | 65579 |
| 28733 | 72489 | 00785 | 25843 | 24613 | 49797 | 85567 | 84471 |
| 65213 | 83927 | 77762 | 03086 | 80742 | 24395 | 68476 | 83792 |
| 65553 | 12678 | 90906 | 90466 | 43670 | 26217 | 69900 | 31205 |
| 05668 | 69080 | 73029 | 85746 | 58332 | 78231 | 45986 | 92998 |
| 39302 | 99718 | 49757 | 79519 | 27387 | 76373 | 47262 | 91612 |
| 64592 | 32254 | 45879 | 29431 | 38320 | 05981 | 18067 | 87137 |
| 07513 | 48792 | 47314 | 83660 | 68907 | 05336 | 82579 | 91582 |
| 86593 | 68501 | 56638 | 99800 | 82839 | 35148 | 56541 | 07232 |
| 83735 | 22599 | 97977 | 81248 | 36838 | 99560 | 32410 | 67614 |
| 08595 | 21826 | 54655 | 08204 | 87990 | 17033 | 56258 | 05384 |
| 41273 | 27149 | 44293 | 69458 | 16828 | 63962 | 15864 | 35431 |
| 00473 | 75908 | 56238 | 12242 | 72631 | 76314 | 47252 | 06347 |
| 86131 | 53789 | 81383 | 07868 | 89132 | 96182 | 07009 | 86432 |
| 33849 | 78359 | 08402 | 03586 | 03176 | 88663 | 08018 | 22546 |
| 61870 | 41657 | 07468 | 08612 | 98083 | 97349 | 20775 | 45091 |
| 43898 | 65923 | 25078 | 86129 | 78491 | 97653 | 91500 | 80786 |
| 29939 | 39123 | 04548 | 45985 | 60952 | 06641 | 28726 | 46473 |
| 38505 | 85555 | 14388 | 55077 | 18657 | 94887 | 67831 | 70819 |
| 31824 | 38431 | 67125 | 25511 | 72044 | 11562 | 53279 | 82268 |
| 91430 | 03767 | 13561 | 15597 | 06750 | 92552 | 02391 | 38753 |
| 38635 | 68976 | 25498 | 97526 | 96458 | 03805 | 04116 | 63514 |

*4th Thousand*

| | | | | | | | |
|---|---|---|---|---|---|---|---|
| 02490 | 54122 | 27944 | 39364 | 94239 | 72074 | 11679 | 54082 |
| 11967 | 36469 | 60627 | 83701 | 09253 | 30208 | 01385 | 37482 |
| 48256 | 83465 | 49699 | 24079 | 05403 | 35154 | 39613 | 03136 |
| 27246 | 73080 | 21481 | 23536 | 04881 | 89977 | 49484 | 93071 |
| 32532 | 77265 | 72430 | 70722 | 86529 | 18457 | 92657 | 10011 |
| 66757 | 98955 | 92375 | 93431 | 43204 | 55825 | 45443 | 69265 |
| 11266 | 34545 | 76505 | 97746 | 34668 | 26999 | 26742 | 97516 |
| 17872 | 39142 | 45561 | 80146 | 93137 | 48924 | 64257 | 59284 |
| 62561 | 30365 | 03408 | 14754 | 51798 | 08133 | 61010 | 97730 |
| 62796 | 30779 | 35497 | 70501 | 30105 | 08133 | 00997 | 91970 |
| 75510 | 21771 | 04339 | 33660 | 42757 | 62223 | 87565 | 48468 |
| 87439 | 01691 | 63517 | 26590 | 44437 | 07217 | 98706 | 39032 |
| 97742 | 02621 | 10748 | 78803 | 38337 | 65226 | 92149 | 59051 |
| 98811 | 06001 | 21571 | 02875 | 21828 | 83912 | 85188 | 61624 |
| 51264 | 01852 | 64607 | 92553 | 29004 | 26695 | 78583 | 62998 |
| 40239 | 93376 | 10419 | 68610 | 49120 | 02941 | 80035 | 99317 |
| 26936 | 59186 | 51667 | 27645 | 46329 | 44681 | 94190 | 66647 |
| 88502 | 11716 | 98299 | 40974 | 42394 | 62200 | 69094 | 81646 |
| 63499 | 38093 | 25593 | 61995 | 79867 | 80569 | 01023 | 38374 |
| 36379 | 81206 | 03317 | 78710 | 73828 | 31083 | 60509 | 44091 |
| 93801 | 22322 | 47479 | 57017 | 59334 | 30647 | 43061 | 26660 |
| 29856 | 87120 | 56311 | 50053 | 25365 | 81265 | 22414 | 02431 |
| 97720 | 87931 | 88265 | 13050 | 71017 | 15177 | 06957 | 92919 |
| 85237 | 09105 | 74601 | 46377 | 59938 | 15647 | 34177 | 92753 |
| 75746 | 75268 | 31727 | 95773 | 72364 | 87324 | 36879 | 06802 |

Fig. 8-1 A Table of 5,000 Random Digits

Reproduced from M. G. Kendall and B. B. Smith. "Randomness and Random Sampling Numbers," Journal of the Royal Statistical Society, 101 (1938), p. 147-66. By permission of the Royal Statistical Society and the authors.

*5th Thousand*

| | | | | | | | |
|---|---|---|---|---|---|---|---|
| 29935 | 06971 | 63175 | 52579 | 10478 | 89379 | 61428 | 21363 |
| 15114 | 07126 | 51890 | 77787 | 75510 | 13103 | 42942 | 48111 |
| 03870 | 43225 | 10589 | 87629 | 22039 | 94124 | 38127 | 65022 |
| 79390 | 39188 | 40756 | 45269 | 65959 | 20640 | 14284 | 22960 |
| 30035 | 06915 | 79196 | 54428 | 64819 | 52314 | 48721 | 81594 |
| 29039 | 99861 | 28759 | 79802 | 68531 | 39198 | 38137 | 24373 |
| 78196 | 08108 | 24107 | 49777 | 09599 | 43569 | 84820 | 94956 |
| 15847 | 85493 | 91442 | 91351 | 80130 | 73752 | 21539 | 10986 |
| 36614 | 62248 | 49194 | 97209 | 92587 | 92053 | 41021 | 80064 |
| 40549 | 54884 | 91465 | 43862 | 35541 | 44466 | 88894 | 74180 |
| 40878 | 08997 | 14286 | 09982 | 90308 | 78007 | 51587 | 16658 |
| 10229 | 49282 | 41173 | 31468 | 59455 | 18756 | 08908 | 06660 |
| 15918 | 76787 | 30624 | 25928 | 44124 | 25088 | 31137 | 71614 |
| 13403 | 18796 | 49909 | 94404 | 64979 | 41462 | 18155 | 98335 |
| 66523 | 94596 | 74908 | 90271 | 10009 | 98648 | 17640 | 68909 |
| 91665 | 36469 | 68343 | 17870 | 25975 | 04662 | 21272 | 50620 |
| 67415 | 87515 | 08207 | 73729 | 73201 | 57593 | 96917 | 69699 |
| 76527 | 96996 | 23724 | 33448 | 63392 | 32394 | 60887 | 90617 |
| 19815 | 47789 | 74348 | 17147 | 10954 | 34355 | 81194 | 54407 |
| 25592 | 53587 | 76384 | 72575 | 84347 | 68918 | 05739 | 57222 |
| 55902 | 45539 | 63646 | 31609 | 95999 | 82887 | 40666 | 66692 |
| 02470 | 58376 | 79794 | 22482 | 42423 | 96162 | 47491 | 17264 |
| 18630 | 53263 | 13319 | 97619 | 35859 | 12350 | 14632 | 87659 |
| 89673 | 38230 | 16063 | 92007 | 59503 | 38402 | 76450 | 33333 |
| 62986 | 67364 | 06595 | 17427 | 84623 | 14565 | 82860 | 57300 |

Fig. 8-1  A Table of 5,000 Random Digits

Reproduced from M. G. Kendall and B. B. Smith. ''Randomness and Random Sampling Numbers,'' Journal of the Royal Statistical Society, 101 (1938), p. 147-66.  By permission of the Royal Statistical Society and the authors.

are two basic rules to follow:

1. A number that is greater than the highest number that may be used must be skipped. Thus, in our example it will be necessary to ignore the numbers 1153-9999. Since the drawer numbering begins with 1, rather than with 0, it will also be necessary to ignore 0000.

2. A number that has already been selected for the sample should be ignored if it is drawn a second time.

Let us say that it is our habit to read the random table as a book, and that the last time we used it we stopped in the 2nd block of 1,000 digits, the 24th row from the top of the block, and the 38th column from the left (with a 7). Now, beginning with the 2nd block of 1,000 numbers, the 24th row from the top, and the 39th column from the left, we begin to read from left to right, starting with 56. Then, moving to the left end of the block and jumping down one row, as in reading a book, we read 08300 92704...... Starting then with 56 and reading groups of four digits, we get 5608 (which we must disregard), 3009 (which we must disregard), 2704 (which we must disregard), 1057 (which is the first member of the sample), and so on. For practice the reader should verify that the sample consists of the following drawers:

| | | | |
|---|---|---|---|
| 1. | 1057 | 11. | 0082 |
| 2. | 0213 | 12. | 1072 |
| 3. | 0059 | 13. | 0676 |
| 4. | 0124 | 14. | 0487 |
| 5. | 0126 | 15. | 0700 |
| 6. | 1106 | 16. | 0840 |
| 7. | 0398 | 17. | 0801 |
| 8. | 0078 | 18. | 0746 |
| 9. | 0312 | 19. | 0839 |
| 10. | 0505 | 20. | 0807 |

In actually recording the number of inches of cards in each sample drawer it is a convenience to have the drawers arranged on the recording sheet in numerical order. Therefore, in practice the sample numbers should not be recorded on a single sheet, as above, but rather on cards or slips, with one number per card.

These slips may then be sorted into numerical order. This procedure is also efficient for catching duplicate numbers. For every duplicate caught, a replacement must be selected from the table.

When forced to bypass most of the numbers read (88.5% in the example), reading the tables can, particularly for larger samples, be tedious and time-consuming, and in any case is wasteful of table numbers. In such cases it is good practice, at least if a desk calculator is handy, or if the number of items in the total group from which the sample is selected is one convenient for mental division (such as 1000, 100, 50, 20, or the like), to adopt an alternative procedure. The digits are grouped as the problem requires and read consecutively. Each group of digits is divided by the total number of items in the group from which the sample is being selected. The remainder represents an item of the sample. In our example the first four digit number read is 5608. Dividing 5608 by 1152, one obtains a remainder of 1000. Drawer number 1000 is then the first member of the sample drawn by this method. The remainder zero represents the number 1152 itself. Since 1152 does not divide evenly into 9999, all table numbers above 9216 (1152 x 8) must be ignored, and also 0000, since the drawer numbering begins with one. However, these numbers represent only 8% of the total possibilities -- as opposed to 88% using the former method. The reader should verify that, using the division method of selection, the sample drawers would be:

| | | | |
|---|---|---|---|
| 1. | 1000 | 11. | 0237 |
| 2. | 0705 | 12. | 1058 |
| 3. | 0400 | 13. | 0058 |
| 4. | 0915 | 14. | 0818 |
| 5. | 0362 | 15. | 0835 |
| 6. | 1057 | 16. | 1047 |
| 7. | 0735 | 17. | 0419 |
| 8. | 1043 | 18. | 0213 |
| 9. | 0814 | 19. | 0028 |
| 10. | 1068 | 20. | 0244 |

Tables of Random Permutations

Tables of random permutations arrange a given group of integers in a random order. There are tables of a wide variety of range available: 0-9, 0-19, 1-9, 1-16, 1-20, 1-30, 1-50, 1-100, 1-200, 1-500, and even 1-1000. As of this writing, the authors know of no generally available random permutation table of more than 1000 numbers. Figure 8-2 is a sample random permutation for the numbers 1-500.

This variety of range can in some instances give a table of random permutations considerable advantage over a decimal table of random numbers for sample selection. Suppose that a sample of 20 card drawers is to be selected from a catalog with a total of 487 drawers. Using a random table, all numbers from 488 to 999 would have to be ignored (plus 000 if the drawer numbers begin with one) -- that is, an average of only 48% of the digits scanned could be used. If, however, the sample was selected by reading the 1-500 permutation of Figure 8-2, it would be necessary to ignore only 488-500 -- less than 3% of the numbers. A second advantage of random permutations is that, unlike the random number table, duplicate numbers have already been eliminated. As an example, let us use Figure 8-2 to select a sample of 20 from 487 drawer. Beginning at the upper left of the permutation and reading it the way a book is read, the sample is:

| | | | |
|---|---|---|---|
| 1. | 428 | 11. | 265 |
| 2. | 083 | 12. | 033 |
| 3. | 293 | 13. | 490 |
| 4. | 004 | 14. | 103 |
| 5. | 234 | 15. | 285 |
| 6. | 200 | 16. | 348 |
| 7. | 257 | 17. | 298 |
| 8. | 397 | 18. | 443 |
| 9. | 095 | 19. | 354 |
| 10. | 114 | 20. | 470 |

Of course the same random permutation cannot be used over and over for the same investigation. Therefore, several permutations of the same range are usually printed together. For example, the book from which Figure 8-2 was taken contains 38 different permutations of 1-500 and (because they require less space per permutation) 960 different permutations of 1-9.

The disadvantage of tables of random permutations is that as of this writing there are no tested tables of over 1-1000 range generally available. Therefore, in selecting a total group of over 1000 items -- such as the 1152 drawers of the major chapter example, they cannot be used.

```
428   83  293    4  234  200  257  397   95  114  265   33  490  103  285
348  298  443  354  470  158  195  223  401  168  391  431  127  336  104
320   39  387   66  496  101   86  380  242  304   46   37  495  295  414
 18   68  375  485  185  343   34   25  358  361  384  402  416   92  125
188   15  473  308  474   16  197   24  329  106  243   54  198   80   28

186  349   79  203   72  413   85   56  107   75  418  425  347  394  269
475  386  253  263  437  193  180  423  405  244  389  251  487    2   43
 11  351  286  306  333  235  214  325  146  468  218  376   82  138   50
381  178  464  162  274   29  372  364  362  196  270  453   73   71  399
481  445  155  328  324  133  340   23  262  211  241  139  282  315  140

228  160  471  489  260  318  478  165   88  484  472  494  232  250  290
 74  153  287  115  254   97   49  132  365  486  310  497  432  393   61
151  458  457  120  322  187  404  137  346  216  499   51   63  182  176
360  411  449  396  410  240  117  281   32  313  118  412  271  488  247
479   10  350  332  126   76  229  296  128  385   99  174  109  288  335

290  119  239  466  424  237  129  398  209  148   94    3  102  316  463
319  433  444  477  121   47  379  202   27  435  179  238    8   98  113
278  201  400   45  420  167  382  170  338  144  455  221  122   52  317
 42  231  450  173  305  206  280   35  123  141  341  339  236   78    7
273  110  112  258   36  307  447  210  177   58  341  246   96  233  374

192  314  355   84   89  227  476  326  377  448  483  134  323  283  469
456  249   26  184  108  311  275  276   53  408  367   40  330  220  226
 44   20  116  145   67  373  465  164  152  194  131   59  267  157  438
 87  199  446  462  142  207  149  409  301  427  452  388  421  368  224
294   30  345   69  136  369  344  417  111  261  439  366    1   91  166

135   81  363  454  392  292    6  383  159  147  422  225  124   21   64
 38  434  334   70   65  156  353  105  171  217  451   93  337   31  492
 17  300  191  259  321  467  430  252   77  248   13  331  352  429  441
491  327  500  143  407  245  302  266  459  181  268  297  371  461  183
436  299  460  312  272    5  498  219  215  163  493  356  205  172  357

426   62  169   14  303  370  100   55   48  378  279  130   60  440  277
406  291  222  403  395  255  342   57  264  415  212   19  256  230   12
213  390  161    9  190  154   90  150  175  309  284  359  204   22  208
482   41  189  442  480
```

Fig. 8-2   A random permutation of the integers 1-500

Determining the Required Sample Size

In the North Carolina Union Catalog example we arbitrarily chose a sample size of 20 card drawers out of a total group of 1152 drawers. Is this sample size sufficient to furnish a reliable average of the inches of cards per drawer? Or do we need to sample 50 drawers or even 100 drawers to obtain an accurate answer? It is obvious that in both this type of problem and in time study problems (where the observations are units of time rather than units of cards) this is a crucial matter. The purpose of the next few sections of this chapter is to show how to answer this question.

Deciding on the Sample Reliability Required

The closer to 100% certainty that an investigator demands that his sample approach, the larger it will have to be. He must therefore begin by making two decisions as to how reliable an answer he needs or desires.

The most common practice is to use a 95% confidence level. This means that the sampler can be confident that his random observations will represent the facts 95% of the time. It also means that 5% of the time they will not. However, these are good odds. Since practicing management people find them satisfactory for most of their purposes, it is recommended that these guidelines also be adopted for library studies. The next most common confidence level is 99%. Since a 99% confidence level is seldom necessary in management practice, and since the 4% increase in certainty may require a substantial increase in sample size, it is recommended that it be used sparingly.

The second decision is with regard to the sample average. This average is obtained by summing the values of the observations and then dividing by the number of observations. In statistics this average is usually called the arithmetic mean and is designated by the symbol $\overline{X}$ (X-bar). In any sample the sample average is expected to vary somewhat from the true average of the total group from which the sample was taken. This variation is expressed in terms of percentage, which is then converted into a range of values.   As an illustration, let us use  $\pm 5\%$ variation and suppose that upon sampling the 20 card drawers, we obtained an average of 722 cards per drawer.   We multiply 722 by 5%: 722 x 0.05 =   36. The true average of the group then lies somewhere between 722 minus 36 and 722 plus 36 -- i.e., between 686 and 758.   The smaller the per cent of variation allowed -- i.e., the closer to the true average of the group that we insist our average sample be -- the larger our sample will have to be.   The most common practice is to insist on a variation of no more than  $\pm 5\%$, and it is recommended that this figure be adopted as a guideline for library studies.

To combine the decision of 95% confidence with that of $\pm 5\%$ variation means that the sampler can be 95% sure and that his sample average is not more than 5% above or less than 5% below the true average of the total group from which the sample was taken.

The Required Sample Size Formula

Once the confidence and variation decisions are made, it is possible to develop a specific formula to show when the sample is large enough to meet these requirements.   Although the formula for any given set of decisions is developed in the same general manner, its constant value will vary with each different set.   Only the formula for 95% confidence and  $\pm 5\%$ variation will be discussed in this text.   By studying the formula's derivation (See the chapter bibliography) the interested reader will learn how to construct comparable formulas for any other set of decision values.

The working formula for 95% confidence and  $\pm 5\%$ variation is:

$$N' = \frac{1600 \left[ N \Sigma x^2 - (\Sigma x)^2 \right]}{(\Sigma x)^2}$$

N' = the number of observations needed to meet the criteria of 95% conficence and  $\pm 5\%$ variation

N = the number of observations actually taken

$\Sigma$ X = the sum of the value of all the observations

$(\Sigma X)^2$ = the square of the sum of the values of all of the observations, i. e., $\Sigma X$ multiplied by itself.

$\Sigma X^2$ = the sum of the squares of the values of all the observations

and 1600 is the square of the constant used at this confidence level.

Recording the Observations

The observations should be recorded in a form convenient for calculation. There should be space for the $X^2$ values adjacent to the X values. If the Xs and $X^2$s are to be added mentally, most persons will find it convenient to record them in columns rather than rows, and with the decimal points in line. The space allowed for recording should be appropriate to the magnitude of the numbers being recorded. Lined paper, graph paper, accounting paper or a homemade form such as is shown in Figure 8-3 will help to keep the work orderly and manageable. The observation sheet should bear all pertinent information about the sampling: Where it was done, When, By whom, and the units in which the observations are recorded. The first 20 observations (the trial sample) of the North Carolina Union Catalog are recorded in Figure 8-3 as observations 1-20.

Determining if Additional Observations Are Necessary

We now have the required data to substitute into and solve the above formula:

(1) $$N' = \frac{1600\left[N\Sigma X^2 - (\Sigma X)^2\right]}{(\Sigma X)^2}$$

(2) $$N' = \frac{1600\left[20(1320.51) - (161.3)^2\right]}{(161.3)^2}$$

(3) $$N' = \frac{1600(392.51)}{26017.69}$$

(4) $$N' = \frac{628016.00}{26017.69}$$

(5) $$N' = 24 \text{ observations}$$

| Observation Work Sheet | | | | | | | | | |
|---|---|---|---|---|---|---|---|---|---|
| Units Sampled: | _Inches of cards/drawer_ | | | | | Name: | | _J. Smith_ | |
| Location: | _North Carolina Union Catalog,_ _UNC Library._ | | | | | Date: | | _9/1/1966_ | |
| Obs no. | Item se-lected | x | $x^2$ | Obs no. | Item se-lected | x | $x^2$ | Obs no. | Item se-lected | x | $x^2$ |
| 1 | 0059 | 9.0 | 81.00 | 15 | 0807 | 8.0 | 64.00 | 27 | 0411 | 5.8 | 33.64 |
| 2 | 0078 | 9.5 | 90.25 | 16 | 0839 | 8.4 | 70.56 | 28 | 0441 | 7.9 | 62.41 |
| 3 | 0082 | 8.9 | 79.21 | 17 | 0840 | 7.8 | 60.48 | 29 | 0454 | 8.3 | 68.89 |
| 4 | 0124 | 9.1 | 82.81 | 18 | 1057 | 5.5 | 30.25 | 30 | 0707 | 8.4 | 70.56 |
| 5 | 0126 | 8.8 | 77.44 | 19 | 1072 | 7.4 | 54.76 | | Total | 240.5 | 1961.75 |
| 6 | 0213 | 8.1 | 65.61 | 20 | 1106 | 8.1 | 65.61 | | | | |
| 7 | 0312 | 8.3 | 68.89 | | Total | 161.3 | 1320.51 | | | | |
| 8 | 0398 | 7.5 | 56.25 | 1-20 | | 161.3 | 1320.51 | | | | |
| 9 | 0487 | 7.6 | 57.76 | 21 | 0109 | 9.2 | 84.64 | | | | |
| 10 | 0505 | 8.5 | 72.25 | 22 | 0138 | 9.8 | 96.04 | | | | |
| 11 | 0676 | 8.5 | 72.25 | 23 | 0208 | 8.9 | 79.21 | | | | |
| 12 | 0700 | 7.8 | 60.84 | 24 | 0239 | 6.6 | 43.56 | | | | |
| 13 | 0746 | 5.7 | 32.49 | 25 | 0340 | 7.3 | 53.29 | | | | |
| 14 | 0801 | 8.8 | 77.44 | 26 | 0365 | 7.0 | 49.00 | | | | |

Fig. 8-3  Sample observation work sheet.

For those readers who prefer to avoid equations, the form illustrated in Figure 8-4 was developed as an aid for the quick solution of the computational form of the sample size equation (See Figure 8-4).

The solution for the 20 drawer sample is recorded in the column headed: Observations 1-20. An explanation of the form is given below:

Step 1.  First, the values must be computed for the $x^2$ column of our observation sheet (See Figure 8-3). In some cases the answer will be known immediately e. g., $(8.0)^2 = 64.0$. In other cases it will not. It is recommended that the squares that cannot be worked mentally be obtained from a table. One common-

| Step | Item | Observations 1-20 | Observations 20-30 | Observations |
|------|------|-------------------|--------------------|--------------|
| 1. | $\Sigma X^2$ | 1320.51 | 1961.75 | |
| 2. | N | 20 | 30 | |
| 3. | $N\Sigma X^2$ | 26410.20 | 58852.50 | |
| 4. | $(\Sigma X)^2$ | 26017.69 | 57840.25 | |
| 5. | $N\Sigma X^2 - (\Sigma X)^2$ | 392.51 | 1012.25 | |
| 6. | $[N\Sigma X^2 - (\Sigma X)^2] \times 1600/(\Sigma X)^2$ | 24 | 28 | |

Fig. 8-4 Determining the required sample size for
the North Carolina Union Catalog Sample.

ly available table (Barlow's) gives square of numbers up to 12,500
and there are many tables giving squares of numbers up to 1000
(See the chapter bibliography). Secondly, the $X^2$ values are added
which give $\Sigma X^2$. This is the value recorded on the form.

Step 2. N is the number of observations.

Step 3. This value is the product of $\Sigma X^2$ and N.

Step 4. First add the X column on the observation sheet
(Figure 8-3) to obtain $\Sigma X$. Then look up the square of this value
in a table. For convenience of subtraction, the $(\Sigma X)^2$ digits
should be lined up directly under the $N\Sigma X^2$ digits.

Step 5. This value is the remainder of $N\Sigma X^2 - (\Sigma X)^2$.

Step 6. This expression is solved quickly on the slide rule.
How to do this is described in the section of this book explaining
the slide rule (See Chapter IX). A desk calculator may be used if
one is handy. The answer is N', the number of observations required.

For the North Carolina Union Catalog sample, we need 24
observations, or 4 more than the 20 we have taken.

Making the Additional Observations

The formula indicates that four additional observations will
be required. However, the formula cannot predict the exact varia-

tion among these additional observations.    If this variation happens to be somewhat greater than that among the observations already made, then, with such a small sample, a few more observations will be needed than stated by the formula.    To be on the safe side we shall therefore make ten additional observations.

The last digit of the 20th observation was taken from the 3rd group of 1000 digits, row 19, column 37 of the random table in this book.    Therefore, we begin to read the table for observations 21-30 with the 3rd 1000 digits, row 19, and column 38.    We continue as before to read the table as a book.    The reader should verify that the following drawer numbers are selected:

| | |
|---|---|
| 21. 0454 | 26. 0208 |
| 22. 0441 | 27. 0138 |
| 23. 0239 | 28. 0707 |
| 24. 0411 | 29. 0365 |
| 25. 0109 | 30. 0340 |

The $X$, $\Sigma X$, $X^2$ and $\Sigma X^2$ values of these observations are recorded in Figure 8-3.    From the cumulative data of observations 1-30 the required sample size (28) is computed using the procedure outlined in Figure 8-4.    We conclude that the 30 observations are sufficient for 95% confidence and $\pm 5\%$ variation, and the problem can now be completed.    Since 30 drawers contain 240.5 inches of cards, the average number of cards per drawer for the sample is 240.5/30, or 8.017 inches.    If this average is within $\pm 5\%$ of the true average for the catalog as a whole, then (since 8.017 x 0.05 = 0.401):

8.017 inches $+ 0.401$ inches = 8.418 inches of cards per drawer

8.017 inches $- 0.401$ inches = 7.616 inches of cards per drawer

The possible range of the average is then from 7.616 inches to 8.418 inches.    At 90 cards per inch, we then determine the total number of cards in the catalog as follows.    Somewhere between: 8.418 inches of cards per drawer x 1152 drawers x 90 cards per inch = 872,778 total cards; and 7.616 inches of cards per drawer x 1152 drawers x 90 cards per inch = 789,627 total cards.    This answer will be quite accurate enough for most management purposes.

The reader who has been struggling along with ideas new to him may at this point be inclined to feel that sampling is more trouble than it is worth.    A comparison of the approximate times involved with and without sampling for the North Carolina Union Catalog example shows clearly that this is not so.    For the sampling technique.

1. Selecting the 20 item sample from the random number table — 15 minutes
2. Arranging the sample drawer numbers in order — 05 minutes
3. Making and recording the observations — 15 minutes
4. Making all additions, calculations for sample size formula (incl. checking) — 15 minutes
5. Selecting and arranging 10 more sample drawers from the random number table — 10 minutes
6. Making and recording these 10 additional observations — 10 minutes
7. Reworking the sample size formula — 10 minutes

Total    1 hr. 20 min.

Now suppose that the sampling technique had been rejected and instead we decided to measure all the drawers.   Since there are 1152 drawers, and making and recording the observations averaged about 3/4 minute per drawer, the work would take about 3/4 x 1152 minutes, or 14-plus hours, not to mention the tediousness of this type of work.   This time does not, of course, include summing the 1152 individual drawer totals.

## Either-or Sampling

So far this chapter has dealt only with quantities that can be arranged among themselves in an order of magnitude-- one card drawer contains more or fewer inches of cards than another drawer.   However, it sometimes becomes necessary to draw samples from universes that can only be classified in terms of either-or. The cataloger either is or is not at his desk.   The book either is or is not returned on time.   The patron either uses or does not use the card catalog.   The book a student borrows is either fiction or non-fiction.   And so on.   In these cases it would be nonsensical to say that one category is greater or less than another in the sense that 3 is greater than 2 but less than 4.

Although this sort of sampling has a variety of applications, its most frequent management use is in work sampling. Work sampling in turn has a variety of uses. One frequent application is the determination of the amount of time per day that an employee is idle or productive. Work sampling can also be used to distinguish different categories of productive work such as typing, filing, stamping, and shelving. The exact times during the day when an employee will be observed are determined by some random process. A random number table is commonly used. Each observation is then recorded simply as a mark under "idle" or "working." (One can of course add other categories if he desires more detailed information.) Just as with sampling quantities with magnitude, the number of observations required for either-or sampling will vary as with the fluctuations in the data, and the confidence level and average variation limits imposed by the analyst. For a confidence level of 95% and $\pm 5\%$ allowable variation, the computational form of the sample size formula (easily solved on the slide rule) is:

$$N' = \frac{1600\ (1-P)}{P}$$

where

N' = the number of observations required

P = the fraction: number of occurrences of the activity or delay being measured/total number of observations. It may be expressed as either a decimal or fraction -- use whichever is most convenient in a specific instance

The final answer may be simply the percentage of the total working day or week an employee or department is idle or productive. Or (for the moment associating P with idle, rather than productive, time), it is possible to convert these percentages into time idle or productive by the formulas:

Hours idle per week = Hours per week on Job x P

Hours productive per week = Hours per week on job x (1-P)

As an example, suppose that the librarian of a sizeable library wants to know the feasibility of utilizing part of the time at the circulation area in other departments of the library at non-peak hours. If there are several employees in circulation, work sampling would be an efficient means of answering this question, for a simultaneous sampling of all of the workers could be made by a single observer. In contrast, a continuous time study, in addition to being more burdensome, would require additional observers -- perhaps even one observer for each employee. The work sampling study may ascertain that some employees are much more idle than others. This may be no fault of the workers, but rather point to poor supervision or poor work procedures or both. The study could also be designed so that the analyst also kept a record of what each individual was doing each time he was observed in order to learn how the circulation people spend their time.

Bibliography

Random Number Tables

Barnes, Ralph M. Work Sampling. 2nd ed. New York: Wiley, 1957.

Contains a 17,500 digit extract from the 1949 table of ICC's Bureau of Transport Economics and Statistics.

Fisher, Ronald, and Yates, Frank. Statistical Tables for Biological, Agricultural, and Medical Research. 6th ed. rev. and enl. New York: Hafner, 1963.

Table XXXIII consists of 7500 two-figure random numbers arranged on six pages. Includes an introduction describing some useful labor-saving devices.

Handbook of Mathematical Tables. 2nd ed. Supplement to Handbook of Chemistry and Physics. Cleveland: The Chemical Rubber Co., 1964.

Contains a 14,000 digit extract from the 1949 table of the ICC's Bureau of Transport Economics and Statistics.

Interstate Commerce Commission. Bureau of Transport Economics and Statistics. Table of 105,000 Random Decimal Digits. Washington, D C., 1949.

Does not seem to be generally available. Not a depository item. It has been identified as Bureau Statement no. 4914, File no. 261-A-1. Rows and columns are spaced in groups of five.

Kendall, Maurice G., and Smith, Babington B. "Randomness and Random Sampling Numbers," Journal of the Royal Statistical Society, CI (1938), p. 147-66.

The table of 5,000 random digits in this chapter was originally printed on pages 164-66.

----, ----, ----. Tables of Random Sampling Numbers. Cambridge: Cambridge University Press, 1946. (Tracts for Computers, no. 24).

One-hundred thousand random digits. Broken into groups of 1,000. Each 1,000 digits are arranged into 25 numbered rows and 10 column groupings of four digits each. The best random table buy around.

Rand Corporation. A Million Random Digits with 100,000 Normal Deviates. Glencoe, Ill.: Free Press, 1955.

A fine table, but a stiff price of $17.50. Twenty-thousand numbered rows; fifty rows and fifty columns per page. Spacing is by fives.

Tippett, L.H.C. ...Random Sampling Numbers. Cambridge: Cambridge University Press, 1927. (Tracts for computers, no. 15).

Forty-one-thousand-six-hundred digits with the columns spaced in fours and the rows in fives. A tried and true table, and an excellent buy.

### Tables of Random Permutations

Cochran, William G., and Cox, Gertrude M. Experimental Designs. 2nd ed. New York: Wiley, 1957.

Chapter 15 contains 1,000 random permutations of 9 numbers and 800 permutations of 16 numbers. Also found in the first edition, 1950.

Fisher, Ronald A., and Yates, Frank. Statistical Tables for Biological, Agricultural, and Medical Research. 6th ed. rev. and enl. New York: Hafner, 1963.

Table XXXIII1 gives 750 random permutations numbered 0-749 of the numbers 0-9. Table XXXIII2 gives 200 similar permutations of the numbers 0-19. Also includes a helpful introduction to the tables.

Moses, Lincoln E., and Oakford, Robert. Tables of Random Permutations. Standard: Stanford University Press, 1963.

Contains permutations of 9, 16, 20, 30, 50, 100, 200, 500 and 1,000 integers. The best tables available as of this writing.

### The Sample Size Formula

Barnes, Ralph M. Motion and Time Study. 5th ed. New York: Wiley, 1963. p. 365-8.

Mundel, Marvin E. Motion and Time Study. 3rd ed. Englewood Cliffs, N.J.: Prentice-Hall, 1960. p. 363-68.

Tables of Squares

Barlow's Tables of Squares, Cubes, Square Roots, Cube Roots and Reciprocals of all Integers up to 12,500. 4th ed. New York: Chemical Publishing Co., 1954.

The most useful such compilation because of its great range. There are many tables giving squares of integers 1-1000. One will be found in any edition of the Handbook of Chemistry and Physics, or in practically any other general collection of mathematical tables. Such tables are also found often in the back of statistics texts.

Work Sampling

Barnes, Ralph M. Work Sampling. 2nd ed. New York: Wiley, 1957.

A substantial portion of this material may be found also in Barnes. Motion and Time Study. 5th ed. Chapter 33.

Hansen, Bertrand L. Work Sampling for Modern Management. Englewood Cliffs, N.J.: Prentice-Hall, 1960.

Heiland, Robert E., and Richardson, Wallace J. Work Sampling. New York: McGraw-Hill, 1957.

Poage, Scott T. 'Work Sampling in Library Administration." Library Quarterly, XXX, No. 3 (July, 1960), p. 213-18.

Sampling

Slonim, Morris J. Sampling in a Nutshell. New York: Simon and Schuster, 1960.

A non-mathematical approach.

Sampling is a part of the broader subject of statistics. Elementary sampling is therefore covered in beginning books on statistics, of which there are dozens in print. Two standard ones are:

Franzblau, Abraham N. A Primer of Statistics for Non-Statisticians. New York: Harcourt, 1958.

Moroney, M.J. Facts from Figures. 3rd and rev. ed. Harmondsworth, Middlesex, Penguin Books, 1956.

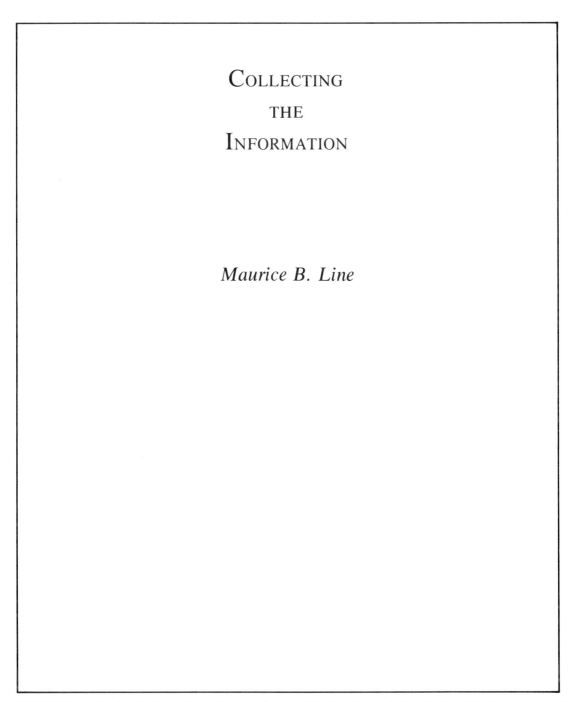

COLLECTING

THE

INFORMATION

*Maurice B. Line*

ONCE THE GENERAL design, sampling, timing, etc have all been determined, the method or methods of collecting data must be decided. Among the factors to be taken into account are, apart from the sort of information required, the usual ones of time and money —this is often the most expensive part of the survey operation— number and experience of staff available, and degree of cooperation required from those contributing information.

It may be possible to obtain a certain amount of the data required from documents already in existence. Reports of chief librarians and library committees, whether printed or duplicated for distribution, or available only locally, are obvious sources which should be exploited to the full before lengthy questionnaires are sent off to busy librarians. Unfortunately these documents are very much less useful than they might be because there has been until recently no standard for the collection and presentation of relevant, detailed, accurate, consistent and comparable figures. (It must be hoped that the recommendations in the American Library Association's *Library statistics* (1966), or preferably something rather more detailed, can be accepted in Britain.) The details given cannot be relied upon unless we know the definitions and criteria on which they are based—and it may require a survey to find out what these are. Similar considerations apply to other published information. Although therefore existing information should never be ignored or neglected, it has very limited value for any survey with pretensions to accuracy. Its main use may be as a starting point and to provide background information.

Where the information required concerns several libraries, and where its nature does not demand that it be obtained direct from library users, it may be collected by a personal tour or series of visits. This method, like the use of published information, is in any case a good way of filling in the background. Whatever sophisticated techniques are used for the main survey, a personal visit to the scene is always desirable, whether at an earlier or a later stage, or both. At an earlier stage it can provide a setting against which other techniques are developed; at a later stage it can help to explain factors which have given rise to odd results.

The survey may, however, be of an impressionistic rather than of a scientific nature, and in this case tours may be the only method used. Surveys of this type are by no means to be despised because they do not require the use of sophisticated methods; they are perhaps of most use in conveying a picture on a large scale, *eg* of library resources in a particular country, but they can be used for more specific areas, such as special collections in university libraries.

The usual procedure is for the investigator, having planned his itinerary, to outline the sort of information he wants, without tying it to too close a schedule, and to give reasonable advance warning to the librarians whose libraries he is to visit, if necessary sending them a note of what he wants to know so that any factual information can be prepared in advance. The visitation is in fact an individual combination of the techniques of observations, interview and questionnaire on a small scale.

Sometimes the information wanted cannot be obtained in any other way. If, for example, the investigator wants to know how well a library is organised, he cannot ask the librarian, and no existing data will tell him, though they may offer broad hints. Or he may want to survey reference services in British libraries—something which cannot easily be explained in words or identified, and which is best studied on the spot. The ability of the human brain to receive large quantities of non statistical information and form from them an accurate and comprehensive picture must not be under-estimated.

If the survey is to cover a large number of libraries, it may be too much for one man. Since the value of an impressionistic survey lies largely in the fact that the impressions all come from one source, and therefore form a consistent picture, the team survey is unusual. If more than one person is involved, such constant and close co-operation is needed between the members of the team that the saving in time may be small.

Some kinds of survey can be carried out by direct observation. Suppose, for example, we are studying catalogue use, and wish to know whether the main use of the catalogue is by users or library staff, we can do this by observing the actual use made, preferably on a random sample of days (assuming that we have learnt to distinguish staff from readers), by counting the number of consultations made by each category and by timing them. If we wish to categorise readers further (in a university library we might want to divide them into staff, students and others) this could probably be done with a high degree of accuracy by one of the library staff. Again, one could learn a lot about use of a classified catalogue by observation: for example, whether users consult a Dewey index first, whether their initial catalogue search is by author or subject, whether they ask library staff, whether they abandon the search almost as soon as looking at the catalogue, and so on. This method could also be used to discover whether, if junior and senior staff are both available at an inquiry desk, there is any clear preference on the part of readers for one or the other.

Reader watching can be as rewarding an occupation as bird watching. Indeed it has some similarity, in that neither readers nor birds should be aware that they are being watched. Observation has much to commend it in the limited areas where it can be used, as it is cheap and not liable to some of the errors which afflict more sophisticated methods. It does however demand concentration, and often considerable knowledge, on the part of the observer; moreover, if several observers are used, it must be quite certain that they are observing exactly the same thing.

One rather special kind of survey method which has particular applications in the library field might be called the self study by the sample, in which the individuals in the sample keep a record of their own activities. The commonest form of this is the reading diary, in which each person records what he reads as he reads it, or, if he is a researcher, what material he uses for what purposes. (Two examples of this method can be found in the *International Conference on Scientific Information proceedings* (Washington, 1959): R M Fishenden 'Methods by which research workers find information' pages 163-180, and I H Hogg and J R Smith 'Information and literature use in a research and development organisation' pages 131-162.) This procedure is open to grave methodological objections. In the first place, the individuals cannot form a random sample. It is certainly possible to draw a random sample and ask all those selected to cooperate, but those who can be induced to do so are likely to be a pretty small self-selected group. This group will become even smaller and less representative if the period over which diaries are kept is long enough to be worthwhile, so that in the end one is left with a few individuals whose persistence is only one of their many nontypical features. Even if they are all honest and conscientious, the likely error is fairly substantial. If the nature of the inquiry allows it, it is much more satisfactory for a large number of individuals to keep diaries for a very short period (say four days). For some inquiries, however, the diary method may be the only way of obtaining data; and even at the worst a few nontypical individuals can yield information of some interest and value.

A method which has been employed in some use studies is the *critical incident technique*. In this, the selected individuals report in detail (either verbally to the investigator or in writing) actual incidents which contributed significantly to the activity or behaviour under investigation (see J C Flanagan 'The critical incident technique' *Psychological bulletin* 51(4) July 1954, 327-358; and, for a study which used it, Auerbach Corporation *DoD User Needs Study: phase I, final report* (AD 615501-2), 1965).

Another method of self investigation is a form of activity sampling, namely ascertaining what proportion of time is devoted to various activities by taking random times and noting what is being done. This has been used in the investigation of scientists in order to find out how much time they occupy in literature use. Cooperation is required here, but the amount of work involved is much less than in the keeping of reading diaries; the period can be shorter and the number of people involved greater. The chief difficulty lies in making sure that those concerned use random times. Much ingenuity has been devoted to this problem. A random alarm which can be carried about by the individuals has been developed for one or two such studies (see M W Martin 'The use of random alarm devices in studying scientists' reading behavior' *IRE transactions on engineering management* EM-9(2) June 1962, 66-71).

The principal survey methods are, however, questionnaires and interviews. They are often used to investigate the same sort of thing and are thus in many cases alternatives to be considered. The purpose of both is to elicit information from human beings which would otherwise be difficult or impossible to obtain.

They are both based on *schedules*, which may be more or less formalised. Most studies of library use which go beyond mere issue statistics, and all studies of attitudes, must be carried out by interviews or questionnaires. Much the greater portion of this chapter will therefore be concerned with these methods, and much of what is said about questionnaires is relevant to interview schedules.

The first step when the use of questionnaires is contemplated is to find out whether any questionnaires previously used for a survey can be repeated in whole or in part. Our knowledge of libraries and their users would be much more complete if the investigations which have been conducted were comparable. Questionnaires are always capable of improvement, but if a relevant questionnaire appears to be on the whole a satisfactory instrument, the urge to improve should be resisted (although the wording of individual questions can of course be clarified if necessary). The gain from having a body of information collected from numerous libraries and communities on a comparable basis would be enormous.

Questionnaires are usually sent out through the mail, or distributed in some other way, to a selected sample. Personal delivery is sometimes feasible; for example, to all the students in a university department, or to all users leaving the library on specified dates. But in most cases the sample will receive the questionnaires in an impersonal way, and it is therefore necessary to make them completely self explanatory. An accompanying letter may, and usually should, be sent with each questionnaire, giving whatever background is necessary (see chapter two, page 38-39), but this will not reduce the necessity for the questions themselves to be absolutely clear.

This means first that the investigator must be absolutely clear what he wants to know. Vague thinking is bound to produce vague or misleading questions. Even when the investigator knows what he wants to know, it may be difficult to put his thoughts into words at all, let alone into a concise form of words that will be readily understood by people to whom his thoughts are quite new. (This difficulty is illustrated by almost any examination paper.) If it is humanly possible to misunderstand anything, there will be those taking part in a survey who will misunderstand it. The attempt to make a statement or question totally unambiguous may result in language of the type associated with legal documents, and for the same reason—it may be precise but incomprehensible. On the other hand, 'plain' English, out of context (and items in a questionnaire can have very little context) is capable of a surprising variety of different interpretations. A subtle balance must thus be struck between imprecision and unintelligibility.

Special attention must be paid to library terminology. It is quite wrong to assume that everyone knows what librarians mean by catalogues, bibliographies, classifications, periodicals, pamphlets or many other terms that we take for granted (not that there is any common agreement among librarians on the definition of a periodical or pamphlet). In any question concerning library or bibliographical matters an explanation should be provided, whether in the wording of the question itself or in a note (*eg* ' The *author catalogue* is the card index to the books in the library arranged alphabetically by author; it is housed in the wooden cabinets to the right of the entrance as you come in. Have you had occasion to consult it during the current month? ').

Clarity and precision in the wording of questions are both necessary. A question can be clearly phrased but mean different things to different people. Efforts must be made to ensure that all respondents are answering the same question. Suppose that one wishes to measure the variation in library use among individual users. The question ' How many times did you use the library last year?', though clear enough as far as its English goes, is open to numerous objections. What, for instance, does ' last year ' mean? During the twelve months ended yesterday, or during the previous calendar year? For the questioner's purposes one period of twelve months may be as good as another, but only so long as all respondents are thinking of the same period; otherwise it will be impossible to attribute with certainty the variations in answers to differences between individuals. Secondly, what is meant by ' use the library '? Some may confine their interpretation to intensive use, some may include the hasty return of an overdue book. The question therefore might be rephrased: ' How many times during the twelve months July 1965-June 1966 did you use or come into the library, for whatever purpose? '

However, it is not really reasonable to expect people to remember accurately how many times they used the library over so long a period. A few conscientious individuals will sit down and try to calculate; and even they will be wrong as often as right, unless their use is very infrequent. Most people faced with such a question will offer a blind guess; if in addition it occurs to them that this is not a very sensible question, they may take less trouble over the remainder of the questionnaire. The period over which they are asked to remember should therefore be reduced, if this can be done consistent with the purpose of the questionnaire. (Recollection tends to fall away sharply in accuracy after a period of about two months.) Another aid to memory, and therefore to accuracy in answers, would be provided by categorising types of use into, *eg,* borrowing, consultation, returning books, and so on. If this is done a further difficulty arises concerning dual purpose visits: is the respondent to count these twice or once or not at all? This particular problem could be solved by phrasing the categories 'for consultation only', 'for borrowing only' . . . 'for two or more of these purposes on the same occasion'.

There is another good reason for categorising types of use, which becomes obvious when the investigator asks himself what sort of answers he is likely to get from the general question. He would have a very crude figure of visits to the library, without any idea of what these visits were for. If the library in question was prompt in sending out overdue notices, a large number of uses might be flying visits, perhaps to return a neighbour's books. Unless these figures are broken down, they mean very little.

It is, incidentally, dangerous to hope that minor errors in interpretation of questions, or in answers, will cancel one another out. They may, but on the other hand they may cumulate and multiply, and cause a serious distortion in the overall results. Moreover, it is not possible to know which is happening—cancellation or cumulation of errors—so that no allowance can be made. There are numerous unavoidable types of error in surveys; any risk of increasing them unnecessarily should be avoided at all costs.

The example above illustrates the difficulty of obtaining a simple piece of information and the complications created in trying to get round the difficulty. Difficulties of another kind arise with questions that cannot be asked directly, either because they might give offence or because they might produce untruthful answers. For example, if you wished to know whether anyone had found any of the library staff difficult to deal with, you would not ask 'Has any member of the staff ever been offensive to you?' since this would immediately arouse suspicion about the motives of the survey. You would probably have to tackle this indirectly by asking such questions as 'Do you normally find the staff helpful/not very helpful to deal with?', or by employing a rating scale which takes away some of the sting by offering a range of alternatives to be checked (see below page 55 ff).

Again, in attempting to estimate the efficiency of a reference service, you could not expect to obtain useful answers to a question 'Do you consider the reference service adequate?' Even if all readers know exactly what is meant by the reference service, their conception of adequacy may be highly variable and extremely crude. Since the direct question will not produce useful answers, the subject must be approached in a more roundabout way by asking several more specific, and, if possible, factual questions. For example:

*If you need help when using the library, do you normally consult the library staff?*

*For what kinds of purpose have you sought their help in the last six months? (For example, wanting a book on a subject, trying to find a particular piece of information, etc.)*

*Were your questions answered to your satisfaction?*

*Were any not satisfactorily answered? (Please state briefly what they were.)*

*If so, what was the reason, in your opinion? (For example, the staff were too busy at the time, they tried but could not find the answer, etc.)*

(Those questions could be substantially improved by offering precoded answers: see below page 53-54.)

Unless something very simple is being asked, it is difficult to formulate questions in such a way that they are both reliable and valid (see page 32). Factual questions have a high reliability, but they cannot be used for anything at all subtle. (The same problem exists with written examinations.) 'Did you go to church last Sunday?' is a very reliable question, but hardly a very valid indication of the respondent's spiritual well-being. 'Are you right with God?' may be a much more valid question, but would be extremely unreliable. To some extent this conflict between reliability and validity can be resolved by asking several factual questions which have been shown (perhaps by factor analysis) to correlate highly with a certain type of personality or attitude. This method is a feature of several of the more common personality tests, where seemingly innocuous factual questions ('Do you go to many parties?') or vague inquiries ('Do you consider yourself to be optimistic?') are often excellent indicators of personality structure. Questionnaires of this kind are usually the result of many years' work and constant refinement, and similar effort would be required to produce really good questionnaires concerning, for example, user needs (this is one of the reasons why the numerous user studies have produced little of value).

All questions should be made as objective as possible, so that the respondent cannot guess what answer the investigator would prefer him to give, or even whether he has any preferences at all. Loaded questions, so worded that a totally uncommitted respondent is more likely to give one answer than another, should at all costs be avoided. Sometimes the loading is deliberate: 'Should obscene books be provided by public libraries?' is little better than asking why children prefer Krapkins as a breakfast food. Both questions are propaganda, not objective inquiries. Loading may be not deliberate, but still obvious: 'What advantages do you think the new borrowing regulations have over the old?' is inviting compliments, and gives no opportunity to state disadvantages (though some respondents may create their own opportunity). Subtle loading is much harder to detect; every individual has his own ideas, and it is very hard for these not to intrude into a questionnaire. 'Do you think public library users should pay a small sum every time they borrow a book?' and 'Do you think authors are entitled to receive some payment for the use of their books by people who do not buy them?' are both apparently honest questions, but they are respectively asking for the answers 'No' and 'Yes' (it is possible experimentally to induce the same respondents to contradict themselves within one questionnaire by the use of loaded questions). Ideally the experimenter should regard his ideas as a hypothesis to be tested, not as a doctrine to be confirmed, however sure he may be of them. In practice the best safeguard is to ask as many people as possible with as many differing views as possible to scrutinise your questions.

The importance of objectivity cannot be overemphasised. Frequently the newspapers contain reports of a 'survey' conducted by some pressure group which purports to prove what it set out to prove. By carefully (or carelessly) worded questions it has been 'proved' that a majority of people in the same town want and do not want comprehensive schools, that they want and do not want nationalisation of industry, and that they will vote labour *and* conservative. This does a great deal of harm to the investigator who wants to conduct a serious and well considered survey. (An interesting example of two surveys on the same topic which gave completely divergent results is given by C A Moser *Survey methods in social investigation* (1958) page 226.)

However objective the investigator may be, he must watch for words which, though innocent to himself, may be emotive to some of the recipients of a questionnaire. Everyone has certain words which evoke a particular emotional response in him (in other words, which act as a conditioned stimulus). While there is no possibility of totally escaping this effect, which is inherent in the nature of language, some words are emotive for particular sections of the community, particularly words which carry racial, religious or social overtones.

If an inquiry clearly emanates from a library (the source of a questionnaire should of course always be stated), however objective the actual questions are, it may be hard to convince respondents that the library is not fishing for compliments, and that it really does want to ascertain the facts, however unfavourable. In such cases a firm statement can be included in the accompanying letter to the effect that the library is concerned to evaluate its performance honestly, and that this will be possible only if everyone gives his completely honest opinion, however uncomplimentary.

The example above concerning reference services illustrated the use of several questions where one direct question would be meaningless. It is often useful, where complex matters are under consideration, to approach them from more than one angle by asking two or more indirect questions designed to produce similar information. These will help to provide a more rounded picture. Moreover, if the answers tally, conclusions can be drawn with more confidence than if only one question is involved. If they are contradictory, it is a clear warning that something is wrong.

For important surveys, a series of pilot studies can be conducted, each asking a large number of questions relating to one aspect of the survey. If a *correlation matrix*, showing all possible relationships between answers (see page 90), is constructed from the results, and a factor analysis carried out, some questions may be found to have a higher validity than others, and these questions can then be used in the main survey. This is much more satisfactory than trying to guess which two or three of many possible questions are likely to be most valid. (This technique is also very valuable for highlighting the crucial areas when new ground is being explored: see A C McKennell ' Correlational analysis of social survey data ' *Sociological review* 13(2) July 1965 157-181.)

Categorisation of questions by breaking them down into possible answers has been mentioned as an aid to the respondent. It not only helps his memory, but provides him with a range of alternatives, some of which might not come to his mind when answering a questionnaire quickly. This advantage is at the same time a disadvantage, in that it may actually limit the range of possible answers. Few investigators can be confident that they have thought of all the possible replies to all their questions, and even if they have there may be so many unlikely ones that it would be absurd to list them all. A reasonable compromise is to follow Dewey's example by providing for all

the main eventualities and adding at the end a category of ' other '
(followed by ' please specify ').

A further advantage of categorisation is that it helps to avoid
ambiguity and misinterpretation by directing the respondent's thought
along the same lines as the investigator. There is no reason why
categories should be mutually exclusive. Overlap between them is
indeed often desirable, provided that the categories are clearly distinguishable one from another; *eg* ' borrowing ' and ' consultation ' are
not mutually exclusive as uses of a library, since a volume consulted
may be borrowed, but they are distinguishable as physical activities.

Questions whose answers are categorised on the form are *precoded*,
as opposed to *open ended* questions, to which the respondent is free
to give any response. In the former case, analysis of the data is
obviously far easier, since the answers can readily be given numbers
or symbols. With open ended questions, the investigator has to
examine all the responses before he can group them into categories. As
an illustration, let us take a fairly simple question about library use.
As open ended version might be ' How often do you use the library
(for whatever purpose)? ' This is too vague to be of much use, and
it would be improved by suggesting examples of the *sort* of answer
required. The question would then become ' How often do you use
the library on average, for whatever purpose? (*eg* once a week, once
a month, etc) '. A precoded version might be:

*How often on average do you use the library?*

| | |
|---|---|
| *more than once a week?* | ☐ |
| *once a week?* | ☐ |
| *once a fortnight?* | ☐ |
| *less frequently?* | ☐ |
| *not at all?* | ☐ |

*(please tick box as appropriate)*

All possibilities are catered for, and the chief difficulty lies in guessing
the distribution of answers. For example, it may turn out that seventy
per cent of the sample use the library more than once a week, and
only two per cent once a fortnight. If this had been anticipated, the
first category could have been broken down more and the third
abolished. You need to guess the answers before you can frame the
questions. This is one of the many factors that make a pilot survey
desirable (see below page 60 and 70).

This is a clear case where precoding is an advantage. A question such as 'What sources of reference do you normally use when searching for material on a topic?' presents more problems. It obviously requires an explanation if the respondent is to understand what is wanted. If examples are used as an explanation, there is the danger that the respondent will look no further than the examples. The same applies to any list of reference sources unless it is exhaustive. It is more likely that a particular kind of source will be marked if it is mentioned, than written down by the respondent if it is not mentioned. On the one hand you do not wish to put ideas into the respondent's head, or to give some ideas prominence at the expense of others; on the other hand, the open ended question carries the risk that he may not be able to think of anything at all. The problem is how to guide the respondent's thought without taking it to any particular destination. (With this particular example, it might be better to ask the recipient how he goes about it when he wants to find out about a subject; he would be asked to explain as clearly as possible in his own words what sorts of books or catalogues he used.)

With mail questionnaires, it is generally wise to have a preponderance of precoded questions. First, the saving in analysis time can be very great. Secondly, with precoded questions the respondent has only to interpret what you mean by the questions. With open questions you have to interpret the respondent's answer to what he interprets your questions as asking, so that the dangers of misunderstanding are multiplied. Thirdly, precoded questions are easier and quicker to answer, and although this may lead to some carelessness, open questions do not necessarily produce fuller answers, simply because some respondents may not bother with them.

It is however often a good plan to include one or two open questions in a questionnaire. Recipients of forms sometimes have hobby horses they wish to ride, or bees in their bonnets waiting for an opportunity to swarm. If provision is not made for this, it may make its presence felt in the rest of the questionnaire; the indignant respondent may ignore all your careful precodings and write an essay of his own, of marginal relevance to your question. In any case, many comments, solicited or unsolicited, are quite reasonable, and the judiciously chosen quotation can make the report of the survey very much more readable.

It should be noted that *considered* opinions are particularly hard to obtain. It might be thought that one of the advantages mail questionnaires have over interviews is that they allow time for reflection.

Unfortunately, you can only surmise whether a given respondent has reflected or not, so that the responses are a mixture of considered and unconsidered opinions without any objective means of knowing which is which. Moreover, while it is true that the longer a person reflects the more considered his opinion, the less personal it tends to be also; he may try to think what opinion would be considered socially acceptable or ' responsible ' and modify his initial reaction accordingly.

The measurement of attitudes has received a great deal of attention during the last fifty years, and research is continually being conducted on it. The following paragraphs are intended merely to indicate the range of possibilities open to the library investigator wishing to measure attitudes. No one should venture on the use of any but the simplest methods without a thorough knowledge; a few of the more approachable and important writings on the subject can be found in the references at the end of the chapter.

It may be said at once that the open question (eg ' How do you like the library? ') is virtually useless for the measurement of attitudes, except perhaps to encourage respondents to air their views generally. Before any serious use could be made of the replies, they would have to be grouped, sorted into some kind of scale of favourability, coded and analysed. The effort is not likely to be worthwhile.

The commonest method used is the *rating scale*. Before discussing this a little more fully, however, mention should be made of two other methods which may be suitable for some special purposes, as they possess high validity and reliability.

The first is the method of *pair comparisons*, whereby several items are paired with one another in as many combinations as necessary (the number of pairs required, if $n$ is the total number of items, is $\frac{n(n-1)}{2}$, so that twelve items would yield sixty six pairs). The respondent chooses between each pair, and from the results the items can be ranked in a single order of preference and given scale values.

The second method is that of *rank order;* here the respondent ranks a list of items in order of preference, and the end product is similar to that of pair comparisons. This method is far more economical in time and effort, on the part of respondents as well as of investigators, particularly when a large number of items is involved. Even so, the computation can be pretty tedious. These methods might be used to compare readers' assessments of several given books, or to assess the relative importance to library users of various aspects of library service.

Rating scales can be of several kinds. The most usual is simply a form of precoded question, *eg*

*Do you think the library's stock in your subject is*:
  *inadequate?*
  *adequate?*
  *good?*
This might with profit be extended:
  *seriously inadequate?*
  *inadequate?*
  *adequate?*
  *good?*
  *excellent?*

The number of categories affects the dispersion of answers. On the longer of the two scales above, there would almost certainly be a lower percentage of 'adequate' responses than on the shorter, even though only a very few marked the two extreme categories. (This incidentally illustrates the danger of comparing the results of two surveys unless the questionnaires used in both are known.) Also, there tends to be a bias towards favourable rather than unfavourable ratings, particularly of people, so that a scale which aimed at an even distribution of ratings would have as its midpoint not a neutral attitude but a slightly favourable one. For example, a scale

  *excellent — good — moderate — poor — bad*
applied to people would tend to produce many more 'excellent' ratings than 'bad', and rather more 'good' than 'poor'. A better scale would be

  *excellent — good — fairly good — moderate — poor.*
Although it has been shown that the order of the categories—from low to high or high to low—does not affect the answers, it should be standardised throughout the questionnaire.

When more than five categories are desired, it may be difficult to find suitable verbal expressions for the small gradations from one category to another. Instead of words or phrases, a numerical scale can be used, respondents being asked to give their attitudes a numerical value on a scale of, say, seven points ('seven' being most favourable). This avoids the risk of verbal bias, but is still highly subjective. Or a sort of emotional thermometer can be used, calibrated in very small sections, on which respondents mark the strength or favourableness of their attitudes.

So far we have assumed that only one question is used for each attitude. Much greater precision and objectivity can be obtained by employing several questions on the same matter. One of the earliest scales, and still a very important one, which uses this principle is the Thurstone scale of equal-appearing intervals. This scale consists of fifteen to twenty five unambiguous and directly relevant assertions on a subject, arranged in random order, with each of which the respondent has to agree or disagree. These assertions are chosen from a very large number (over a hundred) of statements, which are given numerical values on a scale of strength or favourability by numerous expert judges. The judges' values for each assertion are averaged, and the assertions finally selected have values which form a scale calibrated at equal intervals. The score for each respondent is the mean or medium of the values of the statements he endorses.

The need for a large number of expert judges rules the Thurstone scale out of court for all library surveys except the most ambitious. The Likert scale, which is otherwise rather similar, does not have this disadvantage. A number of ordinary individuals grade each statement according to whether they *strongly agree* (1), *agree* (2), are *uncertain* (3), *disagree* (4), or *strongly disagree* (5) (five points are usual but not of course obligatory). The total score of each individual is calculated by adding up his marks, and assertions selected according to their discriminative values (the calculation of which is rather complicated). Thurstone and Likert scales appear to have equal validity and reliability; Likert scales have a higher internal consistency, but their use outside the population in which they were developed has been questioned. Thurstone scales are more subjective in their construction, but may have a wider applicability.

More sophisticated than either is the ' Scalogram ' method developed by Guttman. This method, which uses the concept of a hierarchical scale, is (like several other scaling techniques) too complex to be summarised intelligibly here. When anything of particular difficulty or importance is to be investigated, a thorough exploration of the literature should be made to see whether any technique exists which may be particularly appropriate to the problems in hand.

An attitude scale can of course measure only attitudes to a given object or question; it can reveal nothing about the object or question itself. If you wished to evaluate a library, you would not use an attitude scale (except perhaps to suggest areas requiring more objective investigation). The fact that sixty per cent of users thought the library poor means neither more nor less than that a majority thought the library poor; it does not mean (though it may suggest) that the library is poor. If you wished however to measure the library's standing in the community, an attitude scale would be very appropriate. That people's standards of adequacy or quality of a library differ does not matter, so long as it is clearly realised that it is their *attitudes* that are being measured, not the adequacy or quality of the library.

It is a good idea to include in the questionnaire one or two questions the answers to which can be checked. Occasionally an exact check may be possible, for example by ascertaining how many books the respondents have actually taken out. More often a general check is possible. If, for example, the average number of borrowings per head as indicated by the survey comes to thirty two per annum, and the average number calculated from the library statistics is twenty one, either the sample is biased in some way or, more probably, the respondents are exaggerating. This does not necessarily mean that all their answers display exaggeration; perhaps only their numerical recollection will be at fault, but discrepancies of this kind should lead investigators to be cautious.

The order of the questions needs some thought. If a logical order is adopted, the whole questionnaire becomes much clearer, as each question can be answered in the light of the previous ones which have led up to it. 'Throwaway' questions may be included with the sole function of guiding the thoughts of the respondent in a certain direction; these are especially helpful at the beginning of a questionnaire, or where there is a switch in direction. To a certain extent this can help with the difficulties of avoiding ambiguity. At the same time, it is necessary to be careful that the interpretation of a question is not distorted by the preceding ones. As a crude example, you might wish to ask whether respondents find the hours of opening adequate. If the three previous questions all asked for criticisms of various aspects of the library service, you are more likely to get critical than favourable answers.

The length of the questionnaire must depend on the nature of the inquiry and the educational level of the sample. It is not normally reasonable to expect people to spend more than half an hour except on inquiries of real importance—and not many library surveys come within that category. It should be remembered that open questions take quite a bit longer to answer than precoded questions.

Although some people seem positively to enjoy being interrogated even by questionnaire, and a few may even complain that their favourite questions have not been included, it is usually necessary to prune a questionnaire so that it contains only fruitful items. (The technique mentioned on page 50 can help here.) It may be necessary to leave out supplementary questions which might, as we have seen, shed light. There may have to be a choice between excluding one whole area of interest from the questionnaire, and excluding desirable questions from various parts. Another constraint on the length of the questionnaire may be the method used for processing the data; edge notched cards or eighty-column punched cards can impose very definite limitations on the number of items included. There is a final, and very real reason for limiting the length: the volume of data produced by detailed analysis of a long questionnaire can be overwhelming.

As far as possible, questionnaires should be interesting as well as answerable. This may seem a tall order, and it must of course come a long way after other considerations; but if the curiosity of respondents about the subject of inquiry can be aroused, you are likely to obtain more accurate and thoughtful answers. It is not impossible to construct questionnaires which are imaginative and stimulating, and the attempt to do so will usually result in a better questionnaire.

The physical layout of the questionnaire should be planned with a view to making it absolutely clear. Questionnaires which look like income tax returns are not only uninspiring but hard to complete. The space allocated to answers should bear some relation to the expected length of replies; this would seem an unnecessary point to make if it were not so frequently disregarded (another case where pre-1967 income tax forms can be faulted). Generally forms have to be designed and redesigned several times before an acceptable layout emerges, and even then it should be tested.

Markings on the questionnaires by respondents are usually hasty, and the form must therefore be designed in such a way as to minimise the risk of ambiguous or unintelligible markings. There must be clear instructions on the form explaining exactly what the respondent is supposed to do, and the layout of the questions themselves should leave no scope for confusion. A closely spaced list of categories headed by 'please tick as appropriate' is bound to lead to some ticks which do not clearly relate to any particular item. It is better to provide

' boxes ' to tick, or, particularly if mechanised methods of processing are to be used, numbers to ring. Where codings are very complex, a combination of boxes for the respondent to tick, with numbers in the margin ' for office use only ', can be used. Marginal codings can of course be provided for open questions, so that the coder can mark the appropriate number when he has decided in which category the answer belongs. Since the coding of open questions cannot be finalised until the answers have been studied, a generous number of codes must be provided. If boxes are used, it will help at the processing stage if they are aligned as nearly as possible. An alternative method is to use the instruction ' delete whatever does not apply '; this is reasonably satisfactory, except that positive markings are easier for analysis purposes, and some of the sample are bound to ignore the instruction and use vague ticks.

If printing is an economic possibility, it should certainly be considered, particularly if a questionnaire is long or complex. A good printer can reduce the bulk and improve the clarity of a schedule beyond all recognition: some printers are able to give specially skilled assistance in questionnaire design (see figure 1).

Like other stages in surveys, the physical design of the questionnaire should never be decided until it is known exactly what results are required and how they are to be analysed. Much time can be wasted, and frustration and mistakes caused, by the neglect of quite small details. To put out a questionnaire, even one containing well framed and intelligent questions, and hope to obtain answers with which something can be done, is asking for trouble.

The desirability of a preliminary trial has been mentioned in connection with various points. However many experts and non experts are consulted about a questionnaire (or any other survey instrument), there is no substitute for an actual small scale trial run, or *pilot survey*. This is best carried out with a small random sample, but a selective non random sample will often serve almost as well. A pilot survey can show up faults in the design of the questionnaire and in the framing of questions, and can help to indicate the probable range of answers to open questions (some of which can then be precoded). If the questionnaire appears to be very faulty, a second pilot study may be undertaken .

To some extent this is a counsel of perfection. Often there is not time, or money, to carry out a pilot survey, and sometimes there are enough precedents in a similar field to justify a certain amount of

10    **Nursing Library administration**

38    **Does anyone help you (or the person acting as librarian) with the Library?**

YES  ☐
NO   ☐

If YES
please write in the number of assistants here, including any student or pupil nurse helpers.

KEY: LQ Library Qualifications, FT Full-time, PT Part-time.

STAFF                          PAID    UNPAID

FT/LQ

FT/Non-LQ

PT/LQ

PT/Non-LQ

39    **Is your library collection housed in a room of its own, or is it in a class-room or study-room?**

own room       ☐
class-room     ☐
study-room     ☐
other (specify) ☐

for office use only

38 a

38 b    1 2 3
        4 . .
        . 8 9

39      1 2 3
        4 . .
        . . 9

40    **Please give the times and hours each day and evening that the library is open.** (e.g. Monday 9 am–12.30 pm, 1.30–5 pm, 6–9 pm. Total of 10 hours)

| DAY/EVENING | TIMES OF OPENING | TOTAL HOURS |
|---|---|---|
| Monday | | |
| Tuesday | | |
| Wednesday | | |
| Thursday | | |
| Friday | | |
| Saturday | | |
| Sunday | | |

41    **Can the library be used in the absence of an attendant?**

i    YES, during set hours only          ☐
     YES, during set hours and at some other times  ☐
     YES, any time, day or night         ☐
     NO                                   ☐

ii   If YES
     may books be taken out?             ☐
     books returned only                 ☐
     books returned and taken out        ☐
     NO                                   ☐

42    **Who administers the library in your absence** (e.g. during annual leave)?
     **Please give details:**

for office use only

40 a    Y X 0
        1 2 3
        4 5 6
        7 8 9

40 b    1 2 .
        . . 9

40 c    1 2 3
        . . 9

41      1 2 3
        . 8 9

42      1 2 3
        4 . 9

FIGURE I: *National Book League Survey of nursing libraries.* Reproduced by permission of NBL. This was designed and printed by Pearl Publications & Design Ltd.

confidence (though previous surveys may have been aimed at different types of audience—a survey that has proved satisfactory in a middle class suburb may not be appropriate for a council housing estate).

Every questionnaire should if possible include certain basic questions about the respondents: age, sex, marital status, occupation, and possibly income. The relevance of these factors depends on the nature of the investigation and of the population concerned—only the first two would normally be applicable to students—but in practice there are surprisingly few inquiries where analysis of results by these factors cannot yield information of interest or value. There is little danger that such questions will reduce the response, since any individual can (and some will) refuse to answer any or all of them.

When the time comes for the distribution of the questionnaires, they should if possible all be sent out on the same day. This increases the chance that the questionnaires will be filled in at the same time and will therefore be comparable (in this respect at least), and reduces the danger that later recipients will be influenced by any discussion among earlier recipients. Mercifully few library surveys deal with matters of public controversy, so that serious distortion of later responses is unlikely, but simultaneous distribution is still desirable for the reasons given.

Distributing questionnaires is one thing: getting them back is quite another. With some methods of distribution this presents no problem; if, for example, forms are handed out to individuals as they enter the library, they can be collected on the way out. (This is one of the advantages of this method.) Usually however it is necessary or desirable to distribute forms by mail. Unless return is physically easy (as it might be in a university library where students come regularly), stamped addressed envelopes should always be enclosed with the questionnaires.

Whether or not the proportion of questionnaires returned in response to the first circulation is high, a follow up will always produce more—bringing the total to perhaps half as many again as the initial returns. Two or three weeks is a reasonable interval; a longer period brings the danger (mentioned above) that the situation will have changed between the first and last returns. Since one reason for non response may be that forms have been lost, the follow up letter should explain that further copies can be obtained, or should actually include another copy. Care must be taken to ensure that one person does not return two forms, and that copies do not stray outside the sample.

One way of preventing this is by giving each form a distinctive code number; any additional copy distributed to an individual would bear the same number as the first.

The coding of questionnaires is desirable for other reasons. The sampling frame used may reveal something about the sample itself. If, for example, electoral registers were used, the codes would indicate the area of a town in which a respondent lived. More important, the numbers of forms returned can be checked against the sampling list, so that only those who do not return their forms receive a follow up letter; otherwise the whole sample has to be circulated again.

Although library questionnaires are relatively harmless, anonymity is so important to many that assurances should be given that it will be preserved. If the form carries a code number, there will be some suspicious souls who simply will not believe protestations that it will be used only for follow up purposes, and will either remove the number or tear up the form. Suspicion may be decreased somewhat if the number appears on the envelope for return rather than the actual form, and an assurance is given that the envelope will be scrapped. It is certainly unwise to attempt to conceal the number in some way, since if it is found (as it will be by some) suspicions will be confirmed. A partial solution is to number the original forms, but in the follow up to send out unnumbered forms.

The forms returned during the first three or four weeks should be kept separate from any returned subsequently, especially if there is a follow up after two or three weeks. The two groups may exhibit differences in some respects, and since later respondents are likely to have more in common with non respondents, their special characteristics can help to suggest the directions of bias due to non response. How typical the respondents are of those circulated is obviously a matter of great importance (see chapter five, page 102 ff), particularly when response is small.

How high a response can be expected? Anything over eighty per cent is very good; this is attainable in a compact community like a university, or with surveys where questionnaires are distributed to users as they enter and collected as they go out. With a mail questionnaire distributed to a sample of the public sixty per cent is quite a reasonable response. Any return lower than forty per cent is likely to be prone to very substantial errors and biases due to non response. This is one of the biggest disadvantages of the mail questionnaire.

Interviews in their simplest form are little more than questionnaires administered verbally. The interviewer asks the questions set out on a printed or duplicated schedule (which may be almost identical with a mail questionnaire schedule), and records the answers in the appropriate way. At the other extreme, an interviewer may hold a conversation with the individual, drawing him out and guiding the discussion along certain lines, but aiming to get a personal viewpoint rather than a set of responses; as already mentioned, Ferdynand Zweig is an outstanding practitioner of this kind of interview. The former type of interview is often called *structured* or *formal*, the latter *unstructured* or *informal*.

Even a tightly structured interview has advantages over the mail questionnaire. The response rate may be higher; this is discussed later. The danger of misunderstanding is less, for provided that the interviewer is absolutely clear what is meant, he can where necessary explain it in terms the persons interviewed will understand. There is less danger of hasty or careless answering. The interviewer can help to remove suspicion by explaining the nature and purpose of the survey.

Along with the advantages go several disadvantages. However closely the interviewer sticks to his questionnaire, it has been shown that some interviewers tend to obtain consistently, if marginally, different results from others. The interaction between person and person, even when it consists of formal questions and answers, is an extremely complex and subtle process. The human link between the investigator and the respondent can produce either greater precision or greater distortion than the mere paper link of a questionnaire form, and it is not often possible to be sure which it is doing. Few interviewers can adapt themselves equally well to different social classes and types of individual. Differences of sex, age, colour and class can all make rapport difficult. The way questions are asked, and the way answers are interpreted, are both affected by the individuality of the interviewer. Even if an interviewer is absolutely free of conscious or unconscious prejudice, this will certainly not be true of all those interviewed; and they will react to different interviewers in different ways. They may try to give the interviewer what they think he wants (and they may be quite wrong about what he wants, in ways that the interviewer cannot allow for)—or they may give him what they think he deserves ('Why should I bother with this anyway?').

If these difficulties arise with closely structured interviews, they become much greater as the inherent flexibility of the interview is exploited. The more the interpretation of the questions, or phrasing of them, is left to the interviewer, the less reliable the interviews become (*ie* the greater the chance that different interviewers will achieve different results if repeated with the same person); and the more skilled the interviewers need to be. For these reasons, it is as well to make interviews as formal as possible unless the matters to be investigated are of great complexity or experienced interviewers are available. Most library surveys can probably be carried out by formal interviews.

However, for some purposes unstructured or semistructured interviews are justified. If, for example, one wished to explore within a limited community such as an industrial research institution the needs of researchers for information, it is doubtful whether this could be done by a straightforward questionnaire, since some of the concepts involved would require a great deal of explanation and interpretation; the researchers would have to be induced to think in a way which might be novel to some of them before the questions would make sense. Again, an investigation of the place held by the public library in public esteem, if carried out by questionnaire, would yield superficial results; a more satisfactory picture could be obtained by talking around the question until it became clear what concept each individual interviewed had of the public library and what part it played in his life. For some investigations there is no substitute for the free expression of opinion provided by unstructured interviews. A set question, even if rephrased, compels the respondent to think along similar lines to the investigator if his answer is to be relevant; it does not allow for ways of thinking that may be quite foreign to the investigator. In other words, it is valid only within a certain cultural and conceptual framework. The nearer a question approaches to the heart of the matter, and the more valid it is, the less reliable are the answers likely to be; there may be, as we saw earlier (page 50), a conflict between validity and reliability.

To summarise, there are various degrees of formality possible in interviewing. You can in effect say to the interviewers 'Ask these questions exactly as they are written down', or 'Ask these questions in wording that each individual will comprehend', or 'Ascertain the answers to these questions in any way you like', or 'Find out all you can about what people think on such and such a topic'.

It will be clear that, for interviews of all kinds, the selection and briefing of interviewers are of the first importance. Even for very simple questions, unprejudiced and sympathetic persons should be chosen, and absolutely clear instructions must be given. For less closely structured interviews, long and detailed discussions are necessary. The more straightforward the interviews, the more interviewers can be used, and a large sample covered at a reasonable cost. For very informal interviews, interviewers should not only be very skilled but very few. If possible, one interviewer should carry out all interviews, since so much will depend on the general picture built up in his mind. Impressions of many people obtained by one investigator are hard enough to coordinate, impressions gathered by several different interviewers infinitely harder. A choice must therefore often be made between large sample, simple interviews, unskilled interviewers and superficial results on the one hand, and, on the other, small sample, interviews in depth, and a few interviewers with wide experience.

Interviews of people over an area extending beyond the locality are rarely practicable. The problems of briefing interviewers and coordinating their activities if they are scattered over a large area are formidable. In this respect the questionnaire has a distinct advantage, since once the sampling frame has been established it is not much more difficult to distribute questionnaires over the whole country than in a small area. For similar reasons, methods of drawing the sample may be different. It is not reasonable to spread interviewers over the whole of a large city, nor is it necessary if multi-stage or cluster sampling is used (see chapter two, page 29-30). The procedure in a city might be to divide it into major areas by social or economic or housing factors, to draw a random sample of smaller districts from each major area, a random subsample of streets from each district, and finally a quasi random sample of houses (say every fifth house) in each street.

Interviews score heavily over mail questionnaires with open questions, since respondents are much more ready to speak long answers than to write them down. This imposes on the interviewer the task of summarising the answers. This should be done immediately, if necessary in the form of detailed notes which can be condensed later. If the interviewer considers that a respondent's answer is biased by prejudice (perhaps against himself), this should not be allowed for in the transcription of the answer, but mentioned in a note. Any other personal peculiarities which strike the interviewer should be similarly noted. Where possible and relevant, the actual words of the respondent should be set down if they are likely to be illuminating, though the temptation should be avoided of noting the quotable but untypical remark. A note should be made on every schedule of the date and time of the interview, and the interviewer should sign it. The interviewer's signature serves several purposes. It makes the interviewer conscious of his responsibility for the document; it enables the investigator to pursue with the interviewer any point relating to a particular form; and it makes it possible to allow for interviewer bias by comparing the results of different interviewers.

There is no reason why, with structured interviews, questions should not be precoded in the same way as for mail questionnaires. Any doubts into which category an answer falls should be resolved on the spot. However tempting it may be to pigeonhole it later, it will certainly be no easier to pigeonhole it accurately, since recollection will be distorted by subsequent questions and subsequent interviews.

Whatever editing or tidying up of substance is necessary should be done as soon after each interview as possible—certainly before going on to the next interview. The reasons for this have already been indicated. It can of course cause problems if interviews are being conducted in back streets on a dark night; the most that can be done in such circumstances is to make as full and comprehensive notes as possible at the time. Summarising full notes is much less dangerous than expanding rudimentary ones.

The time which should be allowed for any worthwhile interview is unlikely to be less than fifteen minutes. Even for a fairly simple and straightforward interview thirty minutes is a safer minimum estimate, especially if interviews take place in people's houses. The possible causes of delay are innumerable—the man of the house may be finishing his tea or his bath; the old lady may welcome the first visitor she has had for days; the housewife may insist on making a cup of tea; and anyone may have something much more important on his mind than library problems, which cannot be discussed until the other matter has been aired. To the interview time must be added the time between interviews. If interviewing takes place during the evening, four or five interviews per evening is quite a respectable figure.

The labour involved in interviewing even a smallish sample is, clearly, quite considerable. Ten interviewers, working hard for five evenings a week, can hope to get through a total of about 150 a week, allowing for various delays and frustrations. (The estimates given are for short interviews: loosely structured or unstructured interviews may take very much longer. For example, the interviews used as a basis for P Marris *The experience of higher education* (Routledge, 1964) averaged between two and three hours each.) If voluntary labour is used, expenses may be no more than travel costs, and indeed can work out at less than a mail questionnaire. Except in a university or college or library school, however, it is not easy to obtain ten voluntary interviewers for a week or two. The employment of professional interviewers is not only costly but raises other problems; they may be firstrate at interviewing, but it may be harder for them to understand exactly what is actually wanted than for members of the community that is being studied (*eg* university students), or for librarians.

For the same reasons as it is desirable to send out mail questionnaires at the same time, interviews should take place within as short a period as possible. If interviews are carried out within a week or a fortnight, there is a good chance that all the actual responses will be obtained within a *shorter* period than most mail questionnaires can achieve.

The administration of interviews can sometimes be very simple. For example, everyone leaving a library on a specified day can be interviewed. Within an institution or industrial establishment, appointments can be made. If members of the public are to be interviewed in their homes, advance warning should normally be given by a note through the door. The danger that the prospective interviewee will take special care to be out can be reduced if the proposed time of calling is not given (in any case it cannot be forecast with any accuracy). An advance warning should give some idea of what the survey is about, who is conducting it, and at what time of day (*eg* evening) the interviewer is likely to call; a request to contact the organisers if a particular time of day is not suitable can well be added. Advance notice is an added expense, but is generally preferable to surprise visits, which can increase the number of refusals.

On the first round of interviews, some people will resist, but can be persuaded; some will refuse point blank; and some will be out. No useful purpose is served by pressing the point blank refusals, since in the unlikely event of their yielding to importunity they may be uncooperative in their answers. Reasons for refusal should be set down if possible. The man who will not be interviewed on the public library because he never has any patience with reading is implicitly expressing an opinion on libraries which should be recorded. Sometimes a refusal can be turned into a postponement; an injudicious time (such as pub opening time) may have been chosen. If a selected individual is not in, another member of the household may be able to say when he will be. A second round of calls on absentees will be necessary, and a third desirable. With sufficient interviewers, time and persistence, the response can often be quite high—much higher than mail questionnaires usually achieve.

Interviews and questionnaires are not necessarily alternatives; for some studies both methods can be used together to good effect. Interviews can be employed to follow up in greater depth some of those who have responded to a questionnaire, whether they are selected for interview on a random basis or as exhibiting special characteristics. For example, a sample of library users could, after the response to a mail questionnaire had been partially analysed, be stratified by extent of use, and a few within each category interviewed. Conversely, interviews can be used to explore the ground before distributing a questionnaire, to ascertain which questions are likely to produce the most valid and reliable answers and to indicate problems which may not have appeared in planning. A series of unstructured interviews followed by a mail questionnaire can constitute a powerful research tool.

Because interviews and questionnaires assume such a prominent place in social research, and indeed in this chapter, their deficiencies should not be forgotten. They may be for many purposes the best instruments we have, but they are far from perfect. The more they can be supported by other methods, the more valid the survey results will be. There are numerous combinations: observation *plus* interviews; questionnaires *plus* diaries; personal visits *plus* statistical records; and so on. Two recent examples will illustrate this. In 'The ecology of study', *Library quarterly* 36 (3) July 1966 234-248, Robert Sommer describes the use of observation, experiment, and a questionnaire to investigate the ways in which students found privacy in public reading areas of a university library. In 'The implications of the needs of users for the design of a catalogue' *Journal of documentation* 22 (3)·September 1966 195-207, Laraine Kenney gives the results of a survey conducted by a combination of questionnaires, interviews, and statistics of readers' requests, and thus containing facts, opinions, and impressions. In both these cases, no one method would have given so full a picture or allowed such valid conclusions to be drawn.

When thinking of using any particular method or faced with a particular problem, it is wise to check whether there is any article or paper or report on the topic; for example, the problems of interviewing old people, interviewer turnover, and so on. The Government Social Survey (formerly a department of the Central Office of Information) has issued mimeographed (or occasionally printed) papers on several of these matters, and there are many others in professional literature. The most profitable sources are likely to be found in the literature of sociology, psychology, statistics and education.

The desirability of pilot surveys should be stressed again. Unless there are good precedents for both the subject to be studied and the techniques to be used, a pilot study, even if it involves extra expense and delay, will save much time and trouble later. The conditions observed in a pilot survey should normally be as close as possible to those envisaged for the survey proper, the main difference being the number of individuals involved. There are very few surveys that cannot benefit from a trial; without a trial, at best your results may be slightly less useful than they might have been, at worst they may prove to be quite invalid.

BIBLIOGRAPHY

There are numerous works on the matters covered by this chapter; a few of the more approachable and accessible ones are listed below.

American Library Association *Library statistics: a handbook of concepts, definitions and terminology* (Chicago, American Library Association, 1966).

Central Office of Information—The Social Survey *A handbook for interviewers*. Edited by Muriel Harris (COI 1956). A practical handbook with a wider range than its title would suggest.

L Festinger and D Katz (*eds*) *Research methods in the behavioural sciences* (Staples Press, 1954). Chapter 8 ' The collection of data by interviewing ' by C F Cannell and R L Kahn.

W J Goode and P K Hatt *Methods in social research* (NY, McGraw-Hill, 1952). Chapters 10-13, 14-17.

J P Guilford *Psychometric methods* (McGraw-Hill, second edition, 1954).

P H Mann *Methods of sociological enquiry* (Blackwell, 1968). Chapters 5-8.

C A Moser *Survey methods in social investigation* (Heinemann, 1958), Chapters 9-12.

A N Oppenheim *Questionnaire design and attitude measurement* (Heinemann, 1968).

C Selltiz *and others (eds) Research methods in social relations* (Methuen, revised edition 1965). Chapters 6-8 'Data collection', and Appendix c 'Questionnaire construction and interview procedure'.

P V Young *Scientific social surveys and research* (Englewood Cliffs, Prentice-Hall, fourth edition 1966). Chapters 6-10, 12.

## MEMO ON
## EFFECTIVE LABOR COSTS

*Barton R. Burkhalter*

Labor cost is an important factor in most of the studies. Multiplying the number of hours devoted to a task times the hourly wage rate of an employee is not an adequate means of obtaining the total labor cost for the task, because the hourly wage rate does not represent the real cost to the Library. The Library must pay taxes and fringe benefits of various types over and above direct wages. Furthermore, over an extended period of time any employee has fewer productive hours than paid hours. If a particular task accounts for 10% of an employee's productive hours, then it should also account for 10% of his total cost to the library.

The following calculation provides two factors which, when multiplied by the annual salary of a salaried employee or the hourly wage of an hourly employee respectively, produce the effective hourly cost to the library for this labor. In other words, it gives the total annual cost divided by the number of productive hours per year.

I. Salaried Workers*

   A. Total Cost per Dollar of Salary

   1. Salary                                              $ 1.00
   2. Fringe benefits and taxes (FICA,
         etc.) = 12%                                      .12  Cost $
                                                       -----------
                                                         1.12  Salary $

   B. Productive Hours Per Year

   1. Hours per year (Gross = 52x40)       2,080.0
   2. Two weeks vacation                     -80.0
   3. Twelve days sick leave                 -96.0
   4. Six holidays + one "floating" holiday  -56.0
                                          ----------
      Actual hours in library             1,848.0
   5. Coffee breaks (15 min/4 hours
         worked)                            -115.5
   6. Personal time (3%)+
      Fatigue (3%)+
      Delay allowances (5%) = 11%           -203.3
                                          ----------
         Total annual productive hours =  1,529.2  Productive
                                                   hours per
                                                   year

C. Effective Hourly Cost Per Dollar of Annual Salary

$$= \frac{1.12 \text{ (cost \$/salary \$} \qquad )}{1,529.2 \text{ (productive hours/year)}} = .0007324$$

For example, suppose a full-time employee earned an annual salary of $6,000. Then his effective hourly cost would be:

$$6,000 \text{ (salary\$/year)} \times .0007324 \left(\frac{\text{cost \$ . year}}{\text{salary \$ . prod. hours}}\right) =$$

$$4.39 \left(\frac{\text{cost \$}}{\text{prod. hours}}\right)$$

*Note: The figures for salaried workers appy to clerical positions. Librarians receive 4 weeks of vacation and work 35 hours per week. Thus their effective hourly cost per dollar is .0009418

II.  Hourly

A. Total Cost Per Dollar of Hourly Wage

| | |
|---|---|
| 1. Wage | 1.00 |
| 2. (No fringe or taxes are paid for hourly employees) | - |
| | $\overline{1.00} \quad \frac{\text{cost \$}}{\text{Wage \$}}$ |

B. Average Productive Hours Per Hour Paid

| | |
|---|---|
| 1. Start with one hour paid | 1.00 |
| 2. (No allowances are made for vacation, holidays, or sick time) | - |
| 3. Coffee breaks (25% of time no break; 75% of time 15 min/4 hours) | -.05 |
| 4. Personal time (3%) + Fatigue (3%) + Delay Allowance (5%) = 11% | -.11 |
| | $\overline{.84} \quad \frac{\text{Prod. Hrs.}}{\text{Paid Hr.}}$ |

C. Effective Hourly Cost per Dollar of Hourly Wage

$$= \frac{1.00 \text{ cost \$/wage \$}}{.84 \dfrac{\text{productive hours}}{\text{paid hours}}} = 1.19$$

Thus, a part-time employee earning $1.00 per hour would have an effective hourly cost of $1.19.

# THE GREAT
# GAS BUBBLE PRICK'T;
# OR, COMPUTERS REVEALED

*A Gentleman of Quality*

SOURCE: From Ellsworth Mason, "The Great Gas Bubble Prick't: Or Computers Revealed—by a Gentleman of Quality." *College & Research Libraries*, 32 (May, 1971) pp. 183-196. Reprinted by permission of the author.

On an evaluation visit last spring to a small college (collection 175,000 volumes, peak daily circulation 700), I found the library automating its circulation records, an action tantamount to renting a Boeing 747 to deliver a bonbon across town. Everyone felt great about it; it was a Good Thing! In a college sorely pressed for funds, wasting this amount of money was actually a serious crime against the common weal.

This situation nicely characterizes the fatuousness of one of the most curious periods in our nation's history—the period that began with a rebound off Sputnik, which seemed for a moment to snatch a tip from our crown of world leadership, to strip us of our masculinity, as it were. In this period, which has now passed its peak, money meant nothing, the world of formal education was endowed with magical properties, and technology became an unquestioned God (If we can put a man on the moon we can . . . ). This decade boasted of its technical potency with the false bra-vado of a male virgin, and if the moon rocket in the Sea of Tranquillity was its sexual symbol, the computer, choked in its navel cord of programs, was its abortion.

This fact has yet to be generally absorbed. It has already become painfully clear that technology is a two-edged sword of Damocles. Grave doubt has been raised that the computer has done even major industries much good.[2] But, oblivious to the signs of change, librarians are proceeding in a kind of stunned momentum like a poleaxed steer, because the computer industry and its public handmaidens have polluted our intellects. In one of the most massive public manipulations in history, the computer has been joined to Motherhood, the True, the Good, and the Beautiful. Operational considerations have been stripped to a stark choice between "the old hand-method" (ugh!) and THE COMPUTER. The effect has been to obscure a whole range of machine and machine-manual alternatives.[3]

Technology has been set back many years and intelligence has been uprooted. Any fool who does anything with a computer for any reason (we all know at least one) is automatically a genius; anyone who does not is the last of the dinosaurs.

During a period of study sponsored by a Council on Library Resources (CLR) fellowship which allowed me to study problems in ten major research libraries last spring, my observations convinced me that the high costs of computerization make it unfeasible for library operations and that it will become increasingly expensive in the future.[4] The computer feeds on libraries. We actually devote large amounts of talent and massive amounts of money (perhaps $25 million dollars a year in academic libraries alone) to *diminish* collections and *reduce* services, exactly at a time when libraries are starved for both, by channeling money into extravagant computerization projects which have little or no library benefits. While my original expectations were entirely in the opposite direction, after talking at length with some of the finest computer experts in the library world and probing the thinking behind more than forty computerized library operations, it became clear that the application of computers to library processes is a disaster, and that no one is willing to admit it.

The reasons for its adoption are governed by a range of irresponsible, irrational, and totally unmanaged factors, both within the library and in the university, that cannot fail to disgust anyone seriously concerned about the academic world. This article intends to analyze how we learn to stop thinking and love the machine, and to make possible the return of intellect and managerial methods to an area of library practice from which both have been driven.

## THE ROUGH BEAST WITH THREE BREASTS

Unlike most other machines, the computer is not subject to reasonable surveillance at any level of operation.[5] A college president or the manager of industrial research cannot judge with any reasonable degree of accuracy how much computer capacity is required for his needs, nor can his subordinates. This means that basically he must accept his computer configuration on faith and on the urgings of computer industry representatives.

This condition in which the computer wanders free from quality checks extends right down the line of a computer operation to the head of programming, who cannot judge with any degree of precision the quality of the programs written for him.[6] He can tell whether they run (indeed, the principal struggle is to get them to run trouble-free at all), but he cannot tell how they rate in comparison with the range of other alternatives. This free-form condition of control, which is inherent in the occult nature of the computer, accounts for the great range of loose work and random performance observable in computer operations.

Moreover, a computer operation is incapable of becoming stabilized on its own terms. No matter what level of performance is achieved, if a later generation computer is marketed, it is necessary to shift as soon as possible to the new generation, with all the agonies, dislocations, and setbacks involved in the change, and with no assurance that the same level of results can be achieved. There is no choice of remaining as you are if reasonably satisfied with your results because it is extremely difficult to recruit a systems and programming staff (doubly difficult for libraries, which lack the glamor and loose money that have characterized industry until recently). A good staff will abandon a superseded model computer, since to remain would make them professionally obsolescent.

These two floating conditions make computer operations basically uncontrollable. In managerial terms, these facts alone would argue for discarding out of hand any other machine in existence, until it was amenable to quality control. But we have been conditioned to suspend completely the requirements that apply to all other equipment, and automatically accept the computer as Good, without questioning. We accept the computer as the pot of gold at the end of the rainbow, the touchstone that turns dross into gold. Glittering with spangles, draped seductively in the fluff of unreason, it really has sex appeal, and who applies reason while gulping the lures of a floozie like Myra Breckenridge?[7]

### THE NEW BLOOMUSALEM

When Leopold Bloom, Joyce's common man in *Ulysses*, proclaims, while playing God in an hallucination, "the golden city which is to be," thirty-two workmen wearing rosettes construct "The New Bloomusalem," a megastructure in the shape of a huge pork kidney. Something like this debased miracle happened in library computerization in the decade of the sixties, when computers rode tall in the industrial saddle and librarians flung themselves at the horse's tail. During that decade, our large problems were operational (whereas now they are desperately financial) and we looked for a panacea. Noting us sniffing around the computer, the industry perked up and assured us they were the answer.

A kind of syllogistic thinking followed —we have problems; the computer says it can solve them; therefore, using the computer solves our problems.[8] It's all simple enough and clear enough if you just have Faith, and of course, Reason

is the enemy of Faith; in fact, it gets in the way of certainty. In our awe at the wonders of technology, we forgot the deadly threat of Dr. Strangelove's mechanical hand. Like lemmings moving toward the sea, we surged to get with it, became scientists, became industrialists, and practiced the best that was known and felt in the business world.[9] In the whole range of the academic world, we forgot one of our traditional functions— to suspect the beguilement and evanescence of the moment and "to keep clean our sense of difference between the temporarily and the permanently significant."[10] In short, we embraced with fervor all the sins of the commercial world. Now, look at the commercial world and at the academic world and wonder how it is that student rebels connect the two.

The fascination of the computer, like that of a hooded cobra, lies in its exotic beauty, which fixes its victim for the spurt of poison. On the surface it seems to have many answers. It looks effortless, is pleasantly mysterious, it makes pleasing sounds, it promises great speed, and it has a reputation for performing miracles. Despite its beginnings in 1942 (long before Xerox), it is considered the *latest* technological development. So we got with the new and the technologically best by adopting the computer. We did so to solve simple and clearly defined problems—to save staff (or substitute for staff that we couldn't hire), to speed processing, and to save money. Information retrieval was seen in the distant mist, but these were the clear and central targets.

But when we used the computer, it didn't save staff, and it didn't speed processing, and it cost a great deal more to do the same things we were doing by hand. Our reaction was to computerize more. Although we lost money on every operation we computerized, the theory grew that if you knit enough losses together, obviously you would save money. In Orwellian doublethink, if you waste money in an attempt to save it, save better by wasting more. We still didn't save staff, and we didn't speed processing, and it cost us even more money. Our latest answer is to use newer and bigger and more expensive computers; it still is not saving us staff or speeding processing, and we are now spending extravagant amounts of money. We bombed library problems with the computer, and the strategy didn't work. So we bombed even more problems with the computer and it still didn't work, so we are bombing even more.

## Just Like General Motors

At this point, the third strange fact about the computer becomes clear. It is a half-baked machine. Every other kind of equipment we use is bought for specific purposes, to perform defined tasks, at a known cost. Even highly automated equipment like the MT/ST comes with a simple program to perform known tasks after a modicum of training. A wholly baked computer, nicely browned, would be ordered to specifications, and would come ready to dust off, to insert the program *provided by the manufacturer to do what we wanted to be done*, and to begin our computerized operation. Only under such conditions would we consider any other machine. But we have been brainwashed not to apply the same reasonable standards to the computer. The cobra has us hypnotized.

When it is dumped on your dock, it can do nothing for you; like, Ford delivers you a Continental and deposits it in your yard. You leap with joy and shout to the neighbors who come to admire. You puff with pride, as we do for computers. "Let's go for a ride," they say. Somewhat sheepishly, you explain that it is a new proto-electric Continental, with a wonderful fume-free motor, but that there is no battery known strong enough to power it. When they say,

"Why did you buy it?" do you reply, "Oh, I'll do my own Research and Development to produce the battery"?

Such an answer would be insane, but this is exactly what we do for computers.[11] We assume the responsibility, the elaborate costs, and the human agonies involved in programming to make the machine do what we knew we wanted it to do before we bought it. In one project now underway, it will take a staff of ten, three years to make anything happen. Libraries really are getting important when they can play junior GM (without GM's budget) and launch amateur research and development operations, which is what programming really consists of. No matter how good our systems staff, such research and development must remain amateur. We don't know enough about technology even to know which field we should work in to solve our problems, let alone which machine we should encourage. We haven't the meagerest grasp of the perspective required by industrial R & D. But we have enthusiasm, we have suspended our brains, and we've come to love the computer.

We spend millions making the computer work for library activities, with a guarantee that it will produce a built-in deficit and with only a vague chance that it will improve anything. We simply can't wait for the finished machine, for the one that really works, the one which when it comes will make computers useless. We must develop it ourselves, even if we have to sell our libraries (which we are doing) to do so.

## How We Are Covered With Locusts and How the Invasion Began

How did we get into this mess? There are precedents in human history. The mountebank pulls up at the crossroads and the yokels throng the tailgate to buy snake-oil guaranteed to cure any disorder of libraries. Gullibility accounts for part of it; pressures account for the rest. The physical scientists and mathematicians brought the computer on campus for its computational facility.[12] Engineering, which quickly was seized by electronics specialists, burgeoned later. From these three groups came large demands for computer time in the universities. Administrators, naive and uninformed, began pressures to have all the computer time on campus used because of its heavy cost. They began by offering "free" computer time (an interesting concept at current prices) to any department that would use it. This free offer sprang from the prestige value inherent in using the computer (the industry did supply the prestige) and from a conviction on the part of administrators (also supplied by the industry) that use of the computer saved money for any operation it touched.

As this free time was used, the demand for computer time overran that available, and bigger, better, and much more expensive computers were brought on campus. With even greater increases in expense, administrative pressure (as brainless as all other pressures involved in computerization) intensified, and in some instances became downright nasty to departments that dragged their feet either through lethargy or knowledge. They were joined by the computer engineering faculty, which in recent INTREXed years, has become self-deluded to an extreme degree.[13]

Librarians, most of whom are humanistically trained, are especially sensitive to accusations by technologists and administrators of refusing the best that is known to business and technology. Even when they know better, consistent pressures unsettle their confidence. To cool the hot breath of the president's office, one university made a list of special materials by computer when they knew in advance they could do it considerably

cheaper by more than one noncomputer method. To appease the demands of a renowned and totally impractical engineer, one university went to a computerized circulation system as the least wasteful operation they could run on the computer. The fatuous self-confidence of computer experts is considerably jolted when they have to cope with the demands of library operations, which are far more complex than anything else they tackle in terms of their machine. But so long as they can throw stones from a comfortable theoretical distance their pressures are compelling indeed.

### THE ELECTRONIC CALF

In a time of waning personal confidence, it takes a very strong man to stand up to a university president and tell him he's wrong when he is convinced by technologists that inertia springs from ignorance. There are only a few men left these days. Therefore, with the prod in our rear, or approaching, we adapted to the new campus ecology, now polluted by technologists. Although some librarians seized the computer for its public relations value (Look, mommy: no catalogers!), the more sober members of the fraternity went along with a better conscience by adopting a mystique about the computer that grew partly outside and partly inside librarianship.

This mystique generated, and in turn was generated by, a group of librarians whose livelihood depended on the computer, and whose reason for being depended largely on their ability to believe the computer industry's claims laid out before them. The emergence of this Faith and the band of True Believers have been responsible for the rapidity with which we have gotten into computerization despite all evidence that the fantastic claims for the computer are completely false. This group of the faithful was abetted by enormous sums of government and foundation money that flowed, like Niagaras of champagne (Lucius Beebe's phrase), into computerized projects for a five-year period. With this amplitude of fuel, these neo-Zoroastrians began to burn up the world.

### THE REVELATION

Blazoned across the dark benighted sky of conventional librarianship were the following Truths:

*The First Truth*—Come to the computer all ye who are heavy laden and It will make everything effortless.[14]

The Facts—The computer has involved librarians in greater and more prolonged agonies than anything in recent history short of the Florence flood. Agonies of campus politics (flipped from computer to computer), agonies of financing (since the golden angels have gone), agonies of programming, patching programs, reprogramming, re-debugging programs, agonies of lengthy machine breakdowns, agonies of deception by computer experts (both in industry and in other campus units) have left deep scars on every library computer expert I have known.[15] While I was on campus one university was executing the second major cutback of computer capacity within three years, each causing major upheavals and changes in staffs and procedures and the bitterest kind of infighting to control the nature of the computer configuration. The most efficient road to ulcers on a college campus, short of the president's office, is through library computerization.

*The Second Truth*—Thou shalt do everything with the speed of light, if thou butst computerize.

The Facts—Response time of computers, which is incredibly fast (as fast as the movement of an electron), is not to be confused with the response time of computerized processes.[16] It is common knowledge that computerized class schedules take weeks longer to produce than the old hand method. In librarianship, these are some of the commonplace delays found strewn all along the trail: Circulation, a delay of one day in the ability of the circulation file to account for the location of a charged book (in one case, the costs of *paper* led to updating the file only once every three days). On-line circulation, the alternative to batching, is so astronomically expensive that anyone who adopts it should be summarily condemned as a public malefactor. Acquisitions—consistently slower in placing orders. Acquisitions was so slow the spring that I was on campus that, in one case, 20 percent of their periodical subscriptions were cancelled due to slow placement of orders. Book catalogs—longer and longer delays in cumulations because of the costs involved. In the case of one university, an operation highly touted while in action had left a liberal arts college with its book catalog in four (repeat, four) parts. They were at the point of doing what they were sure would be, forever, their last total cumulation because of its cost, while their future lies in a book catalog always in two and three parts. They would like to go to a card catalog, but at 100,000 volumes, cannot afford to. One circulation operation, where the students were cleared faster than previously at the charge-out point, claimed this advantage, without noting that the new system involved the use of book cards in lieu of user-written cards, and that the computer charging console takes longer than most simple charging machines.

*The Third Truth*—The computer will save you money.

The Facts—Computer experts laughed when I suggested economy as a motive for adopting the computer. No one claimed to have saved any money doing anything by computer, and although the analysis of computer costs is, to be charitable, hair-raisingly casual, estimates of costs of doing by computer exactly the same things that had previously been done manually were extremely high (in one case, five times the cost). We now know there is no clear evidence that the computer has saved industry money "even in routine clerical operations."[17]

*The Fourth Truth*—Well, anyway, once you have done it, thou shalt have economies in future programming by having programs convertible to later generation computers.

The Facts—Absolutely false! About half of the third-generation computers in major industries are in an emulation mode that makes them perform as second-generation computers because industry, having been hooked on the enormous programming costs for the second generation, is unwilling to absorb even higher costs to program for the third generation, which leads to an interesting view of our economy (like our libraries), buying the latest to get with it to avoid losing face.[18]

*The Fifth Truth*—Well, anyway, once someone has done it, programs *can* be converted from location to location, so you save the expense of programming for yourself.

The Facts—This initially was one of the most appealing lures of the computer industry. A few years ago, in a correspondence with Robert Hayes of the University of California, I asked why we all had to make the computer repeat on machine the motions we were doing by hand. Since we all need about the same end products at the same key points in a serials operation, why couldn't one library program it and present the program in modules, each of which could accomplish one thing, for us to choose those we preferred? At length, in a series of letters, I learned

the elaborate and complex reasons why this could not be done. All the library computerators I questioned agree that transferability of programs is completely unfeasible at present and in the future.[19]

*The Sixth Truth*—Thou *shalt* have cheap computerization by sharing computers with others.

The Facts—This, again, was one of the bright promises laid out by the computer industry, but the deeper we get into library computerization, the more evident it becomes that sharing computers to reduce costs is a chimera. Yet within the month, an eminent professor of industrial management who read my CLR report trotted out the old turkey that, with remote access consoles, sharing computers would soon make them economical.

*The Seventh Truth*—Thou shalt save money as you multiply the separate operations that you computerize if you combine them by a systems approach.

The Facts—Though a common belief among the aborigines of Computeria and sustained by a well-developed theology, there is no evidence whatsoever to support this belief.[20]

*The Eighth Truth*—Thou shalt have greater service for the public by computerizing library operations.

The Facts—Most of the libraries computerized seem to have no interest in improving service, as we can see from such things as their average line-staff salaries (mostly at the peonage level), the size of their cataloging backlog (in one case about 300,000 volumes), and the staffing of their campus branch libraries (about half of the staff needed). Money wasted in computerization could greatly improve service if applied to these areas. Also, processes that delay placement of orders, delay accountability of circulation records, and split the card catalog in multiple parts would not seem to be aimed directly at improving service to the public.

## THE CREDO

Throughout the land, the priesthood, with no exception, recited to me "The Credo of Automatic Automators":

I believe in the increasing cost of labor and the decreasing cost of computers.

I believe that in ten years (the time span was standardized) the cost curves will cross in favor of computers.

I believe that even if it isn't cheaper, the by-products of computerization make it worthwhile.

Since my pilgrimage, I have had the same Credo recited by others who were not specifically computerators, so there must be international specifications for its writing. It requires some examination.

First: there really is no "decreasing cost of computers." It is true that, on paper, the unit rental cost of new generations of computers decreases, but in sounding out what actually happens in practical applications, it is evident that the cost of applying the machines has increased due to various factors, one being the difficulty of keeping the computer fed without interruption.[21] But the central fact is that the overwhelming costs in computerization are labor costs (machine costs run about 20 percent of the total), and the salaries of systems analysts and programmers go up even faster than library staff salaries. Even after initial development costs are absorbed, the repeated costs of reprogramming and program adjustment are very high. Since the costs of computerized library operations are far higher than manual alternatives now, and the costs of computer labor are increasing faster than library labor costs, computerization will become increasingly expensive in the future.

Second: we are willing to accept any machine that will save us money at any time,[22] but if that time is ten years from now, then 1981 is the time to adopt the machine. What kind of folly wastes money for ten years on a machine that it hopes will eventually save money? Within ten years new machines, now unseen, will emerge in competition with the computer.

Third: the matter of by-products is the smelliest red herring of all those dragged across our path by computerators. The word is invoked with a kind of awe, as though it descends from heaven to banish all the disabilities of the computer. As Melcher put it: "we find ourselves invited to applaud computer applications that are somewhat in a class with the dog who played the violin—not that it was done well, but rather than it was done at all."[23] I keep having draped before me as accomplishments by-products that either are of no use whatsoever for a library operation, or that have a very low incidence of use, or that can easily be done by hand or by other machines faster and at a lower cost. The questions that are ignored must be asked—what by-products are worthwhile, for what library purposes, at what costs, and for what incidence of use? In sum, I find the Credo, like all matters of dogma, an excuse for suspending the intellect on the part of librarians and managers.

## THE MIRACLES

At the very peak of library computerization we are breeding a group of extremely able librarians, whose otherwise fine intelligence is completely blown when they evaluate their machine. They analyze their daily operations with command and critical brilliance, but when they talk about their future, like a sun-crazed prospector dribbling fool's gold through his fingers, a dull film covers their eyes, and they babble about miracles to come that are just around the corner, with not a shred of evidence to support their beliefs. Their faith is the exact equivalent of a witch's faith in flying ointment. Unfortunately, we have long passed the stage in which we could run a library from a broomstick.

Nevertheless, one can respect the priesthood. It's the acolytes, and at their fringe, the sycophants that make us feel unclean. Here we are in a range of one-upmanship and pretentiousness straight from Madison Avenue.[24] Responses to questionnaires about computerized operations produce amazing answers, if you know what is really going on in libraries. If someone lays down a transistor on a typewriter, the department is likely to respond that it has automated. The computer is used to cover up weaknesses as cowdung was to plaster frontier log cabins. If catalogers are low producers, if circulation is in chaos, the tendency is to computerize instead of reviewing or revising operations, both of which require thinking.

So, the rules of thumb are clear—if you start a library from scratch, computerize and you're fifty years old.[25] If you're upgrading an Ag college, the computer will liberalize cows. If you're a frustrated junior college, computerize and it makes you Ph.D. If your faculty is lousy, computerize and you'll be Harvard. If you're bush league, computerize and you'll win the Series. If you're stupid, computerize and you'll feel great. Instant achievement by machine and cheap attempts to invoke a false sheen of glory have replaced an intelligent confrontation of the problems in a large number of weak libraries.

## Run, Rabbit, Run

In view of the irrationality of the forces that led to library computerization, and the subsequent aggravation of this situation by self-seekers, it should come as no surprise that managerial practice has entirely left this field.[26] Of the forty-odd computer projects re-

viewed on my leave in ten major libraries, not one was begun on the basis of a managerial decision, after carefully reviewing and costing the operation to be converted, costing other machine or machine-manual alternatives (which were obviously available for many of them), or carefully projecting the costs of the computer operation after development costs (which one should be willing to absorb if retrievable over a period of time). Since most of the projects were doing only what had been done manually, price should have been the major factor in making this decision, yet very little cost analysis was applied, although all the libraries were hard pressed for funds. No computerators were surprised when I reported lack of managerial decisions; it was taken for granted that there were none in computerization. Like concupiscence, the desire to computerize simply must be satisfied no matter what the cost, and this at a time when most universities and libraries are bankrupt and facing an even bleaker financial future.

## Downhill All the Way

My discussions of this problem have produced a number of oppositions over the past few months, the most interesting of which is the concept of comparative incompetence advanced by a friend of mine. It makes no difference, the argument goes, that no careful cost comparisons precede computerization, because most librarians do not analyze costs before making other changes in libraries. The premise, I think, is false; but even if it were true, it is almost impossible to make even approximately as large a commitment in any other way in a library as that involved in computerization, where a quarter of a million dollars is meager.

More harrowing than the enormous costs is the fact that *a computerized system is virtually irreversible*, the fourth distinctive disability of this machine.[27] Once you begin a systems approach to computerizing operations, you are hung by the gills on the computer industry's fishstringer for good. Once applied, the computer acts as a *powerful agent against change*. The dynamics here are interesting. One library began to computerize by hiring a systems librarian who hired one programmer when they began to convert their circulation operation. Two years later, when the agonies of this conversion had subsided (and the circulation costs were fantastically more expensive than the manual system, and they were cumulating circulation records only every three days), the staff of this department was five, and, having been blooded, was eager to begin computerizing another operation. Even if you could prove that further computerization was diabolically evil, you still could not stop this momentum.

In addition, once computerization begins, the campus pressures on the library to get with it have been assuaged, the operation has been tapped for its public relations value, and personal and institutional egos are heavily invested in ploughing ahead to disaster. This is especially true if the computer project is the librarian's baby. One highly touted serials project began on "free" computer time, then later was charged for the campus computer costs (which hurt, but were not disabling). When the campus changed its computer and this operation had to use commercial firms for the computer configuration necessary to run its program, the cost more than doubled previous costs. After reprogramming for over a year, this serials operation is still processed partly off and partly on campus. It is known as a disaster area among computer experts, but this librarian stated recently that he thought it had done his library a lot of good.

Inertia also results from sheer moral exhaustion. The prolonged agonies inflicted on any sane person during the process of converting to computerization push him to the extremes of human endurance. After all the bugs are exterminated and the system is running, it is virtually impossible for a survivor of the process to summon up the moral strength to rethink, reorganize, redevise processes, and restaff. In one case, where superficial cost comparisons convinced an acquistions operation it was saving money, more sober thought made clear that it was losing money and taking longer by computer. But the department head was very indignant when I proposed that they could return to their former system—"After going through all of *that?*" Another department head refused even to reconsider and attempt costs comparisons when, after three years, her computer system was finally working.

Then, of course, there are enormous inflexibilities imposed against change by finances. Development costs in one case seem to be running over a million and a half dollars. You can be sure that it will take quite a jolt to make a library abandon that large an investment. In other instances, the costs of changing to an alternative system require large amounts of money not in hand, as in the college with the four-part book catalog that would prefer a card catalog. Until the totality of waste in operating by computer becomes so large that the figure really appalls, the library is not likely to make the sensible move, especially in the face of the beneficent connotation that (in libraries, at least) is still attached to the computer.

### THE BRAVE NEW WORLD

Anyone who computerizes at this point in time is hitching his wagon to a falling star. The honeymoon is over, if our seduction by the computer can be so termed. We have been sucked in by one of the most potent information control powers in recent history. Computerizing library operations at present and projected costs, and with foreseeable results, is intellectually and fiscally irresponsible and managerially incompetent. The proper answer to idiots who beamingly dangle their computerized projects for our admiration is, "Why don't you do something useful, instead."

The shrinking financial support of the academic world will drive us to sense even against our will. On the campus where I found forty-nine computers (four of them IBM 360's; one, the largest capacity known), the president gave the bloodiest state-of-the-university speech to date—dropping three academic programs, cutting back the current budget forthwith a million dollars, forecasting a further rollback of 3.5 million over the next three years, and even this predicated on unusual success in fund raising.

This is no temporary condition tied to the recession. More than two years ago, it was apparent that the public had become disillusioned with technology and education. They expected miracles of both; yet it is clear that each is the answer to only a part of our national problems. Public support for technology, a keystone in education's expansion, will continue to decline. Alumni disillusionment with campus products has seriously diminished alumni support. Foundations have been turning from the academy to other social agencies. The production of bachelor's, master's, and doctorate degrees has overrun the market for their products. Elementary and secondary school populations continue to decline. Education has costed itself out of sight, either in tuition costs or in the total costs of public institutions. All of these factors guarantee us future curtailment of programs in higher education and a continual decline in financial support, except for those programs immediately responsive to immediate problems that enjoy public favor. Make no mistake, we are about to shake out the men from the boys, and the future in libraries (as in other areas of university services) lies with the managers, who can make the most out of every cent available. The computer is the machine that evaporates money the fastest.[28]

In sum, our experience with the computer in library operations has been one more replay of *The Emperor's New Clothes,* and what we were led to believe were distant mountains laden with gold, available merely by boring a drift in the slope, turn out, upon close inspection, to be the hairy buttocks of the well-fed computer industry. And from such a source we have gotten exactly what we should expect.

## REFERENCES

1. Ellsworth Mason, Director of Library Services, Hofstra University.
2. Quite obviously, this kind of view is not encouraged by industry, but when it emerges, it is extremely revealing. In "Computers Can't Solve Everything," *Fortune* (Oct. 1969, p.126–29+, Tom Alexander reports the principal findings of a highly disenchanting survey by the Research Institute of America of computerization in 2,500 leading U.S. industrial companies. In "Automation: Rosy Prospects and Cold Facts," *Library Journal* (15 March 1968), p.1105–09, Daniel Melcher, president of the R. R. Bowker Company, indicates in detail that, although computerization is costing the publishing industry more than former processes, its effect has been to diminish performance.

Alexander contends: "But now, after buying or leasing some 60,000 computers during the past fifteen years, businessmen are less and less able to state with assurance that it's all worth it." (p.126) "As the Research Institute of America survey revealed, most companies are unsure that there is a payoff from computers even in supposedly routine operations." (p.128) "Relatively few companies have yet succeeded in devising nonclerical applications for the computer (because) programming and equipment costs are so high." (p.127)

Melcher contends: "To be candid about it, however, I think we could have done all this if anybody had wanted it, even before the invention of the computer." (p.1109) "Anything can be done (by computer), I guess, but that isn't the issue. What matters is whether anyone in his right mind would choose that way of doing it." (p.1109) "They all hope for tangible economies in the future—though it is a bit puzzling to note that the $5 million companies seem to expect those economies when they reach $10 million, and the $10 million companies think there might be economies when they reach $20 million, etc." (p.1105) "Computers have unmistakably lengthened the time it takes to fill an order, and have made it almost impossible to understand a royalty statement or get an intelligent answer to a complaint or a query." (p.1105) "The near-term result often seems to be that information formerly available by means of a phone call to the order department is reported as unknowable until the computer makes its next periodic report." (p.1106) "Batch processing . . . can delay your orders, delay your deliveries, delay your payments, and cut you off from ready access to your own data." (p.1109)

Victor Strauss, a consultant for printing management and contributing editor of *Publisher's Weekly*, states: "The computerization offered neither price advantages nor delivery advantages to book publishers." "The New Composition Technology: Promises and Realities," *Publishers' Weekly* 195:62 (5 May 1969).

3. Circulation is one operation in which librarians seem to see nothing between a manual and a computerized system, whereas in reality, a large range of alternatives exists. See also Melcher, p.1106: "Other machines also cost less or do more. The cost of offset printing plates drops from $1.50 a page to $1 a page to ten cents a page, even to five cents a page—in an almost unbelievable series of technical breakthroughs."

4. "The old idea that an automated system could be operated at a new lower cost than a manual system is dead, indeed." [Allen Veaner, "The Application of Computers to Library Technical Processing," *CRL* 31:36 (Jan. 1970).]

"Wishful thinking about present and future costs may give us librarians a black eye with the very administrators who are urging us to 'get with computers.'" [William Locke, "Computer Costs for Large Libraries," *Datamation* (Feb. 1970), p.74.]

"I talked to one wholesaler who had really made his automation work, but who had wound up with costs a good deal higher than a competitor's. I asked whether he really thought he could get his costs down. He said: 'No, but I think the other fellow's costs will rise—he's automating, too.'" (Melcher, p.1107)

See details of the high costs of computers in educational processes in Anthony G. Oettinger (of the Harvard University Program on Technology and Society), *Run, Computer, Run* (Cambridge, Mass.: Harvard University Press, 1969), p.189-200. This is the most penetrating analysis to date of the application of various technologies that are "force-fed, oversold, and prematurely applied." See also the frank statement on the costs of computers, including limitations on the cost reductions possible in the longterm future, in Frederick G. Withington (of Arthur D. Little, Inc.), *The Real Computer* (Reading, Mass.: Addison-Wesley Publishing Co., 1969), p.37-41.

The literature is riddled with irresponsible accounts of project costs that make no real attempt to include the full range of costs of computerization.

5. This fact was called to my attention by a manager of an aerospace satellite systems division.

6. "Programming is still very much an art and one in which there seem to be no standards of performance." (Alexander, p.171)

7. Just compare. We are lured by the frills of computerization and forget its enormous basic costs. Myra, with her six-foot-seven escort, proposes to forget the six feet and concentrate on the inches. "The glamor, let's face it, is in the computers, but the breakthroughs are elsewhere." (Melcher, p.1107)

8. "They (computers) are creatures of their time, and they come because they are

needed." (Melcher, p.1106) Melcher makes the common mistake of assuming that, because we needed something to help us in volume operations, the computer is what we needed. I contend that it is not. He states later, "It must be noted, however, that as yet the utilization of the computer to meet those changing needs has been massively disappointing." (p.1107)

9. "The service bureau put out cards through its computer instead of through the far simpler card lister formerly used. The result was no different, and they charged us three times as much—but it made us feel kind of big league." (Melcher, p.1109)

10. A phrase by one of our best poets of the 1930s, now reemerging, Laura Riding.

11. "In effect, each new task for a computer entails the design, development, and fabrication of a unique machine, assembled partly out of the boxful of hardware, partly out of software." (Alexander, p.171)

12. I still accept on faith the remarkable computational facility of the computer, though cautioned by friends in industry that unless the computer is checked at each permutation point in a computation, they cannot be sure that the results are right, because of possible disorders in the machine. Since checking takes too much time, technologists accept the computer's results, fully aware that often they are working with unreliable data!

13. This may have eased somewhat, since the extravagantly financed and well-publicized grunts of INTREX at M.I.T. have brought forth a mouse. The self-delusion of electronics engineers is demonstrated in the fact that, since they have taken over control of the engineering schools, "insignificant" courses, such as Power and Illumination, have been dropped from all of them. Maybe if we don't look, need for such knowledge will go away.

"When the new specialists were asked to understand before they criticized, some of them were outraged. 'We should learn from you? You've got to be kidding. Should we, the Knights of Systems Analysis, soil our anointed hands with that old rubbish? Learn about it? We will simply sweep it away in no time with our electronic broom. We'll put you out of business!' " [Victor Strauss, "Betwixt Cup and Lip," Publishers' Weekly (26 Jan. 1970), p.263.]

14. Similarities to Christian doctrine are due to the fact that Computer theology is vaguely Christian in orientation.

15. The most extreme deception encountered involved a campus computer unit which contracted with the central library (apparently to get access to its grant money) to handle a library operation, one of whose basic requirements was the integrity of the information stored in the computer (an IBM 360/67). Months after programming began, the library discovered that the chances of this machine wiping out its storage file are high, a fact known to the experts from the beginning.

16. See Melcher's statements in footnotes two and four.

17. Alexander, p.126.

18. "One knowledgeable consultant estimates that about half the System/360's now installed are still operating in the 'emulation' mode (i.e., are acting as second instead of third generation computers) . . . at least a billion dollars worth of new machine capacity is, in effect, wasted." (Alexander, p.129) If we can brainwash people to be so stupid, why can't we brainwash them to be virtuous?

19. Allen Veaner discusses this problem in full in "Major Decision Points in Library Automation," CRL 31:308-09 (Sept. 1970).

20. "Within limits, the more of our processes we get computerized, the better chance we have of matching the costs of the manual system." [William Locke, "Computer Costs for Large Libraries," Datamation (Feb. 1970), p.72.] When pursued by mail, Locke admitted that he has no evidence to support this contention.

21. "Despite the fact that, on a capacity basis, the IBM System/360, RCA Spectra 70, and GE 600 series are cheaper to lease or buy, they have been the hardest put to show a demonstrable payoff . . . they are too costly to be sitting idle, but they also need more highly qualified—and more highly paid—personnel to operate effectively." (Alexander, p.129)

22. Hofstra is now running final cost estimates on an MT/ST card production system despite warnings against it. But we began with careful cost control of our manual production and will be able to compare costs to decide whether or not to continue.

23. (Melcher, p.1107) His figure, of course, is stolen from Samuel Johnson.

24. "The rules of the computer game are that you talk only about what you are going to do, never about how it turned out. This is a science in which you publish the results of your experiments before you make them." (Melcher, p.1105)

25. For what happens when computerization

begins with the library, see Dan Mather, "Data Processing in an Academic Library," *PNLA Quarterly* 32:4–21 (July 1968).

26. "In companies everywhere the reasons for buying computers were not thought out. From the top, the attitude was that you can't let the competition get ahead of you; if they buy computers we've got to buy computers. The result was great euphoria." (Alexander, p.126, quoting a GE internal consultant on computer usage.) "According to the survey, the majority of computer users believe they themselves were too precipitous in acquiring their machines." (Alexander, p.127)

27. "Yet once in the grasp of an automated system, there is no turning back. Entering upon an automated system in any enterprise is practically an irreversible step." [Veaner, "The Application of Computers to Library Technical Processing," *CRL* 31:37 (Jan. 1970).]

28. "But do people only want to save money?" plaintively writes a computerator to me. If at no other time, certainly when they are bankrupt.

NOTE

An extension of these remarks can be found in my article, "Automation? or Russian Roulette?," published in the Proceedings of the 9th Annual Clinic on Library Applications of Data Processing, 1972, by the University of Illinois.

Since the first publication of this article, two other significant articles have come to my attention. Dan Smith, "The Accident-prone Miracle; a Survey of the Computer Industry," *Economist* 238: 39 pages in Roman numerals following p.50 (Feb. 27, 1971). This is a brilliant comprehensive article on reasons for the poor state of the art; and Robert C. Kenagy, "Why is Data Processing Such a Mystery?" *Publishers Weekly* 202:30-32 (Sept. 11, 1972). This is the only article I have found by anyone seriously committed to the computer that shows as deep a concern as mine about the lack of managerial control over computerization. Kenagy defines six steps required in the development of any computer project and explains how they can be controlled by non-technician managers.

# LIBRARY COST ANALYSIS:
## A RECIPE

*John Kountz*

IN ECONOMICS there is an axiom ascribed to Jody Todd, an obscure but erudite economist of the last century: "They ain't no g*d d**m free lunch!" (My apologies, Jody, but they get the gist.) And to set this axiom in context, libraries are no exception. Everything done in libraries represents a cost, with everything meaning *everything*. However, in a manner apparently contradicting the Todd Axiom, we as librarians are wont to be unaware of the existence of certain activities and their affiliated cost.

The result of this peculiar *Weltanschauung* is that actions occur for us without anyone doing them and are, therefore, without cost. Examples of this phenomena include screening and sorting LC proofslips ("Get rid of the stuff in Hindoo, Marge"), or specially preparing books for patron use *after* they have been processed ("Jay reads 'em all and puts the green dot on the spine for new fiction...the red dot means, well—those books are risqué..."). Perhaps our outlook stems from logical causes ("Gotta keep Marge busy; besides, she likes it"), but the activities do not happen spontaneously and they are not free.

Inadvertently, the Todd Axiom has been acknowledged by librarians. A benchmark example of this acknowledgement is easily Fremont Rider, who in 1936 reported cost accounting data in the *Library Quarterly* for the Olin Memorial Library. From this beginning, a very meager trickle of information has resulted. Implied from the late beginning librarians made in their acknowledgement of the Todd Axiom, and the subsequent unpopularity of the subject, is a lack of interest which, at best, has an odor of doom about it.

Unforgivably, time has passed since the days when the library's patron was the local monarch and cost was no deterrent. Time's passage has replaced the monarch with taxpayers or stockholders, and, concurrently, sensitivity to cost has attained stellar importance.

The causes for being unaware of costs may stem from a variety of reasons, but they cannot, in all fairness to the profession, belie an inability to perform the simple arithmetic of cost accounting. What is suspected is a lack of the few simple ground rules and the logical operations that bind them together, in short—a recipe for cost accounting and analysis.

In the following is outlined one such set of ground rules and their related procedural requirements, which have evolved and been applied with success over the past few years. It is stressed that since this set represents the findings of one library, it may not fully satisfy the specific requirements of your own shop. Therefore, feel free to adopt the ground rules to your immediate requirements.

With regard to discipline, it is pretty much summed up in the six steps and five resource requirements which follow. In addition to identifying steps, requirements, and the mysterious ways of cost analysis, these ingredients are blended together in a manner which will be meaningful for your internal operations and may be significant for your library's future. Finally, rest assured that some discipline and an adding machine are the only prerequisites for entry into the 20th Century.

*The Steps*

1. *Identify and Quantify Process (What's In A Name):* In essence, the process to be analyzed must be named in a manner which will permit isolation of this particular item throughout the accounting processes. Invariably, a process does not stand alone; rather, it is preceded and followed by processes which may or may not be within the same organization. For example, the process of receiving books depends on a vendor's shipping them.

As a function of the scope of your review, the precision with which these names are developed may permit the identification of similar or identical processes elsewhere in the organization. In this way, a process which is recurring need be costed but once. For example, once you have costed one bibliographic search, the same figures will probably apply wherever bibliographic search is performed—unless the specific search radically differs from your "standard."

Last, but not least, avoid the natural tendency to become overly general in naming activities or functions, e.g., "Oh, that's part of circulation,"; or overly specific, "...that's slipping." Similarly, avoid jargon, e.g., for "...she's discharging books," perhaps "clearing circulation records for books returned by patrons" is suspected to be less anatomical and more to the point.

*Performance Indicators*

Once the process has been identified, quantifiable performance indicators for it are required. Examples of performance indicators include number of requests handled per week, number of books processed per day, number of boxes opened per hour, etc. Understand that for the item to be quantifiable, it must relate to a designated period of time, since work is performed in time. Once a unit of time is set, it should be used throughout. Don't set yourself up with conversions of "millenium to light years" or "angstroms to feet."

Finally, relate the performance indicators to cost, e.g., filling a request should cost "X" dollars. The use of standard times will facilitate the development of such cost equivalents. As we proceed, you may be astounded at the difference between what you think should be (your performance indicator) and what is (your cost accounting total).

2. *Shatter the Process into Its Component Steps:* After a process has been identified and performance indicators designated, look the process over to determine and document the activities comprising it, and the special materials and equipment it requires. This "looking it over" is analysis; yea, systems analysis, and there are entire libraries stuffed with information on how to do it, so we needn't dwell.

Of importance at this point is the determination of the process' component activities (its requisite motions and their logical sequence) beginning with an input and yielding an output. The documentation alluded to can consist of, but is not limited to, flow charts, narrative, tables, or whatever you are most comfortable with. The secret to good documentation

for cost analysis is the ease with which data can be tracked to its source. This track, often referred to as an audit trail, is vitally important since the results of cost analysis represent, without exception, totals. The individual costs for the activities represented in those totals must be identified before the totals are significant or can be used for comparison.

3. *Determine Each Step's Resrouce Requirements:* Analyzing and documenting the activities internal to a process are vital to understanding its "how" and "why." Equally vital, and usually only hinted at in analysis and documentation, will be the resource mix necessary to sustain the process. The more glaring resources, and those suspected to dominate in the bulk of published cost recapitulations, are manpower and supplies. However, manpower and supplies are but two of the four resources almost invariably affiliated with an activity (manpower, supplies, supervision, and environment). In addition, a fifth resource (equipment) may also be encountered.

Since each resource type has a distinct set of characteristics, it is valuable to examine the types to delineate their peculiarities and their impact on the cost of a process activity.

Finally, and before proceeding with an examination of each: note that each resource type is indicated parenthetically as either direct or prorated. Remember "...no free lunch"? Cost accounting, which is the name of this game, yields a summarization of all expenses affiliated with performance of a process. The indicated direct or prorated category will come in handy when it comes time to determine "how much of what" before adding up the tab, as you'll soon see.

A. *Manpower (direct):* For each process activity you have documented, a human being will be involved somewhere, somehow. Manpower means how much involvement and by what level of person. Involvement simply stated is the amount of time spent on this activity per unit produced by it. Level of person relates to the cost to the organization for the person who is involved. Let's review these two terms, involvement and cost, in greater detail as their application to cost accounting is linked to our understanding of them.

Involvement, in a finer sense, means the direct, hands-on participation in the activity by a staff member. For example, that Lulu searches the author catalog to fill requests delineates Lulu's involvement in a step of the request handling process. That Lulu is one of four people employed for this activity is relevant in that she costs the organization so much per hour, produces so many "searched requests" per hour, and represents one of four individuals functioning in that area. That a fifth person supervises Lulu, and the other three searchers, and perhaps ten others who are performing tasks not related to the "search" is important, but it is not relevant. Involvement thus limits manpower resources to mean only the direct participation in the activity, the time dedicated to it. Lulu's supervisor is a special case (considered later) and cannot be included in the accounting for direct contribution to the activities cost.

Cost, as hinted above, means, "How much does Lulu cost us to keep her As such it means much more than the wage she receives once a week for her services. Included in the cost to keep Lulu are expenses for vacations with pay, sick leaves, medical plans, retirement, workman's compensation, but not the cost of handling the paperwork affiliated with both Lulu and her fringe benefits.

*"The secret to good documentation is the ease with which data can be tracked to its source"—using the track or "audit trail"*

Manpower cost, then, means wages and fringe benefits. Lulu's wages are probably a matter of record, in terms of range and step, as will be her fringes; the cost of handling Lulu's paperwork is liable to be something else and is examined separately under Overhead.

B. *Supplies (direct)*: As Lulu works, she consumes pencils, erasers, "P" slips, flags, paper clips, paper, certain janitorial items, and the like. These are supplies. Identify them and the quantity consumed over a period of time. "Paper clips?" Yes, and pencils, too. If supplies are handled casually in your library, it may be time to exercise control. You may not forestall Lulu opening her stationery store, but you'll have a better idea of the cost of handling a request, and ammunition for meaningful administrative questions. Expendable they may be, but gratis they are not; and the expense represented by the supplies needed to perform an activity contribute to the cost of that process, regardless of how seemingly miniscule they may be.

"*How much does Lulu cost us to keep her!*"

C. *Equipment (prorated)*: If supplies are the raw materials consumed by most office functions, then equipment are the tools. Certain rules for costing must be applied, however, else each new machine purchased will have a devastating effect on the allied activities cost. In addition, for libraries, not all equipment will plug into an electrical outlet. NUC, for example, represents a sizeable outlay and is not expendable. Thus, a general rule of thumb might state that equipment cost more than $100 and not be considered expendable.

Obviously, not all activities would have equipment affiliated with them (Lulu doesn't type), and in view of their cost, a second rule comes into play: It would not be wise to attempt to charge the outlay they represent in one time period. Therefore, the cost for equipment should be prorated (amortized) over the useful life of the device. As a result of this division of the total cost into "pennies per day," the cost contribution represented by a piece of machinery to the total process cost can be determined. For each organization, usually, a schedule of useful life for amortization has been developed. In the case of NUC and similar bibliographic tools, you may have to determine the amortization schedule, if you haven't done so already.

D. *Supervision (prorated)*: Lulu may be the world's best searcher, but. . .the amount of time and the level of supervisor required to guide Lulu's efforts is also important. Who supervises Lulu is no mystery: Martha does! However, without using a strictly enforced system of time cards, the precise amount of time Martha spends with Lulu would be extremely difficult to track. Also, Lulu is one of four searchers. Two of those four were

hired last week, and Martha has had to spend extra time bringing them up to speed. This latter factor, that learning periods require closer supervision, must be isolated for the individuals performing the activity. Conversely, as the trainee becomes a swan, the amount of supervision should become minimal (and sometimes does).

Because of these aspects of supervision, it is best to prorate the cost of the supervisor and apply the prorated amount to the activity. The proration indicated must consider the rate of turnover for the activity being costed, and it should be adjusted periodically to reflect current staffing.

E. *Space (prorated)*: It is understood that most library facilities are owned, not leased. Still, for the purpose of costing a process step, it is reasonable to assume that the process could be performed anywhere if suitably housed. The cost of space, then, is the cost of the suitable housing, e.g., how much floor area is required for the process step; does it require special treatment, such as cool air, soundproofing, immediate availability of water, and so on. The utilities affiliated with the process step (e.g., electricity, water, gas. . .) enter into the equation. Remember, Lulu doesn't live by bread alone.

These, then are the resource types in concert with hints at their handling. They form the basic menu for cost accounting; but they do not complete the cost accounting meal. The final ingredient is more spice than staple and as such cannot be called a resource. Rather, it is an expense which applies to the direct cost of doing things. This expense is:

*Overhead (a percentage usually)*: Earlier, under Manpower, there was mention of handling Lulu's paperwork. Implied was a cost lurking behind Lulu, and perhaps others. Well, it is a cost, but it is neither mysterious nor lurking. Overhead is very real, and briefly, consists of a summary of all expenses incurred during the operation of your library which are not related to a specific process step. To further clarify, let's examine the handling of Lulu's paperwork. Somewhere in the library or its mother organization, someone keeps records relating to Lulu. In fact, and directly related to the size of the library, there may be a platoon of record keepers, supervisors of the record keepers, supplies of yet to be filled out records, yea, even unto supply clerks. Further, this group not only sees that Lulu gets paid, it keeps the air conditioning blowing, the flush toilets flushing, the lights lit, the parking spaces assigned, and the wastebaskets empty. All that without mention of catching hell whenever these and other things aren't done right now! Well, this noble yet ignoble group represents essentially the same resource mix as an activity. Indeed, it is unto itself an activity called administration, and the cost of administration is characterized as overhead or burden. In addition, since this administration contacts and serves practically all the activities performed in the library, it would be extremely difficult to document and determine the dollar value of each. Therefore, overhead is usually seen as a percentage of the operating budget, e.g., the total cost for administration (in some instances, including the librarian himself) divided by the operating cost for the library, or the mother organization. For formulators, this looks like:

$$A = \frac{BA}{BT - BA}$$

A = Overhead or Burden Percentage
BA = Administrative Budget
BT = Total Budget

Once determined, the overhead percentage is applied to the direct and prorated costs for a particular activity to complete the cost picture.

4. *Identify Times and Volumes and Relate to Cost*: Everything mentioned above can be quantified and costed. Specifically: the human (Lulu's rate of pay per unit time); the material (the forms, erasers, paper, etc. used by Lulu, including lights and air conditioning); the supervision (how much time the boss has to spend to "make 'it' right"), and the overhead (Lulu's sick leave balance) are blended to their dollar equivalent per "it" performed.

The arithmetic of cost accounting is quite simple:

1) Direct labor, materials, equipment, and supervision are quantified and each reduced to its cost;

2) The identified costs are added together; and multiplied by the overhead percentage. This figure is then added to;

3) The sum of identified costs themselves to yield a grand total, into which;

4) The number of "it" performed is divided, and;

5) The answer written on a sheet of documentation (for future reference).

For those with formulitis, this might be shown as:

$$Icost = \frac{A(\Sigma D, S, E, M) + (\Sigma D, S, E, M)}{I}$$

Icost = Cost per "it," of item, handled during the sample period.

A = The overhead rate in its original equivalent, e.g., 25 percent is 0.25 (here's where employee newsletters, water coolers, and so forth go).

D = The total cost for direct labor used during the sample period (how much did Lulu cost. . .)

S = The total cost of supervision during the sample period ("We never underline *those* in red. . .").

E = The total cost of equipment used (except water coolers, etc.).

M = The total cost of materials consumed during the sample period (including stuff "ripped-off").

I = Number of "it," or items, handled during the sample period (how many did Lulu do. . .).

Notice time is only a factor here, insofar as work occurs in time (Lulu's rate of pay, or the lease of a machine).

Implied, and with reason, are a group of activities which do not occur in mechanical time and cannot be equated to dollars per unit; e.g., cataloging and book selection. Further, while a goal of the library may be an increased number of "it" handled per unit time, the goal of cost analysis is more "it" for less bucks . . .the two goals may not mesh.

Many mechanical contraptions probably can do Lulu's job in one thousandth the time—but the cost per item is liable to make Lulu look damn effective. By this same token, the converse is just as apt to be true, which leads us to cost analysis and alternatives.

5. *Cost Analysis and Alternates*: There are those who would call everything done to date cost analysis. This may coincide with their experience and, therefore, be true for them. However, finding how much it costs is less than half the job. For any significant result to evolve, it is necessary to know where you presently stand, of course, but there's more to it if results are desired. To produce results demands alternates and cost analysis. To understand the interplay of these factors, let's briefly review the effort so far.

Tersely, the discipline and documentation of cost accounting will have produced three things:

a) a detailed knowledge of the resources, practices, and procedures of each activity requisite to a particular process;

b) quantitative data unto cost concerning each activity, and;

c) alternate ways of doing each activity as they came under scrutiny.

It is precisely at the point you realize the madness of your previously established performance indicator(s) that one or another of those alternate ways of doing "it" begins to look good. It is also at this point that cost analysis comes into play.

But before you become analytical it is necessary to cost account the alternate. Yep, same old stuff all over again. There is a difference, however: the existing way of doing "it" can be documented "a posteriori," so to say; the alternate must be documented "a priori," which implies three considerations.

The first is honesty—which is what the Todd Axiom is all about. Beware of the blue sky, the under-capitalized gadget manufacturer, the convention demonstrator unit, and the sympathetic salesman. It is true that in our technological age most of the processes or functions which we, as librarians, have traditionally performed manually can now be performed mechanically. But it is true that many devices are but gleams in the eyes of prospective parents who are searching someone to pay for the delivery. Don't document and cost alternates which exist on paper only (unless you've an unlimited budget—in which case the exercise is academic).

The second consideration grows out of the first: don't re-invent the wheel. Certainly we would all like to pioneer something: but the school of hard knocks reveals that the demands of pioneering are stringently enforced and curiously unique (asbestos trouser seats, the tenacity of a bull pup, and cast iron attitudes, for example). It is far better to use a proven alternative, one that is post partum, operating, documented, and available. The cost accounting for alternates that exist somewhere (and not just in the mind) will not be quite as exciting, it is true; nor will there be the excitement of budget and schedule setbacks, service disruptions, staff alienations, and patron complaints should the brainchild be implemented.

The third and last consideration rests on higher mathematical principles: the alternate must do everything the existing activity does. If it doesn't, it is not an alternate, and the fun has just begun as you set about revising the entire library to accommodate what appeared to be a better way of gluing pockets into books. In addition, this consideration applied will lead you away from enviously trying to compete costwise with another shop's quick and dirty methods. For example, Orange County, California, creates card sets, pockets, and spine labels on receipt of the book for less than a nickel per volume, but for this technique to be used elsewhere would necessitate the automation of the entire library, so that precataloging, mechanized receiving procedures, and book catalogs, with cumulative supplements were integral features. (Reference here is to Orange County Total System BIBLIOS, which is documented in the February 1970 issue of *Datamation*, and the *6th Annual Proceedings of the American Society for Information Science*, 1969.)

Cost analysis occurs once the existing and alternate ways of doing "it" have been cost accounted. Oddly enough, it consists of analyzing the costs attributed to the various ways of doing "it," with the goal of determining what, if any, benefit might be derived by replacing the existing with an alternate. The scope of cost analysis is not limited to the comparison of operation and maintenance costs: it includes the cost of converting from the existing to the alternate as well. This conversion cost is, in itself, a composite of expenses which, depending on the magnitude of the alternate, may include retraining of personnel, new forms, new equipment, file conversion (both manual and mechanical), test runs of the alternate, and parallel operation of both the existing and the alternate.

Cost analysis thus will not only confirm suspicions, but will hint at how much it will cost to get there from here. And hint only! It can be postulated unequivocally that the cost of conversion or implementation will not be known with precision until those activities are completed. But the hint can be damned accurate if you have done your homework.

The value of cost analysis rests in its ability to let you do it on paper before you do it. Realism and detail are the keys, quicker and cheaper ways of doing things the products.

6. *Review*: Now then! The alternate not only works, it was implemented without Lulu's protesting (too viciously) and didn't interrupt patron service (too thoroughly). Are you satisfied. . .happy. . .complacent? Nope! Now it's time to start at ground zero and undergo the discipline a third (or fourth, or. . .) time. This seemingly masochistic bent is called review, and it is as important as the other ingredients in the recipe.

Cost accounting of existing operations and alternates, analysis and comparison of the results, selection or rejection of proposed alternates: this is the stuff of cost analysis and good administration. Such efforts are not, however, one shot. Kindly understand that while knowledge of all the facts may eliminate the need for decisions, knowledge must reflect current reality. Material, labor, supervision, space, and equipment costs all change in time, and the modulating pattern which they represent can impact directly on the way you are doing things.

Thus, a "running" review may steer you to the adaptation of alternatives which can genuinely enhance the performance indicator (cost) figures for your particular shop with a result in overall savings. It is not uncommon, in certain instances, that such a review yields an abandonment of a function or an alternative which has grown too expensive for continued use.

Finally, and a direct product of review (which implies accounting and analysis), you will be in a much stronger position to evaluate the mechanical gadgets with which the market is flooded today. In fact, it may be you who calls the salesman, rather than the other way around. And you'll have the spirit of Jody Todd at your right hand.

## ZENS & KELVINS

*An Open-Minded Bibliography
on Cost Techniques for Libraries*

The following brief bibliography is not intended as a comprehensive survey of literature of library cost efforts, but rather a selection of the type of literature which has sprung up around the subject. The goal of the literature review was to discern a general methodology which could be applied in any library and which encompassed the entire spectra of costs, not just wages and salaries. Oddly, the most applicable references are not in the library literature, and so a single recent representation from a management periodical is cited as an example of the type information sought. Once underway, it became apparent that while no general methodology was discovered, two camps were being represented. This polarization had been suspected, but it is most evident in the literature of budgets, cost accounting, and cost analysis.

The first faction is far and away the most popularly disseminated, having been and being currently promulgated by library schools and practicing librarians alike. It is an approach to fiscal matters most appropriately called "Zen," in which the subject is dismissed as: too complex; too outlandishly beneath the dignity of the profession; or, so mundane that it is best mentioned and forgotten. A principal characteristic of Zen writers is their understanding that budget must be rationalized even though it is last year's stew warmed over and padded out with spuds for an additional guest at the table. Terrifyingly, the Zen approach continues to draw zealous adherents, who, lemming-like, are hell-bent to chase their phantom into a sea of oblivion.

The second faction recognizes the detail and complexity of cost accounting and analytical techniques as well as their rewards. This approach perhaps is best called "Kelvin" after an expression attributed to Lord Kelvin which in paraphrase goes, "if you can't express it numerically, you don't know very damned much about it." Characteristically the Kelvin followers realize the value of planning; appreciate the necessity of control; exercise reporting techniques; and luxuriate in the visibility available to those who "know something about it." Hopefully, the Kelvin adherents, while few in number, are sufficiently vociferous to develop librarianship into a profession that will be perpetuated into the future, rather than shelved as an anachronistic approach to earning a living.

The citations are arranged by date. The individual works represent thought during a particular period, and the number of citations in each date bracket is representative of the interest which occurred during the periods listed. It is curious to note the periods of proliferation with clusters happening in the middle 30's, early 40's, and late 60's. Perhaps exterior pressures have caused these outbursts (dollar squeezes known commonly as the Depression, World War II, and the current "crisis" might be implied, but this is for a higher power to determine). It is also curious to note that in each of these periods, the Zens match the Kelvins almost tit for tat. Originally, it was hoped that representative texts in librarianship could be included. Not so! Texts for library school administration courses were reviewed and found unhappily to represent Zen not Kelvin. Thus, recent texts are not well represented since they do not contribute to the corpus of information relating to library cost activities. Implied here is that the followers of Kelvin will have to be either self-made or those who had to work for a living; for certainly, without appropriate text material, it is clear that library schools cannot be their source of inspiration.

Finally, a series of documents produced at Purdue University under funding by the National Science Foundation was reviewed. These materials had titles as "Comparison of the Effects of Library Control Systems" and "Mathematical Models for Library

*Zens and Kelvins: ". . .it became apparent that while no general methodology was discovered, two camps were being represented"*

Systems Analysis." It was suspected they might be relevant. They weren't. In fact, they resemble the Morse volume cited which is an adequate representative of the statistical extrapolations engineering courses are liable to degenerate into from time to time.

Pre-1900—RHEES, William J. *Manual of Public Libraries, Institutions, and Societies, in the United States and British Provinces of North America.* Univ. of Illinois, Graduate School of Library Science. Monograph No. 7. (Reprint of 1859 ed.)
An interesting volume compiled for the Smithsonian in 1859. The statistics gathered (collection size, collection growth, circulation…) indicate the "traditional" nature of the measurements applied to libraries. Implied is either a lack of interest or the absence of motivation to determine an empirical measurement of service, factors which continue to impede the development of standards. The cost of book catalogs should impress those currently contemplating this approach.

1925—THOMSON, I. R. Howard. *Reasonable Budgets for Public Libraries.* ALA, 1925, 44p.
A curious volume representative of the nolocontendere "tradition" in which, on p. 8, Thomson condemns the past by stating, "Just because there were no standards to which a library must attain, or of the service it must render, there were no standards as to what its expenditures should be." Implied was the possibility of a solution elaborated in the remainder of his book—the solution does not evolve; rather, there ensues a vague enumeration of items, jobs, services, and the like. Budgeting, in Thomson's eyes, remains little more than the sagacity of hindsight warmed over. It is this view which dominated the literature of the period and finds a continuing voice to the present.

1928—LOWE, John Adams. *Public Library Administration.* ALA, 1928, 175p.
Again, the "tradition" is carried forward in which cost accounting and analysis were apparently neither known nor required. The only attempt to know how much does it cost was predicated on the requirement that some rational basis be used to prepare a budget. As such, Lowe states on p. 50, "The necessary first step in adoption of a budget system is the determination of the items which shall be separately recognized." With this as a start, Lowe then examines various categories of costs and finally exhorts that, "Too many separate divisions may be a source of unnecessary intricacy…" and "If the items are too few, the budget will fail in its purposes…(of tracking expenditures) to permit intelligent analysis." Thus, while recognizing that cost analysis requires detail neither method nor categories are outlined (Lowe suggests the ALA Revised Form for Public Library Statistics—Finance, and refers to Thomson—Ref. 1925).

1933—TOLMAN, Frank L. "Shrinkage of Public Revenues" in *Current Problems in Public Library Finance.* ed. by Carl Vitz. ALA, 1933, p. 21-41.
Recognizes the need for a marriage of control and administration through statements such as, "Cost accounting has had little application in library administration" and "Next must come objective measurements of each service in terms of the cost and nature and quantity of the product." In addition, Tolman hints at need for performance indicators. However, there is no mention of how to get there from here.

WRIGHT, Ida F. "Measuring the Results and Informing the Community" in *Current Problems in Public Library Finance.* ed. by Carl Vitz. ALA, 1933, p. 76-96.
What possibly could follow the realization that, "Analysis of costs by working units, such as the Reference Department, the librarian realizes, involves an elaborate technique…" Why the seminal article for use of circulation as an indicator of efficiency—what else? Mrs. Wright continues to rationalize that cost analysis (cost accounting) is too difficult for librarians. She does list categories of functions to be accounted for and divides them into

permanent and nonpermanent expenditures. Object of her presentation seems to be rationalizing a budget.

1936—RIDER, Fremont. "Library Cost Accounting," *Library Quarterly,* October 1936, p. 331-81.
A monumental application of cost accounting techniques to library operations yielding unit costs for labor is represented here. The work reflecting operations at the Olin Memorial Library is, to the best of the author's knowledge, the first such work to appear in library literature, and presents not only a detailed listing of the functions internal to a diversity of library processes, but also an implied methodology for the reduction of the costs attributed to each process step to a unit cost for a handled item. Unfortunately, many of Rider's sample sizes and clear definitions of the work unit are not included. Still the technique represented is as modern as today and could well act as a model for libraries.

1937—MILLER, Robert A. "Cost Accounting for Libraries; Acquisitions and Cataloging," *Library Quarterly,* October 1937, p. 511-36.
Both the components of cost analysis and a fine-grained outline of specific tasks are delineated by Miller. In addition, timesheet layout and an overall methodology for determining personnel times for specified functions are covered. Reduction of this information and a methodology which could be applied generally, however, is not included in Miller's work. What is of value is Miller's delineation of the detail necessary to fully account for a function.

1941—BALDWIN, Emma V. & William E. Marcus. *Library Costs and Budgets: A Study of Cost Accounting in Public Libraries.* Bowker, 1941.
A report on the application of cost accounting techniques by 37 participant libraries. Good for identification of tasks (worksheets included). Does not actually outline steps but details reporting of labor and reduction to cost for developing labor cost per item handled. Volume is valuable for definitions of service types, e.g., cataloging, reference, etc.

1943—BRAY, Helen E. *The Library's Financial Records.* Bowker, 1943. 58p.
Relates to components of accounting and budget content rather than cost accounting. Touches subject briefly in two very short sections: Securing Budget Estimates (p. 47), and; Job Analysis (p. 50). However, the touch is so soft as to vaguely imply a requirement without specific detail on what the requirements actually might be.

McDIARMID, Errett Weir & John McDiarmid. *The Administration of the American Public Library.* ALA, 1943, 250p.
A latter day perpetuator of the intuitive librarianship advocated in the 1920s. As stated on p. 1, "No effort has been made to analyze library processes and routines…" There it rests, a curious counterpoint to Wight (below) and Baldwin.

WIGHT, Edward A. *Public Library Finance and Accounting.* ALA, 1943, 137p.
An excellent section is devoted to cost accounting (p. 95-111). With surprising candor Wight recognizes that, since public libraries lack a profit motive and occupy a monopoly-like role, there has been little use of cost data for their operations; but the times are liable to change as, "Pressure for increasing efficiency is exerted by the taxpayer." He then recapitulates the cost accounting efforts reported in the literature and a rationalization of the use of cost data. Wight also recognizes direct and indirect costs; the work unit and standards.

1952—BRYAN, Alice I. *The Public Librarian.* Columbia Univ. Pr., 1952, 474p.
The single allusion to cost and budget analysis is as an entry in a string of things to be included in library school curricula.

1962—WHEELER, Joseph L. & Herbert Goldhor. *Practical Administration of Public Libraries.* Harper. 1962, 571p.
Well-indexed, extensive references, this volume "covered the waterfront." However, Wheeler's comment on Wight's *Public Library Finance and Accounting* now applies to Wheeler, i.e., "An up-

dated edition would be helpful indeed." The "waterfront" covered here was upset with performance budgeting, surveys by nonlibrarians, the vocabulary of industrial engineers, who's professional and how to deal with people. Needless to say, very few pages are relevant (p. 121-22, and 194-97 or less than one percent).

1967—FASANA, Paul J. "Determining the Cost of Library Automation," *ALA Bulletin,* June 1967, p. 656-61.
Nine factors affecting the cost of mechanizing a library process are described. These factors are: design, machine; programming; input-output; conversion; implementation; building (space); personnel, and; supplies. Curiously, all factors in one degree or another, relate to practically any cost effort. The author also recognizes that for two alternates to be compared they must perform the identical same function, e.g., additional services, etc., can only be recognized qualitatively but not entered into the equation.

1968—MORSE, Philip McCord. *Library Effectiveness: a Systems Approach.* M.I.T. Pr., 1968, 207p.
An extensive reduction of circulation and collection maintenance functions to a set of probabilistic models. Unfortunately, this chrystomathy of Operations Research with a library flavor overlooks cost. If only they had tinkered with linear programming instead of Markov.

U.S. NATIONAL ADVISORY COMMISSION ON LIBRARIES. *Library Services for the Nation's Needs.* U.S. Office of Education, 1968, 63p.
A broad survey of the state of the art with related objectives responding to a series of questions posed in the Statement and Executive Order establishing the National Advisory Commission. Of importance here is the question, "Are we getting the most benefit for the taxpayer's dollar spent?" It is noteworthy that the Commission responded by disclosing the worthlessness of library statistics and the lack of overall expenditure figures from which to work coupled with a lack of means by which to reduce any data to meaningful terms. While the Commission did fund a study by Mathematica entitled, "On the Economics of Library Operations," it did not appear to relate the findings of that study in its report.

MATHEMATICA. *On the Economics of Library Operation. Final Report Submitted to the National Advisory Commission on Libraries.* N.J.
With an intriguing a title as this, how can one go wrong? Ho Hum! Pie charts, tables, conjecture about growth rates and lots of obsolete figures. The lack of relevance of this work is matched only by an introductory disclaimer revealing the inadequate time allotted by the funding agency. One is led to believe that with more time a technique might have evolved to reduce the data to meaningful terms; but assumptions are always dangerous. Perhaps the commission was Wight, oops, right. Still, this is a good example of how to present piles of data in a variety of ways.

WYNAR, Bohdan S., *Library Acquisitions.* Libraries Unlimited, N.Y., 1968, 275p.
This annotated classified bibliography, while focussed on acquisitions, contains a mingled selection of budget, work measurements, and cost accounting materials representative of the late '50s and mid-'60s (p. 45-49).

1971—NANCE, Harold W. "6 Factors to Weigh When Buying Equipment," *Administrative Management.* September 1971, p. 30-35.
This is an example of a genre of cost articles which appears from time to time in the nonlibrary literature. It is a presentation of cost comparison (including spread sheets) for decisions entailing capital outlay. The six factors: machine cost; space; maintenance; labor; supplies; and power—while directed to analysis of two (or more) machines—is fully applicable to the reduction of alternate methods in libraries to tenable terms for decisions. Nance also details the components of his six factors and recognizes that much of the effort invested in cost analysis will yield data valuable both to immediate decisions as well as during operation and maintenance after the decision has been made.

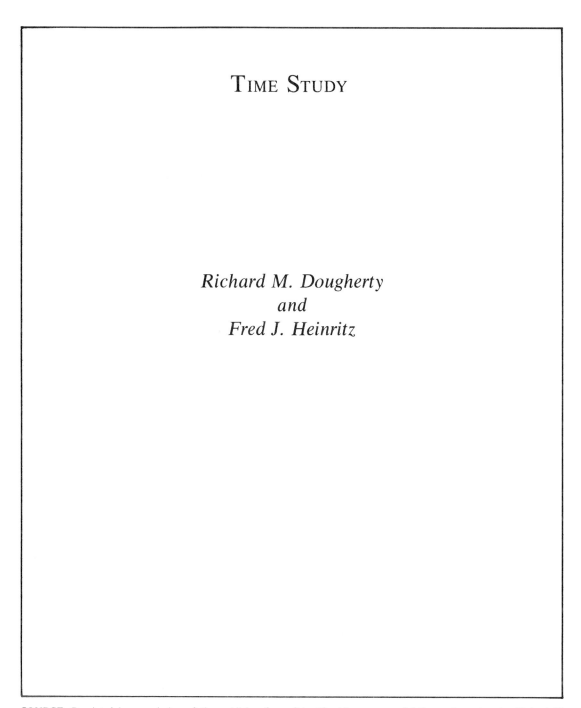

Time Study

Richard M. Dougherty
*and*
Fred J. Heinritz

SOURCE: Reprinted by permission of the publisher from *Scientific Management of Library Operations* by Richard M. Dougherty and Fred J. Heinritz (Scarecrow Press, 1966). Copyright © 1966 by Richard M. Dougherty and Fred J. Heinritz.

The Value of Time Study

Time study is a technique for determining the time required by a qualified, well-trained person working at a normal pace
to do a specified job.  It is a form of work measurement.  Time
study is an activity which follows logically after an analysis of the
existing job and the elimination of unnecessary steps (motion study).
The standard methods for making motion analysis have been covered in previous chapters.

Before plunging into the details of time study, some of its
uses should be considered.  Time study data can be used to determine time standards.  These figures can then be used to establish
fair and reasonable performance standards.  Performance standards,
their construction and usefulness, are discussed in detail in Chapter XI.  Secondly, time study data are necessary to calculate the
costs of a system.  A budget with its cost estimates based on work
measurement data will be a more accurate and supportable document than one with cost estimates based on guesswork.  Finally,
time study, through standards, enables administrators to gauge employee performance -- to reward above standard performance, to
give those who need it such additional training as required to
achieve the standard and, when necessary, to take disciplinary action.

Making a Time Study

Time Study Equipment

For library time studies an ordinary wrist watch is usually
all that is needed.  The watch should have a clearly visible second
hand.  In industry, where many jobs require more precise timing
than do many library operations, it is standard procedure to use a
stop watch.  A stop watch should, for convenience' of computation,
have a decimally divided face (See Figure 7-1).  It need not be a

heavy investment.    As with ordinary watches, there is a wide price
range.    A good quality watch that reads to 1/100 minute and will
total up to 60 minutes may be obtained for as little as $25.00 to
$30.00.    There are numerous stop watch manufacturers listed in
Thomas' Register of American Manufacturers.

Fig. 7-1 Decimal-Minute Stop Watch

Time study men in industry sometimes record times by
means of motion picture cameras and even more sophisticated devices.    The choice of equipment depends on the nature of the job
and the degree of accuracy required.    The accuracy required depends to a large extent on the frequency with which a job is performed.    A small error in a task requiring a small amount of
time, projected a million times will result in a substantial miscalculation; the same job performed only a few thousand times does
not require the same degree of timing accuracy.    It is worth em-
phasizing that most library time studies can be performed successfully with an ordinary watch with a sweep second hand.

Professional time study people generally use decimal stop
watches and record their times to the nearest hundredth of a minute.    A librarian may have available only an ordinary watch with a

sexagesimal face, or even a stop watch so divided (perhaps borrowed or begged from the Athletic Department).  It is not wrong to time in minutes and seconds.  It is, however less convenient for decimal computation.  In any case, it is an easy matter to convert from seconds to hundredths of a minute and vice versa by using the following formulas:

Hundredths of a minute = 5/3 x Seconds

Seconds = 3/5 x Hundredths of a minute

For example, 16.2 seconds would be equivalent to 5/3 x 16.2, or 27 hundredths of a minute; 27 hundredths of a minute would be equivalent to 3/5 x 27, or 16.2 seconds.

If many such computations must be made, a conversion table is a convenience.  Such tables are at hand in any library.  For example, in the mathematical tables section of any edition of the ubiquitous Handbook of Chemistry and Physics one will find together a table of "Minutes and seconds to decimal parts of a degree" and one of "Decimal parts of a degree to minutes and seconds."  Or conversions may be made conveniently on the slide rule.

In addition to a timing device, the person making the study will want some sort of clipboard to hold the paper on which he records his observations.  Special observation boards that are designed to hold both paper and stop watch in a convenient position can be purchased for about $10.00 (See Figure 7-2).  Special time study forms may also be purchased; or the analyst may devise his own (See Figure 7-4).  Slide rules and calculating machines are useful for lessening the time and drudgery of computation.  Even if a librarian does not happen to have a calculator in his department, there is almost always one he can use in the library office or in some other nearby office.  A librarian can learn all he needs to know about the slide rule -- namely, how to multiply and divide -- in only a few minutes.  The basic instructions for using a slide rule are covered in Chapter IX.

Developing a Standard Method

The customary procedure for an industrial methods analyst is to develop a standard procedure before undertaking a time study.

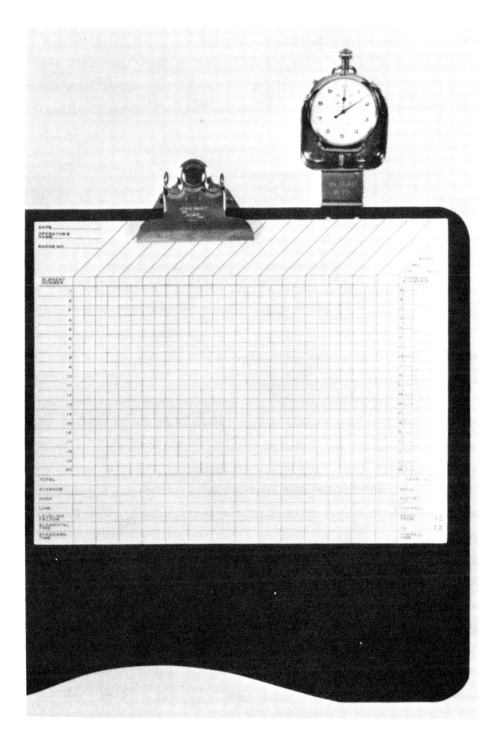

Fig. 7-2   Observation Board With Stop Watch
and Observation Sheet for Recording Times

The reason for this is that if the original procedure is later modified, the timing operations will have to be repeated.    Of course, developing a standard method is one of the purposes of motion study and accounts for the fact that motion study usually preceeds time study.

It is usually necessary, since libraries do not generally know their costs, to chart the present procedure and collect times so as to provide a basis for comparison when new methods are studied.

Dividing the Job to be Timed Into Elements

The standard job description must next be broken into observable elements which may be easily timed.    In industrial situations experience has shown that observations more than 30 seconds in duration tend to become inaccurate.    For most library studies, however, such a high degree of precision is not necessary.    This is fortunate because in many library processes the most natural and convenient elements to time can run as long as several minutes in duration.    However, every effort should be made to keep the time of each element as short as possible.

For ease of timing, the elements should be divided in such a way that the analyst will be able to tell readily when one element ends and another begins, either by sight or sound.    An analyst should also separate out those elements that are not controlled by the operator.    For example, a study of a Xerox 914 in operation would require two types of elements:    The cycle time of the machine, and the handling time of the operator.

Finally, constant elements should be distinguished from variable elements; that is, those elements which do not experience time fluctuations from cycle to cycle should be separated from those elements that experience large cyclical fluctuations.    Although it is very common in the metal industries for processes to include predominately constant elements, the reverse is true in libraries.    In industrial environs it is customary to define a constant element as one that is free from the influences of size, weight, length, and shape.    To these four factors, an additional variable must be added, that of human interaction.   Numerous library routines involve dialogues between two or more individuals -- for example, registering a new reader or collecting fine money.   Such interactions produce cyclical variations.   This is not a serious problem; all that is required are additional observations to offset the extra variation.   The number of observations required is discussed in Chapter VIII.

## Selecting the Person to be Timed

There may be two or more persons performing the same operation either simultaneously or at different times of the day.   As an example of the former, several persons are likely to be typing or filing at the same time; as an example of the latter, the person manning a charging machine during the evening hours may not be the same person who worked the afternoon shift.   In choosing an operator to be timed, it usually makes sense to time one who regularly works at a normal pace, rather than either the high producer or the low producer.   If the fastest worker is timed, there is a risk that other employees may view the study findings with suspicion.

## Taking the Times

There are two common ways to record time data.   One is to let the watch run continuously.   The person doing the timing merely records the time reading at the end of each element in the appropriate space on his data sheet.   The actual time for each element is determined later by subtraction (See Figure 7-4).   This method does not require a stop watch.   It has the disadvantage of requiring a good deal of subtraction.   The second method, commonly known as "snapback," requires a stop watch.   Stop watches are so constructed that, at any time, by pressing the stem of the watch, the hands may be snapped back to zero.   At the end of each element, the observer reads his watch, snaps the watch hand back to zero, and then records the time for the element he has just read (See Figure 7-3).   The advantage to this method is that the times are recorded directly, thus eliminating the need for subtraction, and allowing the person timing to see more easily variations in time values while he is still on the job.    There are other less commonly used techniques beyond the two described in which two, and even three, stop watches are used.    The interested reader will find descriptions of these methods in the books listed in the chapter bibliography.

Recording the Readings

To a beginner it might not seem possible to observe an operator, read a watch, and record time readings almost simultaneously; but with practice he will discover that timing a worker is not at all difficult.    The first rule in timing a job is to observe and record each phase of the job.    Occasionally the analyst will observe an element that occurs only infrequently.    For example, a circulation discharging routine does not ordinarily include collecting fine money for overdue books, but the time required to collect fines should be recorded.    Additional watch readings will have to be collected for infrequently performed elements in order to have sufficient data. In addition, it will be necessary to establish the frequency of occurrence of the elements that do not appear in every cycle so that their times can be prorated.

An analyst will also be confronted with a third type of element -- the foreign element.    A foreign element is one that does not occur regularly in the cycle, such as a reader dropping his ID card while presenting it to the desk attendant.    Times for foreign elements should be recorded during the observations.    They may be ignored later if they happen so infrequently and involve so little total time as to be insignificant.

Time Study of a Binding Preparation Process

A study of a library binding preparation process will illustrate the essentials of making a time study.    The standard procedure was divided into six elements (See Figure 7-4).    1) The bindery clerk pulled the binding specification card for each title to be bound.    2) Next, using the specification card as a guide, he recorded on the bindery slip such information as the volume number, index and TPI (title page index) placement, disposition of covers and ads and so forth.    This was done by hand, so most of the recording consisted

| no. | Element | 1 | 2 | 3 | 4 | 5 | 6 | 7 | 8 | 9 | 10 |
|---|---|---|---|---|---|---|---|---|---|---|---|
| 1 | Pull bindery spec. card | .65 | .80 | .60 | .83 | .82 | .87 | .68 | .60 | .65 | .73 |
| 2 | Write info. on bindery slip | 1.40 | 1.52 | 1.67 | 1.72 | 1.45 | 1.26 | 1.34 | 1.52 | 1.67 | 1.15 |
| 3 | Type info. on bindery slip | .92 | .73 | .87 | .92 | .78 | 1.15 | .85 | .68 | .73 | .72 |
| 4 | Inspection of typing | .45 | .20 | .33 | .31 | .40 | .27 | .38 | .47 | .48 | .53 |
| 5 | Stamp T-no. on bindery slip | .10 | .07 | .08 | .09 | .10 | .08 | .10 | .08 | .09 | .10 |
| 6 | Refile spec. card | .55 | .48 | .57 | .63 | .70 | .67 | .68 | .55 | .56 | .67 |

Fig. 7-3 Example of repetitive or snap-back method of timing.

merely of checking boxes.    3) Then he typed the title and call number on the bindery slip.    4) Next, he inspected his typing. [1]    5) Then, using a Bates numbering machine, he stamped a transaction number onto the slip.    6) Finally, he refiled the specification card into the master file.    The completed slip was then put with the material to be bound and sent to a commercial binder. [2]

The time observation sheet (Figure 7-4) summarizes the time observations for ten cycles.    The continuous method of timing was employed.    The procedure for calculating the times for each element per cycle, the total elemental times for each of the ten cycles, and the average observed time for each element are calculated in the following manner.    The analyst observes the operator until the end of an element is reached.    He then reads his watch and records the time on the time study observation sheet in the box labelled "R" ('Reading') for the first element in cycle one (in the binding example this reading is 0.65 minutes).    At the end of the second element the analyst records the watch reading in the "R" box for element two, cycle one (2.05 minutes).    This procedure is continued until the last element in the cycle has been observed and the time recorded (4.07 minutes).    The stop watch hands are snapped back to zero and the observation process started again for cycle two.

Once the observations for all of the cycles have been recorded, the elemental times in each cycle are computed by a process of subtraction.    The time for the first element is correct as recorded on the observation sheet and thus is written without subtraction in the "T" ('Time') box.    In the illustration this value of 0.65 minutes for element one is subtracted from the reading for element two (2.05 - 0.65 = 1.40 minutes).    To obtain the elemental time for three, the reading for element two is subtracted from the reading for three (2.97 - 2.05 = 0.92 minutes).    This subtraction procedure is continued until the elemental times for all cycles have been calculated.    The next step is to add the times for each element and record the sums in the TOT (total) column.    Finally, the totals are divided by the number of observations -- which yields the average elemental times.    This is recorded in the AVG (aver-

TIME STUDY OBSERVATION SHEET

| Operation | Preparing bindery slips for periodicals. | Allowances | 12 % |
|---|---|---|---|
| | | Std Time | 5.84 min. |
| Time started 8:30 | Time finished 11:10 | Units/hr. | 82.4 |
| Observer J. Jones | Date 10/10/66 | | |

| no. | Elements | | 1 | 2 | 3 | 4 | 5 | 6 | 7 | 8 | 9 | 10 | ΣT | T̄ | RF | NT |
|---|---|---|---|---|---|---|---|---|---|---|---|---|---|---|---|---|
| 1 | Pull bindery spec. card from file. | T | .65 | .80 | .68 | .83 | .82 | .87 | .68 | .60 | .65 | .73 | 7.31 | .73 | | |
| | | R | .65 | .80 | .68 | .83 | .82 | .87 | .68 | .60 | .65 | .73 | | | | |
| 2 | Write info. on bindery slip. | T | 1.40 | 1.52 | 1.67 | 1.72 | 1.45 | 1.26 | 1.34 | 1.52 | 1.67 | 1.15 | 14.70 | 1.47 | | |
| | | R | 2.05 | 2.32 | 2.35 | 2.55 | 2.27 | 2.13 | 2.02 | 2.12 | 2.32 | 1.88 | | | | |
| 3 | Type info. on bindery slip. | T | .92 | .73 | .87 | .92 | .78 | 1.15 | .85 | .68 | .73 | .72 | 8.35 | .83 | | |
| | | R | 2.97 | 3.05 | 3.22 | 3.47 | 3.05 | 3.28 | 2.87 | 2.80 | 3.05 | 2.60 | | | | |
| 4 | Inspection of typing. | T | .45 | .20 | .33 | .31 | .40 | .27 | .38 | .47 | .48 | .53 | 3.82 | .38 | | |
| | | R | 3.42 | 3.25 | 3.55 | 3.78 | 3.45 | 3.55 | 3.25 | 3.27 | 3.53 | 3.13 | | | | |
| 5 | Stamp T-no. on bindery slip. | T | .10 | .07 | .08 | .09 | .10 | .08 | .10 | .08 | .09 | .10 | .89 | .09 | | |
| | | R | 3.52 | 3.32 | 3.63 | 3.87 | 3.55 | 3.63 | 3.35 | 3.35 | 3.62 | 3.23 | | | | |
| 6 | Refile spec. card in file. | T | .55 | .48 | .57 | .63 | .70 | .67 | .68 | .55 | .56 | .67 | 6.06 | .61 | | |
| | | R | 4.07 | 3.80 | 4.20 | 4.50 | 4.25 | 4.30 | 4.03 | 3.90 | 4.18 | 3.90 | | | | |
| 7 | | T | | | | | | | | | | | | | | |
| | | R | | | | | | | | | | | | | | |
| 8 | | T | | | | | | | | | | | | | | |
| | | R | | | | | | | | | | | | | | |
| 9 | | T | | | | | | | | | | | | | | |
| | | R | | | | | | | | | | | | | | |
| 10 | | T | | | | | | | | | | | | | | |
| | | R | | | | | | | | | | | | | | |
| 11 | | T | | | | | | | | | | | | | | |
| | | R | | | | | | | | | | | | | | |
| 12 | | T | | | | | | | | | | | | | | |
| | | R | | | | | | | | | | | | | | |
| 13 | | T | | | | | | | | | | | | | | |
| | | R | | | | | | | | | | | | | | |
| 14 | | T | | | | | | | | | | | | | | |
| | | R | | | | | | | | | | | | | | |
| 15 | | T | | | | | | | | | | | | | | |
| | | R | | | | | | | | | | | | | | |
| 16 | | T | | | | | | | | | | | | | | |
| | | R | | | | | | | | | | | | | | |
| 17 | | T | | | | | | | | | | | | | | |
| | | R | | | | | | | | | | | | | | |
| 18 | | T | | | | | | | | | | | | | | |
| | | R | | | | | | | | | | | | | | |
| 19 | | T | | | | | | | | | | | | | | |
| | | R | | | | | | | | | | | | | | |
| 20 | | T | | | | | | | | | | | | | | |
| | | R | | | | | | | | | | | | | | |
| | TOTALS | | | | | | | | | | | | | 4.11 | 1.25 | 5.14 |

Fig. 7-4  Sample time study observation sheet.

age) column. For example, for element one the total time is 7.31 minutes and the average time is 7.31/10, or 0.73 minutes.

Determining How Many Observations Are Necessary

One of the first questions a neophyte time analyst is likely to ask is: "How many observations should I make: two, five, ten, twenty-five, one hundred; just when can I stop?" The question is pertinent because if too few observations are taken, the data collected is likely to be unreliable; on the other hand, if too many observations are taken, study costs will be increased needlessly. Therefore, what the analyst is interested in determining is the minimum number of observations to produce data with sufficient reliability to satisfy his study requirements. As long as his needs are met, the method employed to gather the data is of secondary importance.

The magnitude of variation in the times from cycle to cycle correlates directly with the number of readings that will have to be gathered. As variation increases so will be the number of readings required. For example, if an analyst were to time a call number labelling process and the times (in minutes) for just ten cycles proved to be .55, 1.22, .32, 1.13, 1.30, .44, .72, .43, .82, and .62, could the analyst state with certainty that the average of .75 minutes is truly representative? If the answer is no, then additional readings would be required. On the other hand, if the observed times had been .69, .81, .75, .85, .68, .72, .76, .84, .64, and .89 which yields the same average .75, the analyst could have been more confident that .75 minutes actually was a representative average even though he still did not know the exact number of readings he should record.

Statisticians have developed a simple formula for calculating an adequate sample size. This procedure is described in Chapter VIII. When this approach is used, an analyst will have a solid foundation for substantiating the validity of the study. However, there will be occasions when the precision that statistical technique can provide is unwarranted, and in such cases a simpler common sense approach will prove equally satisfactory. For example, ten observations of a charging process might yield the following times: .10, .08, .07, .08, .09, .10, .09, .11, .12, and .11.  The average for these times is 0.095 minutes.  Suppose we had some mystic power and learned later that the correct average time for charging out a book was 0.11 minutes.  Our study is 0.02 minutes in error, but would an error of this magnitude have proven critical?  For the vast majority of libraries the answer would be no.

Although the example in the preceding paragraph was not contrived, it was at least convenient.  An analyst will not always be so fortunate as to have a definite pattern emerge so quickly.  In an actual study, what must be done is to continue recording times until one is satisfied that a truly representative pattern has emerged.  This prgmatic approach can be checked by the formula given in Chapter VIII.

## Rating the Worker

If a time study analyst were to calculate work standards and costs on the basis of observed times, the results would be completely misleading.  Costs would be underestimated and production overestimated.  It is necessary to adjust observed times so that they reflect reality.  Industrial time analysts have devised elaborate methods for this purpose.  This procedure is called "rating the worker."  A worker under observation tends to work at a faster pace than he would normally work.  The chief reasons for this are nervousness and job motivation, both of which stem from being singled out for special attention.

At this point it is natural to ask what is "normal?"  Normal time is the time it will take a qualified person working at a normal pace and using a standardized method to perform a job.  The rating process itself consists of adjusting the observed speed to the time analyst's conception of normal speed.  This adjustment is based on the judgment of the analyst and is therefore subjective.  Consequently, regardless of the elaborate systems employed, rating cannot be carried out with mechanical precision.  The accuracy of rating depends almost entirely upon the skill and judgment of the rater and on what appears fair to the employee.

There are several rating methods now in use. Here only the commonest will be described. The reader is referred to the chapter bibliography for information on the others. While the analyst is collecting time observations, he is also evaluating the operator's skill, effort, and speed. If the analyst judges the operator's speed to be equal to the "Theoretical normal," he assigns a rating of 100%; if the observed speed is thought to be less than normal, a rating of less than 100% is given; or if the worker's pace is judged faster than normal, a rating of more than 100% is assigned. The percentage rating factor, expressed as a decimal or fraction, is multiplied by the average observed time to obtain the normal time.

The formula for computing normal time is:

Normal Time $(N_T)$ = Observed Time $(O_T)$ x Rating Factor (RF)

The binding preparation time study will help illustrate the use of this formula. In that study the analyst judged that the worker was working much faster than his normal pace. Overall he estimated that the worker was producing approximately 25 per cent above normal; therefore, he assigned a rating factor of 1.25. The normal time was then calculated as follows:

$$N_T = O_T \text{ x } RF$$
$$N_T = 4.11 \text{ x } 1.25$$
$$N_T = 5.14 \text{ minutes}$$

The preceding paragraphs illustrate in capsule form the essential steps of rating a worker's observed speed. Rating a person with precision requires a great deal of skill which can be acquired only through long years of practice and experience. Most industrial analysts attend special classes where they are taught how to rate normal time. They spend hours analyzing motion picture films and conduct many time studies under close supervision. In view of the training that these analysts receive, it would be folly to expect inexperienced library analysts to duplicate their consistency or accuracy. The best that can be expected is for the analyst to exercise his best judgment, fully aware that his rating factor may be in error by twenty-five per cent or more.

Adjusting Normal Time to Standard Time

If workers produced without interruption or break for eight straight hours, normal times could be used to calculate standard times. However, workers must be granted time off for personal needs, rest periods, and for events which are beyond their control. These adjustments are termed worker allowances and are added to normal time in order to produce a standard time for performing a job.

Barnes states that "for light work, where the operator works eight hours per day without rest periods, 2 to 5% (10 to 24 minutes) per day is all that the average worker will use for personal time."[3] This allowance time would appear to be appropriate for library studies. In most libraries, either officially or tacitly, workers are allowed personal time; but in addition, they are also allowed rest periods, which have become more or less synonymous with "coffee breaks." Usually an employee is allowed one break around mid-morning and another in mid-afternoon. Breaks seem to run anywhere from ten to twenty minutes each. Some people are still reluctant to accept the value of breaks, primarily because most library tasks do not involve physical exertion. However, studies tend to support the contention that a worker's output increases sharply immediately after a break, and that mental fatigue can produce the same effects on a worker as physical fatigue.

Delays may be either avoidable or unavoidable. Delays caused by the operator should be disregarded and not included in allowances. However, unavoidable delays must be considered. Two examples of unavoidable delays are machine break downs (charging machine, photo-reproduction equipment, etc.), or waiting for an elevator (e.g., vertical movement of book trucks in a multi-stack library). In order to ascertain reliably the amount of delay associated with a particular job, it is necessary to resort to all-day time study, or work sampling. Fortunately, most library oriented jobs do not involve much unavoidable delay time.

In order to illustrate how standard times are calculated, we can again refer to the binding preparation example. The library allows each employee two rest periods of twenty minutes each and twenty minutes personal time per eight-hour day. Delay time for the binding routine as observed is negligible and no allowance is made. The normal work day consists of 480 minutes (8 hours x 60 minutes/hour). Thus the total worker allowances are:

Personal needs allowance $= \dfrac{20}{480}$ minutes/day $= 4\%$

Fatigue allowance $= \dfrac{40}{480}$ minutes/day $= 8\%$

Delay allowance $= 0$

Percentage of total working day permitted for allowances

$\overline{12\%}$

The standard time can now be computed by substitution into the following formula:

Standard time $(S_T)$ = Normal time $(N_T)$ x $\dfrac{100}{100 \text{ - Percentage of total working day permitted for allowances } (A)}$

$S_T = N_T \text{ x } \dfrac{100}{100 \text{ - A}}$

$S_T = 5.14 \text{ x } \dfrac{100}{100 - 12}$

$S_T = 5.84$

The standard time for the binding process is 5.84 minutes per volume prepared. Thus, if a worker devoted his entire day to this process, a supervisor could expect an output of 82 volumes $(480/5.84 = 82)$.

## A Method for Approximating Standard Times

The reason for converting observed times to standard times should be obvious. It should be equally obvious that the study technique itself introduces several types of errors. There will be timing errors, particularly if a regular watch is used for timing; rating errors are virtually assured -- the only question is their magnitude; and sampling will also introduce some errors. This seeming preoccupation with errors is not intended to belittle the value of time study, but to underscore that the results of time studies, particularly in libraries, will not be as precise as one might expect.

Fortunately, as mentioned previously, very few library studies require a high degree of precision.  If for no other reason, job frequencies are not sufficiently great for tenths of seconds or seconds to become a critical factor.  One method for approximating standard times is to add 50% to the observed times.  For other time costs, and added non-time costs, please see Chapter X.

### Notes

1. In a minority of cases, correction or retyping due to mistakes was necessary.  This would be an example of an infrequent element.  It is not shown here.

2. It should be noted that the entire process could and probably would be done in batches, pulling several specification cards at once, and so on.  In this case, we would work with smaller elements in order to keep the individual timings relatively shorter.

3. Ralph M. Barnes, Motion and Time Study, Design and Measurement of Work, (New York, Wiley, 1963), p. 401.

### Bibliography

#### General

Each of the two books listed below contains an extensive bibliography:

> Barnes, Ralph M.   Motion and Time Study. 5th ed.   New York:  Wiley, 1963.   Chapters 24-26.

> Mundel, Marvin E.   Motion and Time Study.   3rd ed. Englewood Cliffs,  N. J. :  Prentice-Hall,  1960. Chapters 17-24.

#### Libraries

For up to 1954 consult:

> Logsden, Richard.   "Time and Motion Studies in Libraries," Library Trends, II, No. 3 (January, 1954), p. 401-09.

For studies since 1954 look in Library Literature under the headings:

> Time and Motion Study and Time and Cost Studies.

ANALYSIS PHASE

OF THE

SYSTEMS STUDY-DETERMINATION

AND

SURVEY OF REQUIREMENTS

*Edward A. Chapman, Paul L. St. Pierre,*
*and*
*John Lubans, Jr.*

SOURCE: From Edward A. Chapman *et al, Library Systems Analysis Guidelines,* N.Y. John Wiley & Sons, Inc., 1970.
Reprinted by permission of the publisher. Copyright © 1970 by John Wiley & Sons.

GENERAL OVERVIEW

In the analysis of systems, the initial phase of any study, the "analyst" or "analysts" on the study staff should acquire a complete understanding of the system to be surveyed and become thoroughly identified and conversant with all of its components. It is equally important that the analyst determine the interactions, the interrelations, existing between the system under investigation and all other systems that place demands on it. The analysis phase, as briefly outlined in Chapter 2, consists of four distinct but interacting surveys.

1. *The Survey of Requirements.* The requirements of the system are the results of the *demands* (for information, reports, and action) placed on the system from all sources. The analyst should determine the demands in the light of stated goals and also where they originate and how they are satisfied by the system. Are there requirements that are not satisfied and if so, why not?

2. *The Survey of Current Operating Conditions.* The analyst should obtain a working knowledge of operations required in the system, the sequence in which they are performed, the functions, decisions, and actions required for the performance of each operation, and of what the inputs and outputs of each operation are. He should make a survey of equipment to establish the utilization and capabilities of available equipment and make a detailed survey, including work measurement, of the individual jobs in the system to learn whether staff job descriptions match the functions required.

3. *The Survey of Outputs.* The outputs of a system consist of the reports, records, and actions that are prepared or performed in the system to satisfy the demands being placed against it. The analyst should determine why, how, when, and by whom the output is prepared, what information it contains, and what functions and actions are required for preparing this output.

4. *The Survey of Inputs.* An input to the system is the information that must be used in order to generate the necessary outputs needed to satisfy the system's requirements. The analyst should determine what inputs are received; how many, how often, and where they originate; what information they contain; and what functions, decisions, and actions are required in order to convert this information into the desired outputs.

## REQUIREMENTS VERSUS DEMANDS

Requirements to be met by a system are based on requests or "demands" for specified information and action originating with management, the library user, and other sources both within the outside of the library. The questions needing to be asked are: What is required of the system? What must it do to satisfy the demands and needs of the library user, of management, and of the library's other systems? Exogenous demands must be identified and evaluated for their necessity and impact on the system and for their possible elimination if they are unnecessary and hamper fulfillment of the system's operational objectives.

Until requirements are known in detail, it is impossible to proceed with the analysis phase of the systems study. Understanding of present procedures, considered in Chapter 4, cannot occur without knowledge of the precise requirements a system is supposed to meet. Although the outputs of a system are derived from requirements, the two should not be confused. The outputs of a system are considered in Chapter 5. Outputs (information and action) flow from the demands with which the system is concerned.

We shall consider a few examples illustrating what requirements are in the operation of a system. Let us say that a fundamental demand of the serials system of the library is to provide the user with an accurate and current catalog or listing of the library's serials holdings. This demand obviously places a requirement on the serials system. We shall assume further that the demand is placed on the serials system to provide, within a reasonable time, special listings or special reports concerning the serials holdings felt to be necessary by the library's users and by the library staff itself. The complexity of this type of demand indicates that any serials system capable of handling it must be flexible enough to provide an analysis or report concerning the serials collection without prior knowledge of what specifically will be asked. This means that the serials system requires records that can be manipulated in various ways to satisfy multiple demands. Internal requirements of the serials system include the prompt claim of issues on a definite schedule, precise control of renewals of subscriptions when due, and improved accuracy and speed in the check-in of issues as received.

Another example of a demand is the preparation by the library's management of annual statistical reports for national data gathering agencies. This demand commits the director of the library to preparing the report and causes him to place demands on each system within the library to provide the data for this report. Thus this demand placed on the director causes the placement of multiple demands on systems throughout the library and the records required in each system must provide the information needed for the director's report.

Although the terms "demands" and "requirements" sometimes are used interchangeably, there is a difference in the meaning of these terms until the action taken translates a demand into a requirement. This fact may have become evident in the foregoing example of the demand for an annual statistical report placed against the library's management.

Whether demands originate from within the library or from outside, the reactive process explained above occurs. This complex of demands and requirements can be the source of a system's breakdown if it is not thoroughly controlled and analyzed with respect to the need, duplication, or overlapping of each element of the set of demands leading to requirements placed on the systems. Establishment of the need and correlation of demands and requirements is particularly important at the level of so-called "routine" operations. Here the library clerk with limited responsibility, an incomplete understanding of the process to which his work contributes, and no defined limitation of his own contribution may impose demands, or requirements, or both on a fellow worker resulting in a gradual deterioration of a process as it was originally designed.

The origin of the demand resulting in a requirement against a system has bearing on any judgement of whether the demand is unnecessary or must be accepted as a bonafide requirement of the system. Again alluding to the annual statistical report "demanded" by a national agency, this represents a demand from outside of the library over which the library has no direct control. The administration has two courses open. Either it can comply with all of the items of statistical information requested, placing additional requirements on the various systems of the library, or it can decide to comply partially using whatever statistical information is available. If this demand for a statistical report were an internal one and, for example, had originated with the library's administration itself, the management of the other systems of the library would be able to question elements of the demand with respect to the library's needs to be served and with respect to the applicability of the same information being made available in a different form from that requested by the library's administration.

Some demands placed on the library from outside sources may be questioned as to their being legitimate requirements acceptable by a system. This is illustrated by the instance of a comptroller's office of a parent organization demanding the continuance of manual procedures in the clearance of book invoices by the acquisitions department whose operations are to be automated. It may be pointed out that the computer system could furnish and summarize payment and accounting information in almost any manner required by the comptroller. The comptroller's demand may be shown to be unnecessary and uneconomical in the operations of both the library's acquisitions system and the comptroller's system. The comptroller may take the position that the library would create an exception in his commonly applied procedures that he is not readily inclined to change. Whatever the resolution with the comptroller of this outside demand, it serves as an illustration of why it is necessary to examine the validity of each demand.

Some demands from outside the library are patently arbitrary and create requirements deleterious to a system's operation. In a university library, for example, a department chairman may demand that his record of the books he has requested for purchase be kept up to date with the acquisitions department's record of action and ordering—that is, when each book is ordered, reports of delay, date received, and date cataloged. This requirement obviously would hamper the efficiency of the acquisitions system and create added costs if indeed the total demand could be met. This demand represents an unreasonable extension of an acceptable requirement; that is, notification to the requestor when the book is available for his use. Thus demands coming from any and all sources must be critically reviewed and analyzed in relation to the impact of the resulting requirements on the efficiency and mission of systems.

In analyzing or designing any system requirements cannot be established by asking what is wanted and blindly accepting stated requirements. Intrinsic requirements in the operation of a system frequently are not recognized or identifiable by the personnel responsible for fulfilling the purposes of a system. Explanation should be made of the *concept of requirements* if workers are to understand the meaning of their jobs and be able to identify for themselves the subsidiary requirements they fulfill in satisfying the primary demands of a system. A member of the acquisitions staff, if asked, might reply that the requirement of the acquisitions system is to order books, not realizing that many other requirements must be fulfilled to meet the basic demand of the ordering of books. Also a number of requirements are entailed if the serials system is to meet the primary demand of maintaining accurate and up-to-date records. Thus not only the primary demand of a system needs to be known but also all requirements contributing to the execution of the primary demand.

## DETERMINATION OF REQUIREMENTS

The first step in the analysis of a system is that of determining what requirements are being placed on the system. The study staff must translate the stated goals of the library into the demands that the goals place on each system and, in addition, must identify all requirements and their sources as listed below.

1. From outside the local organization there are ALA rules for filing; rules for main entry; reports required by governmental agencies and professional groups.

2. From outside the library locally accounting information is required by central purchasing; user requests are required for information and services.

3. From within the library there are systems depending on other systems for information; the director of libraries requiring certain reports and statistics. Illustrative of the interdependence of systems the cataloging department may wish to receive from the acquisitions system certain information contributing to the cataloging of a book; in addition it may expect that the acquisitions department will have ordered the Library of Congress cards at the time the book was ordered; or that the acquisitions department

will supply the proof card or copy of the cataloging information from the *National Union Catalog* if available.

4. From within a system information is required by one subsystem from another; information required within a subsystem. For example, the ordering subsystem within the acquisitions department may place the requirement on the preorder search subsystem that it furnish the correct information about the availability of the publication, the publisher, the date of publication, edition, cost, author, and title.

The need for identifying sources arises in connection with the opportunities the study staff may or may not have in suggesting helpful modification, change, or elimination of existing requirements. Typical sources of requirements are represented by the following, whom the study staff should interview regarding the requirements each one places on the system: (a) the director of libraries; (b) the users of the library; (c) the heads of departments within the parent organization affecting library operations, such as purchasing, accounting, and so on; (d) the head of each major operating system within the library; (e) the head of the system being surveyed; and (f) the personnel within the system.

In analyzing the requirements of the system it is necessary to sift out those invalid requirements arising from artificial organizational separations as well as those of a traditional character serving vestigial needs. The magnitude of the need served by the requirement should be observed. If, for example, a record is maintained to answer a need that may or may not arise, it will be necessary to prove that this record is necessary. In many instances requirements have been perpetuated in a system and with the growth of the system those that are obsolete have not been eliminated. When the "why" of a requirement is explored, it may be found to be spurious. A simple relocation of records can lead to the elimination of a so-called "requirement." For example, an "official catalog," a main entry catalog of each title in the library and on order, was deemed essential in a technical processing area because of the distance of the public card catalog from the cataloging and acquisitions personnel. This requirement to maintain an official catalog, entailing a considerable drain on staff time, could be eliminated by relocating the public catalog closer to the acquisitions and cataloging departments and by filing open-order records directly into the public catalog. Thus a primary phase of the preorder searching operations of the acquisitions department could be satisfied in one place—the public card catalog.

In analyzing requirements the analyst plays a rigorously objective role finding out how the system operates and inevitably in the process evaluating the validity of the requirements. If, for example, the library's management wants a report every month of every account maintained by the library, the question "why" arises: Is this report needed every month? Would a report simply on those accounts that are running low at a particular time be more pertinent? Just how much information does management need? How is this information going to be used? In the matter of reports it will be found many times that various reports will present the same information and that several records within the library will duplicate this information. The obvious question is can one or more of these records/files be eliminated or combined? This is typical of what must be investigated in order to determine the practicability of retention of existing records and files.

This point is belabored in order to emphasize that *no* requirement, formal or informal, is too small to investigate and evaluate. In addition no requirement can be retained for any reason except that it is necessary to meet the valid demands made on the system.

The analyst must probe and question until he knows why the information flowing from a given requirement is needed, how it is used and where, whether used elsewhere or filed, and if so, why, until he knows every use and disposition of the information being generated by each requirement. In order to do this he must learn the content of specific jobs in depth and the purpose each is intended to serve. It is not a question of his being able to do the job being analyzed but rather to know what the job is about; to know the requirements the work is intended to meet; and to know the processes through which personnel attempt to satisfy the requirements. The extent of the analyst's responsibility is more precisely indicated by the questions that must be answered in the form, Worksheet for the Survey of Requirements, Figure 3-1.

Each system looked at in a cursory way may reveal apparent requirements but such requirements cannot actually be accepted unless it is found what the demands are that cause these requirements. For example, the primary demand placed on the acquisitions system is the ordering of books. This, of course, could be done by simply accepting a request and ordering the item desired. However, all requirements of the acquisitions system would not be fulfilled by this action that satisfies the primary demand. Such action would not satisfy the requirement of "no duplication"; it would not satisfy the requirement that if a book is on order another copy should not be ordered; and it would not satisfy the requirement that before ordering a book verified bibliographic information should be supplied. Without identifying within a given system all requirements both modifying and enforcing established primary demands, the system cannot be properly analyzed or designed.

Requirements are determined on the basis of actual need rather than on desire without any demonstrable reason. Otherwise an administrator who states his requirement as being the need for information about *all aspects* of an operation rather than for the *critical elements* of it will only find that his decisions affecting the maintenance of the effectiveness and efficiency of that operation are more difficult to make. Such a requirement ignores the principle of "management by exception," which is knowing what has not occurred as planned in an operation rather than all that has occurred. This principle will be reviewed in Chapter 7.

Knowledge of all requirements in specific detail is vital to determining the work force capable of satisfying these requirements. Awareness of the requirements will uncover the extent of duplication of work, multiplication of the same reports from various sources, and the actual need for the information being supplied. It is not infrequently the case that staff members receive reports serving no useful purpose to them but merely perpetuating a traditional referring of information based on defunct requirements. Such unnecessary requirements should be ferreted out and eliminated.

## SURVEY OF REQUIREMENTS

In the beginning of this chapter were listed the people to be interviewed by the analyst, beginning with the director of the library, to determine the specific demands being placed on systems by the officers, workers, and users of the library and by those officers of the library's governing organization whose demands affect the library's operations.

The results of the interviews should be written if the analyst is to analyze systematically and to synthesize the information gathered. Availability of this information in a correlated condition is prerequisite and critical in the determinaton and survey of the "outputs" of systems, a step in the analysis phase of a systems study, subsequently treated in Chapter 5.

As a guide in interviewing and in maintaining a record of the information given by each interviewee, the form Worksheet for the Survey of Requirements is used (Figure 3-1). Before considering the use of this worksheet in illustrative detail, it is used to record the answers to questions such as the following:

1. What bibliographic, statistical and account records or reports are needed?

2. What system(s) generates the requested records or reports?

3. What information must these reports and records contain?

4. Why and how is this information used?

FIGURE 3-1

| WORKSHEET FOR THE SURVEY OF REQUIREMENTS (PART 1) |
|---|

1. Prepare a copy of this worksheet for each of your requirements.
2. Attach a completed sample of each record or report.

System:          Accounting and          Analyst Name:    Date:
    Acquisitions   Reporting Subsystem

Name of Person Interviewed:                 Position:
    Mrs. Jones                                  Acquisitions assistant

Name of Supervisor:                          Position:
    Mrs. Brown                                  Acquisitions librarian

Describe the requirement:
    To maintain accurate and current balances on each fund and school account and expenditures and encumbrances by department and also to provide a monthly report on each of these accounts.

Describe decisions you make in satisfying this requirement:
    What is the account number to which the book is charged? What is the variance between the list and delivered price of each book? Has the account balance fallen below the minimum that was established?

Describe what action you may take as a result of your decisions:
    In the case of the major decision regarding the status of the account balance: notify the acquisitions head for authorization to transfer additional funds into a depleted account or discontinue ordering; or if the fund is not exhausted, continue charging against it.

Describe the functions you perform in satisfying this requirement:
    By department, post the estimated price for each book to proper account. Post the actual cost, from invoice, for each book to account. Calculate the variance between estimated and actual cost. By school, post total expenditures and variance to school account and calculate balance. Prepare monthly financial report.

5. Is the information received in usable or final form; if not, what functions must be performed to adapt it for use?

6. Is the proper information for making necessary decisions furnished; and what are these decisions and the basis of each?

7. What actions are normally taken as the result of these decisions?

The information obtained in the survey of requirements serves the purpose of supplying the basis for the analyst to arrive at the following determinations:

1. The information unconditionally required by each staff member at each level of the organization in order for him to meet his responsibilities; the origin of the stated requirements and their pertinence to the operation for which he is responsible;

2. The reports unconditionally needed by management, including their frequency and content;

3. The statistics that are unquestionably required for analysis of the workloads and efficiency of a system;

4. The information generated at each level of operation actually required by management for the making of valid decisions and the taking of actions actually contributing to the objectives of the system's functions.

Analysis and survey procedures can be somewhat confusing and overwhelming unless undertaken *seriatim,* proceeding gradually and systematically from the greater to the lesser factors being determined, analyzed, and surveyed. This technique should be followed in the analysis of requirements as well as in the analysis and understanding of current procedures and in the determination of outputs and of inputs of the system under study.

The first step is to analyze the system broadly in order to gain a general idea of the demands placed on it and the operations that the system must undertake if it is to satisfy its objectives. Having gained this general understanding and come to deductive conclusions, the analyst probes to a more specific level. This process is continued until he possesses sufficient detailed knowledge to allow him to evaluate the effectiveness and efficiency of the system and suggest an improved design that will best satisfy the overall requirements and goals of the library. Again, using an acquisitions system as an example, its major functions may be broadly typified by the designated subsystems: preorder search, order, accounting and reporting, and receiving and checking, all being subsystems in the model acquisitions system illustrated in Figure 1-1. With a knowledge of the system's primary demand and major functions, subsidiary or supporting requirements can now be sought and deduced. This step-building process continues until all of the operations, functions, decisions, and actions of the system are determined and synthesized in preparation for the factual systems' evaluation and design.

Referring to the Worksheet for the Survey of Requirements (Figure 3-1), the descriptions of requirements and of decisions, actions, and functions performed in satisfying each requirement are approached concurrently because of their close interdependence and are treated in this manner. The acquisitions system is taken as an example in an effort to clarify what information is being sought in the four sections of the form. Although an acquisitions system has many requirements placed on it, only one placed against the accounting and reporting subsystem (Figure 1-3) is used in illustration.

The requirement placed on the accounting and reporting subsystem is that of maintaining accurate and current balances for each book fund including a record of actual expenditures from each fund, estimated outstanding encumbrances, and the supply of monthly reports of the status of each fund. There are many minor decisions entering into this requirement: What is the fund to which each book will be charged? Is the list price of the book correct? What is the estimated variance between the list and delivered price of each book? The major decision is has a given account balance fallen to or below the minimum established for administrative action? The action triggered by the affirmative may be the transfer of funds to bolster the account or the suspension of further purchasing.

The functions performed in satisfying the requirement of accurate and timely balances include the posting of estimated prices of books at the time of order, posting of the actual costs on receipt of invoices, calculating the variances between estimated and actual charges, and the posting of total expenditures and estimated outstanding charges against each fund. The preparation of the monthly financial report for each book account is based on the execution of these functions.

Referring to Part 2 of the Worksheet for the Survey of Requirements (Figure 3-2), the description of each record or report said to be needed for satisfying given requirements should be described and justified with respect to its actual need and application. Each person preparing and maintaining a record or a report should describe each record or report and the role each plays in satisfying specified requirements. As illustrated in Figure 3-2 two records and one report are maintained and prepared by the accounting and reporting subsystem: departmental control records, school or account records, and the monthly financial report.

In determining requirements it is necessary to gain an understanding of a given system before moving to another. The analyst, however, will

FIGURE 3-2

| WORKSHEET FOR THE SURVEY OF REQUIREMENTS (PART 2) |
|---|

Describe below each bibliographical, statistical, and/or accounting record or report that you prepare or maintain to satisfy this requirement. (Use additional forms if necessary.) *Please attach a completed sample* of each record/report (on sample underline in *black* information you use; underline in *red* information you add; underline in *blue* any unnecessary information).

Departmental Control Record: An accounting control is maintained for each academic department as well as certain library accounts, such as reference, gifts, and so forth. Each book purchased is posted to its respective record. The record contains the estimated price, actual cost, variance, and total expenditures for each book. Periodically the information from this record is summarized and the totals posted to the proper school or major account in the school or account control record.

School or Account Control Record: A summary accounting record is maintained for each school or account. The purchases for each department are summarized and posted to the appropriate control record. The record contains the estimated cost, variance, and balance. This record is the basis for preparing the monthly financial report.

Financial Report: This report is prepared and distributed monthly. It contains the distribution of library funds as well as the breakdown of expenditures this month and year, to date, by department and school. It also presents the balance for each school and major account.

For each input received by you which in any way affects your fulfillment of this requirement, please complete a *Worksheet for the Survey of Inputs.*

inevitably gain knowledge of relationships between the system under study and other systems in the library. In studying the acquisitions system, for example, it may become evident that demands are placed on the acquisitions system by the reference system as well as the cataloging system. The detailing of such demands from other systems must be made in order to reach an understanding of the requirements placed on the acquisitions system. Thus in studying a system the analyst must determine factors outside of that system affecting or causing the requirements to be met by the system under study.

At whatever level in the organization requirements are being investigated much the same questions are asked: What is wanted? What is actually required? How is it used? In summary, the survey of requirements is intended to yield knowledge in the following areas: (a) information definitely required from each person at each level in order for him to fulfill his functions within the system; (b) reports required by management—their frequency and content; (c) statistics required about operations; (d) information at each level clearly required for making decisions; (e) actions resulting from such decisions; (f) relevant requirements; and (g) unnecessary requirements.

In order to evaluate and correlate the findings of the survey of requirements of each system and its components, findings should be systematically summarized for study. Preparation of the Summary Worksheet for the Survey of Requirements (Figure 5-4) is discussed in Chapter 5. The reports, information, and records purported to be needed to satisfy the requirements placed against the system will have to be evaluated as valid or invalid *outputs* in fulfilling the system's requirements.

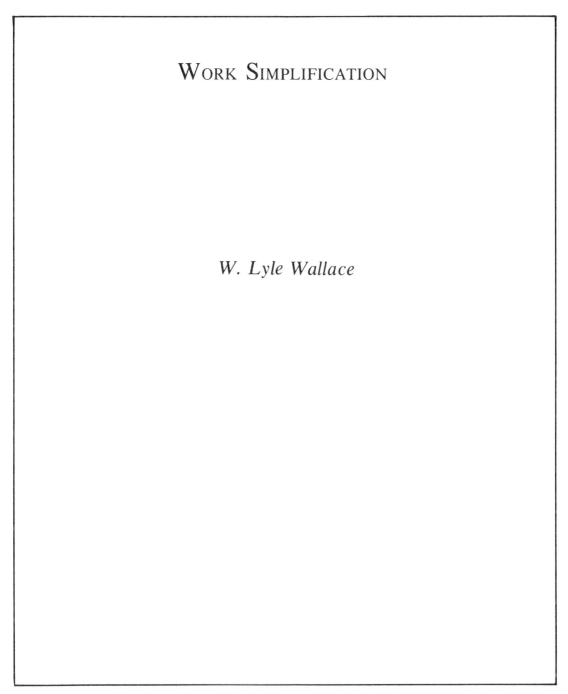

WORK SIMPLIFICATION

W. Lyle Wallace

SOURCE: From *Work Simplification* (Systems Education Monograph No. 1) edited by W. Lyle Wallace, Detroit, Michigan, Systems and Procedures Association, 1962. Reprinted by permission of the publisher. Copyright © 1962 by Systems and Procedures Association. (now Association for Systems Management)

## SECTION I

## INTRODUCTION TO WORK SIMPLIFICATION

### 1. Definition

A definition of work simplification should be one that in itself employs the techniques associated with this subject. Expressed simply, therefore, work simplification is systematically eliminating unnecessary work and streamlining that remaining so that it will move faster and better.

### 2. Why Work Simplification?

During the first half of the twentieth century, the total American working force multiplied two and one-half times. In the same period office workers multiplied 30 times. During the third decade of this century, one office worker was able to handle the paper work for 15 production workers. In the fifth decade of this century, calculated national averages indicate that six office employees were required to handle the paper work for the same 15 production workers. These statistics are strongly indicative of the trend which work simplification is intended to check. Since office work is rarely productive, the salaries of such employees usually result in a direct charge against the profits of a company. If one assumes that normal operations of a profit-making organization result in $6 of profit for each $100 of sales, it follows that $6 saved in paper work operations has the same impact on profits as $100 of new sales. Furthermore, with good follow-up it can be expected that savings in paper work costs will repeat year after year whereas sales must be consummated each year. It is important that the reasons for this trend are understood. It is not that individual productivity of the office worker is less now than it was in the 1930's. To the contrary, the productivity per capita of today's office worker is much higher than it was 30 years ago. Furthermore, it is steadily increasing with the spectacular improvements in office machines and data processing equipment.

Instead, it should be emphasized that the opportunities for simplification of paperwork processes have multiplied many times in the past 30 years, simply because the requirements for paperwork have multiplied many times.

### 3.  Effect on the Quality of Paper Work

Complicated paper work procedures are inherently costly because they require time-consuming training periods, they generally require more skilled and more highly paid personnel, and usually require a large staff.  These procedures are also more likely to generate mistakes, misunderstandings, bottlenecks, and delays.  It may be expected, therefore, that application of work simplification techniques will make it possible to train office workers more quickly and effectively.  The resulting simplified work routines will promote greater accuracy because error exposure will be reduced.  Simplified work routines should become faster as needless operations are eliminated and the remaining ones become simpler to perform.  In addition, improvement in the quality of work will invariably result in improving the morale of the employees performing the work.

### 4.  Effect on Supervisors

Management personnel are seldom effective when they are required to spend considerable time in detailed work.  To be a successful supervisor means that time must be set aside for planning.  Eliminating and simplifying paper work procedures enables the supervisor to assign detailed work to the workers.  It reduces the supervisory time necessary to correct errors and enables him to meet deadlines.  Most importantly, it provides the supervisor time for effective planning.

### 5.  Effect on Employees

It must be clearly understood what work simplification does not mean.  First, it does not mean that the individual worker is expected to "speed up" or work harder.  Second, in most companies it does not imply the intent to dismiss employees if the application of work simplification principles results in fewer people being needed to accomplish the job. Usually personnel can be transferred to other operations or, over a period of time, normal attrition will take care of the excess. Furthermore, most companies are enlarging their operations in our expanding economy and more often than not ways and means must be found to accomplish more work with the same work force. Generally speaking, simplified and efficient paper work provides more job security because it adds to the profitability of the enterprise.

### 6. Effect on Top Management

The statistics mentioned earlier in this letter are of grave concern to top management of American business. Any intelligent approach to the solution of the problems of sharply mounting paper work costs will invariably receive a sympathetic hearing by top management. The decisions of management are almost invariably based on the results of paper work. Any improvement in the timing, completeness, and accuracy of the paper work will certainly improve the validity of management decisions.

SECTION II

APPROACH TO WORK SIMPLIFICATION

This section discusses the "work simplification" approach to a procedure under

four general headings: (1) Select the Procedure, (2) Break it into Component Parts,

(3) Chart the Process, and (4) Eliminate and Streamline.

### 1. Select the Procedure

Be specific in selecting an office procedure as the target for a "work simplification"

effort.  The selected procedure should be one that can be clearly defined in scope, and

which presents one, or more, of the following problems:

    a.  Demands an excessive amount of supervisory attention,

    b.  Seems to be a "bottleneck" characterized by excessive delays,

    c.  Involves personnel requirements which are obviously out of line with

       the importance or value of the results,

    d.  Involves overtime work on a chronic basis, or

    e.  Is under fire as a procedure which produces results of questionable value.

Any business enterprise usually includes a number of procedures which contains one or

more of the elements listed above.  For example, the procedures incidental to recording

receipts and issues, or distributing the costs, of stationery stores are frequently much too

expensive to be justified by the results obtained.  Another example might be payroll pro-

cedures.  These are frequently the cause of chronic overtime.

### 2. Break It Into Component Parts

The initial breakdown of the selected procedure should consider the major components.

Here, care must be exercised to describe adequately what is done in each component.  For

example, it is not sufficient to say "Personnel clerk prepares a folder to hold employee records". Instead, it is most desirable to go to the person who actually does the job and ask him these questions: (1) How do you prepare the employee record folder? and (2) What specific operations do you perform in preparing it?

In the example cited, several other specific questions might be asked.

a. How does the employee record get to the personnel clerk?

b. What examination must be made of the record before any further action can be taken? Possibly different action might be necessary, depending on whether the record is for a new employee, or one who previously worked for the company.

c. Where are the empty folders stored? Are they easily accessible in his desk, or must he walk to a file cabinet or storage closet each time?

d. Is it necessary for the clerk to copy an excessive amount of information on the folder?

It is clear that the work simplification analysis must include the diligent application of the questions, what, when, where, how, and why, to every component of the procedure.

### 3. Chart the Process

A convenient method of recording the detailed breakdown of a procedure is the procedure analysis worksheet, a sample of which is included with this lesson (Exhibit 1). It is difficult for most people to work with a lengthy written description of a procedure. This difficulty is compounded when a discussion of the procedure is undertaken with others. By charting the procedure, symbols can be used which are readily understandable and universally recognized.

Exhibit 1

**PROCEDURE ANALYSIS WORK SHEET**

| | SUMMARY | | | | | | PROCEDURE CHARTED |
|---|---|---|---|---|---|---|---|
| | PRESENT | | PROPOSED | | SAVINGS | | |
| | NO. | HRS. | NO. | HRS. | NO. | HRS. | |
| ◯ OPERATIONS | | | | | | | |
| ▭ TRANSPORTATIONS | | | | | | | CHART BEGINS          CHART ENDS |
| ▢ INSPECTIONS | | | | | | | |
| ▯ DELAYS | | | | | | | CHARTED BY          DATE |
| ▽ STORAGES | | | | | | | |
| DISTANCE TRAVELED | | | | | | | ☐ PRESENT    ☐ PROPOSED |

| LINE NO. | STEPS IN PROCEDURE | OPERATIONS | | | | | DISTANCE IN FEET | TIME | ACTION | | | | | | |
|---|---|---|---|---|---|---|---|---|---|---|---|---|---|---|---|
| | | OPER. | TRANSP. | INSPECT | DELAY | STORE | | | ELIMINATE | COMBINE | SEQUENCE | PLACE | PERSON | IMPROVE | |
| 1 | | ◯ | ⇩ | ▢ | D | ▽ | | | | | | | | | |
| 2 | | ◯ | ⇩ | ▢ | D | ▽ | | | | | | | | | |
| 3 | | ◯ | ⇩ | ▢ | D | ▽ | | | | | | | | | |
| 4 | | ◯ | ⇩ | ▢ | D | ▽ | | | | | | | | | |
| 5 | | ◯ | ⇩ | ▢ | D | ▽ | | | | | | | | | |
| 6 | | ◯ | ⇩ | ▢ | D | ▽ | | | | | | | | | |
| 7 | | ◯ | ⇩ | ▢ | D | ▽ | | | | | | | | | |
| 8 | | ◯ | ⇩ | ▢ | D | ▽ | | | | | | | | | |
| 9 | | ◯ | ⇩ | ▢ | D | ▽ | | | | | | | | | |
| 10 | | ◯ | ⇩ | ▢ | D | ▽ | | | | | | | | | |
| 11 | | ◯ | ⇩ | ▢ | D | ▽ | | | | | | | | | |
| 12 | | ◯ | ⇩ | ▢ | D | ▽ | | | | | | | | | |
| 13 | | ◯ | ⇩ | ▢ | D | ▽ | | | | | | | | | |
| 14 | | ◯ | ⇩ | ▢ | D | ▽ | | | | | | | | | |
| 15 | | ◯ | ⇩ | ▢ | D | ▽ | | | | | | | | | |
| 16 | | ◯ | ⇩ | ▢ | D | ▽ | | | | | | | | | |
| 17 | | ◯ | ⇩ | ▢ | D | ▽ | | | | | | | | | |
| 18 | | ◯ | ⇩ | ▢ | D | ▽ | | | | | | | | | |
| 19 | | ◯ | ⇩ | ▢ | D | ▽ | | | | | | | | | |
| 20 | | ◯ | ⇩ | ▢ | D | ▽ | | | | | | | | | |
| 21 | | ◯ | ⇩ | ▢ | D | ▽ | | | | | | | | | |
| 22 | | ◯ | ⇩ | ▢ | D | ▽ | | | | | | | | | |
| 23 | | ◯ | ⇩ | ▢ | D | ▽ | | | | | | | | | |
| 24 | | ◯ | ⇩ | ▢ | D | ▽ | | | | | | | | | |

APPROVED BY          DATE          TOTALS

PAGE    OF    PAGES

These are five of these symbols:

Circle—The circle indicates an operation which might be the creation of a

form, adding information to a form, collating the pages of a proposal,

or rearranging the sequence of a series of pages.

Arrow—The arrow indicates transportation, or the movement of paper work

from one person to another, or from one work station to another.

Square—The square represents an inspection. It indicates that paper work is

being proofread or that the paper work is being scrutinized to see that

it meets certain standards of quality.

Elongated D—The symbol of an elongated D is used to indicate that a delay

occurs in handling of paper work. Such delays might occur when one

piece of paper must await the completion of a batch before being pro-

cessed, or when paper work must await a scheduled time for processing.

Triangle—A triangle indicates that paper work is filed.. Be careful to make the

distinction between a delay and filing.

The process analysis worksheet is a valuable tool in work simplification. Sometimes the

worksheet is never completed because the examination of the procedure to prepare the

worksheet discloses the fact that the whole procedure is unnecessary. Frequently many

improvements can be incorporated in a procedure during the study necessary to prepare the

analysis sheet. When the analysis sheet is completed, the most cursory inspection will

sometimes disclose so many symbols indicating delays and transportations that the source

of trouble is readily apparent.

### 4. Eliminate and Streamline

The complete elimination of an unnecessary step in a procedure is obviously the most effective simplification.   To eliminate, ask these questions:

   a.   What is the purpose of the whole procedure, and what is the specific

        purpose of each part of the procedure?

   b.   Is the purpose important enough to justify the cost?

In examining the purpose of a procedure, be sure that all copies of a letter or form
are really necessary.   Who is checking the form and what is the purpose of the check?
In asking these questions and in finding the answers, important savings can frequently
be realized.

Streamlining the parts of the procedure that are left after elimination of the unnecessary
steps is a logical progression.   There are several frequently used methods of streamlining
procedures.

   a.   If examination of the operations discloses frequent recopying of the same

        information, the analyst may consider combining several forms into one form.

   b.   Scrutiny of the distance the paper work travels and of the delays incidental

        to the travel, may point out the possibility of rearrangement of the sequence

        of steps.

   c.   Reassignment of similar steps of a procedure, improved work aids or

        business machines, and replanning of office layouts should be considered to

        minimize the work needed to complete the essential steps of a procedure.

## SECTION III

## SIMPLIFYING A SPECIFIC PROCEDURE

Sections number I and II have described in general terms the approach to a work

simplification program.  Section III begins the discussion of simplifying a specific

procedure.  Before beginning this section the student should have selected the procedure

or process which he will analyse as a practice assignment.

This section will present pointers on breaking down a procedure for step by step

scrutiny on a process flow chart.  A sample of a typical flow chart is included.  (Exhibit 2).

POINTER # 1.  An explanation of what will be done should be given to all personnel in-

volved in the process to be analyzed.  This may be done by calling a group meeting which

may be started by a discussion of why this particular process has been selected for a

simplification study.  Each member of the group should be encouraged to express his

ideas, and you should make a point of writing out each idea in your notes.  There is no

greater stimulus to the adoption of procedure imporvements than having had those who

must carry them out participate creatively in their development.

POINTER # 2.  The analysis of the selected procedure must be based on accurate data,

preferably obtained through direct observation.  There are four distinct and different

versions of the way any process is accomplished: First, the way the Procedures Manual

says it should be done; second, the way others think management wants it done; third, the

way the supervisor thinks it is being done; fourth, the way it is actually being done.  The

fourth way, of course, is the one that must be charted and analyzed.  In order to be sure

the process is correctly charted, the analyst should observe the people who are actually

performing the operation, detail by detail.

Before                                    Exhibit 2  (Page 1)

**PROCEDURE CHARTED: Withdrawing Stationery from Central Plant Supplies**

CHART BEGINS: In Ordering Office   CHART ENDS: Distribution of Charges

CHARTED BY:                  DATE:

| SUMMARY | PRESENT NO. | PRESENT HRS. | PROPOSED NO. | PROPOSED HRS. | SAVINGS NO. | SAVINGS HRS. |
|---|---|---|---|---|---|---|
| ◯ OPERATIONS | 21 | .57396 | | | | |
| ▷ TRANSPORTATIONS | 8 | .19128 | | | | |
| ☐ INSPECTIONS | 3 | .167 | | | | |
| D DELAYS | 1 | .167 | | | | |
| ▽ STORAGES | 2 | – | | | | |
| DISTANCE TRAVELED | 741 ft. | | | | | |

☒ PRESENT   ☐ PROPOSED

| LINE NO. | STEPS IN PROCEDURE | OPER. | TRANSP. | INSPECT | DELAY | STORE | DISTANCE IN FEET | TIME | ELIMINATE | COMBINE | SEQUENCE | PLACE | PERSON | IMPROVE |
|---|---|---|---|---|---|---|---|---|---|---|---|---|---|---|
| 1 | A. Section Clerk inspects stock | ◯ | ▷ | ■ | D | ▽ | 5 | .167 | | | | | | |
| 2 | A. Writes one copy of Form 1772 | ● | ▷ | ☐ | D | ▽ | | .083 | | | | | | |
| 3 | A. Carries Form 1772 to Supervisor | ◯ | ▶ | ☐ | D | ▽ | 40 | .0167 | | | | | | |
| 4 | B. Supervisor signs Form 1772 | ● | ▷ | ☐ | D | ▽ | | .033 | | | | | | |
| 5 | A. Section Clerk takes Form 1772 to Stationery Stores | ◯ | ■ | ☐ | D | ▽ | 225 | .083 | | | | | | |
| 6 | C. Plant Stationery Clerk fills requisition | ● | ▷ | ☐ | D | ▽ | | .167 | | | | | | |
| 7 | A. Section Clerk waits for Stationery | ◯ | ▷ | ☐ | ◆ | ▽ | | .167 | | | | | | |
| 8 | A. Takes Stationery back to section | ◯ | ■ | ☐ | D | ▽ | 225 | .083 | | | | | | |
| 9 | C. Checks each item as it is filled | ● | ▷ | ☐ | D | ▽ | | .0167 | | | | | | |
| 10 | C. Initials Form 1772 | ● | ▷ | ☐ | D | ▽ | | .0042 | | | | | | |
| 11 | C. Delivers Form 1772 to Stationery Storekeeper | ◯ | ▷ | ☐ | D | ▽ | 75 | .0083 | | | | | | |
| 12 | D. Stationery Storekeeper checks each item against Kardex file | ● | ▷ | ☐ | D | ▽ | | | | | | | | |
| 13 | D. Determines if Stationery was orginially charged to M&R or Engr. from Kardex | ◯ | ▷ | ■ | D | ▽ | | .167 | | | | | | |
| 14 | D. Prices each item if credit must be developed and change made | ● | ▷ | ☐ | D | ▽ | | | | | | | | |
| 15 | D. Places red check opposite each item to be credited to Engineering | ● | ▷ | ☐ | D | ▽ | | | | | | | | |
| 16 | D. Extends Quantity x Unit Price | ● | ▷ | ☐ | D | ▽ | | .083 | | | | | | |
| 17 | D. Totals cost for each requisition | ● | ▷ | ☐ | D | ▽ | | .0167 | | | | | | |
| 18 | D. Sorts Form 1772 to Requisition Dept. | ● | ▷ | ☐ | D | ▽ | | .00056 | | | | | | |
| 19 | D. Runs adding machine total by Dept. | ● | ▷ | ☐ | D | ▽ | | .00028 | | | | | | |
| 20 | D. Runs adding machine total for All Engineering Credits | ● | ▷ | ☐ | D | ▽ | | .0014 | | | | | | |
| 21 | D. Runs adding machine total of Dept. totals | ● | ▷ | ☐ | D | ▽ | | – | | | | | | |
| 22 | D. Writes out distribution of charges and credits | ● | ▷ | ☐ | D | ▽ | | .00056 | | | | | | |
| 23 | D. Types letter - three copies - to Accounting Dept. to make entry | ● | ▷ | ☐ | D | ▽ | | .00056 | | | | | | |
| 24 | D. Attaches bundle of requisitions to letter | ● | ▷ | ☐ | D | ▽ | | – | | | | | | |

APPROVED BY                    DATE

| | TOTALS | 17 | 4 | 2 | 1 | | 570 | 1.09896 | | | | | | |

PAGE 1 OF 2 PAGES

Exhibit - 2  (Page 2)

| | SUMMARY | | | | | | PROCEDURE CHARTED | | | |
|---|---|---|---|---|---|---|---|---|---|---|
| | PRESENT | | PROPOSED | | SAVINGS | | | | | |
| | NO. | HRS. | NO. | HRS. | NO. | HRS. | | | | |
| ○ OPERATIONS | | | | | | | | | | |
| ⇨ TRANSPORTATIONS | | | | | | | CHART BEGINS | | CHART ENDS | |
| ☐ INSPECTIONS | | | | | | | | | | |
| ◗ DELAYS | | | | | | | CHARTED BY | | | DATE |
| ▽ STORAGES | | | | | | | | | | |
| DISTANCE TRAVELED | | | | | | | ☒ PRESENT        ☐ PROPOSED | | | |

| LINE NO. | STEPS IN PROCEDURE | OPER. | TRANSP. | INSPECT. | DELAY | STORE | DISTANCE IN FEET | TIME | ELIMINATE | COMBINE | SEQUENCE | PLACE | PERSON | IMPROVE |
|---|---|---|---|---|---|---|---|---|---|---|---|---|---|---|
| 1 | D.  Delivers letter and requisitions to Stationery Supervisor | ○ | ■ | ☐ | ◗ | ▽ | 5 | - | | | | | | |
| 2 | E.  Stationery Supervisor signs two copies of letter | ● | ⇨ | ☐ | ◗ | ▽ | | - | | | | | | |
| 3 | E.  Gets folder from file cabinet | ○ | ■ | ☐ | ◗ | ▽ | 8 | - | | | | | | |
| 4 | E.  Files copy of distribution letter in folder | ○ | ⇨ | ☐ | ◗ | ▼ | | - | | | | | | |
| 5 | E.  Replaces folder in file cabinet | ○ | ■ | ☐ | ◗ | ▽ | 8 | - | | | | | | |
| 6 | E.  Takes requisitions and two copies of letter to Dept. Staff Supvr. | ○ | ■ | ☐ | ◗ | ▽ | 150 | .00028 | | | | | | |
| 7 | F.  Dept. Supvr. initials original copy of letter | ● | ⇨ | ☐ | ◗ | ▽ | | - | | | | | | |
| 8 | F.  Mails original copy and requisitions to Accounting Dept. | ● | ⇨ | ☐ | ◗ | ▽ | | - | | | | | | |
| 9 | F.  Takes file of distribution charges from desk. | ● | ⇨ | ☐ | ◗ | ▽ | | - | | | | | | |
| 10 | F.  Compares current month with past month. | ○ | ⇨ | ■ | ◗ | ▽ | | - | | | | | | |
| 11 | F.  Files copy of letter in distribution file | ○ | ⇨ | ☐ | ◗ | ▼ | | | | | | | | |
| 12 | | ○ | ⇨ | ☐ | ◗ | ▽ | | | | | | | | |
| 13 | | ○ | ⇨ | ☐ | ◗ | ▽ | | | | | | | | |
| 14 | 35 steps | ○ | ⇨ | ☐ | ◗ | ▽ | | | | | | | | |
| 15 | | ○ | ⇨ | ☐ | ◗ | ▽ | | | | | | | | |
| 16 | | ○ | ⇨ | ☐ | ◗ | ▽ | | | | | | | | |
| 17 | A.  Section Clerk / B.  Section Supervisor | ○ | ⇨ | ☐ | ◗ | ▽ | | | | | | | | |
| 18 | C.  Plant Stationery Clerk / D.  Stationery Storekeeper | ○ | ⇨ | ☐ | ◗ | ▽ | | | | | | | | |
| 19 | E.  Stationery Supervisor / F.  Dept. Staff Supervisor | ○ | ⇨ | ☐ | ◗ | ▽ | | | | | | | | |
| 20 | | ○ | ⇨ | ☐ | ◗ | ▽ | | | | | | | | |
| 21 | | ○ | ⇨ | ☐ | ◗ | ▽ | | | | | | | | |
| 22 | | ○ | ⇨ | ☐ | ◗ | ▽ | | | | | | | | |
| 23 | | ○ | ⇨ | ☐ | ◗ | ▽ | | | | | | | | |
| 24 | | ○ | ⇨ | ☐ | ◗ | ▽ | | | | | | | | |

| APPROVED BY | DATE | TOTALS | 21 | 8 | 3 | 1 | 2 | 741 | 1.09924 | PAGE 2 OF 2 PAGES |
|---|---|---|---|---|---|---|---|---|---|---|

POINTER # 3.  There are three major sequences of most operations: First, make ready; second, do; third, clean-up.  The analysis of an operation must cover all three sequences. In Lesson II reference was made to a procedure described as "personnel clerk prepares a folder to hold employee records".  In this example, the "make ready" steps might include sorting the employee records from the incoming mail, obtaining a supply of blank folders, and getting out any of the rubber stamps which might be used.  The "do" steps would include the actual transcription of information to the folder.  The "clean-up" steps might consist of putting away the various stamps used, and placing the completed folders in the file.

POINTER # 4.  Section II described the five symbols to be used in charting the steps of a process.  A circle should be used to represent any kind of an operation, whether involving the do, make ready, or clean-up steps.  An arrow should be used to represent any kind of a transportation.  A square represents any kind of an inspection.  An elongated D represents a delay which occurs when work is being held pending some other operation, or temporarily stored in a work tray.  The triangle represents a step of filing, or storage.  Description of the content of each step that is represented on the analysis sheet should be written in the space provided to the left of the preprinted symbol area.

POINTER # 5.  When a process involves a multi-part form, a separate analysis sheet should be used to follow each copy.  Each analysis sheet should begin with the receipt of the form copy by the first person who handles it and should follow it through to the ultimate disposition.

POINTER # 6.  The name of the person who performs each charted step should be shown on the left side of the chart and should be underlined for emphasis.  This will simplify a later analysis of whether a given task has been allocated to the right person, that is, the person who can do it best in the least time.  The finished chart will show if there are too many people handling the same document.  Each handling of a piece of paper represents a definite cost and warrants careful consideration by the analyst.

POINTER #7.  Record key distances and times involved in the performance of each step.  These need not be minutely accurate measurements from stop watches and tape measures, but they should be thoughtful estimates made in cooperation with the office workers, and then doublechecked by direct observation.  The results should be recorded in the columns provided for such data to the right of the symbol area on the work chart.

POINTER #8.  After the flow chart is completed, tha analyst should mentally challenge each step.  A number of key questions should be asked:  What is the contribution of the step?  Is it worth what it costs?  Is the step necessary in view of its duplication at some other point?  Can the step be done better and faster by someone else?  Could the step be performed at an earlier stage in the process to avoid re-sorting?  Could the form be re-designed so that one form would do the job of several?  Would a cross reference index eliminate all but one file?  Is each inspection necessary or are some of them only done as a matter of habit?  Is it possible to reduce some of the make ready and clean-up operations?

POINTER #9.  Indicate steps on a flow chart by shading the applicable symbol and by drawing from it a line leading to the applicable symbol for the next step in the sequence of process flow.

WORK MEASUREMENT
APPLIED TO LIBRARIES

*Elaine Woodruff*

SOURCE: Reprinted with permission from Woodruff, E. L., Work Measurement Applied to Libraries, *Special Libraries* 48 (No. 4) p. 139–144 (April, 1957). Copyright © 1957 by Special Libraries Association.

ALTHOUGH work measurement has been widely utilized in industrial and technical operations as an effective means of production and quality control, it has not had the same widespread application in administrative and professional areas. It is with this latter area, that of the application of work measurement to libraries, that this article is concerned.

Since the author's experience with work measurement is limited to its use in the library of the United States Civil Service Commission, this discussion relates primarily to the objectives, procedures and terminology common to the Commission's work measurement program. The program in the Commission emphasizes cost—total cost and unit cost of each activity. In other organizations the emphasis may well be on time. The emphasis would depend entirely on the use to be made of the figures obtained and the objectives of the program. Since the Commission's cost figures are used for budget purposes, the tendency is to be very cost-conscious.

## Definition

Before discussing what work measurement is, what it is *not* should be explained. Work measurement is *not* a standardized, arbitrary system for checking up on the hour-by-hour performance of a librarian and his staff. It is

Adapted from an address presented at the Technical Library Conference sponsored by the Bureau of Ordnance and Bureau of Ships, U. S. Department of the Navy, Washington, D.C., June 8, 1955.

*not* a new system superimposed on present management functions. If properly administered, it should be a part of the fabric of day-to-day operations. It enters into everything done, into every major decision made.

At this point, it should be emphasized that although time and attendance reports may be kept and although some sort of production statistics may be recorded in every library, until the two are related and until the relationship between time and cost and production can be shown, the processes are ones of recording—not measuring.

Briefly defined, then, work measurement is a tool, a tool to be developed and used, sharpened and re-used, and refitted to a particular situation and need. No one can tailor a ready-made system and say, "Here it is—use it." It has to be worked out individually and changed as needed.

The Bureau of the Budget defines work measurement as a management tool designed to establish "an equitable relationship between work performed and manpower used."[1] In other words, how much time and money does it cost to do a job—not only the whole job but each of its components?

## Balancing Workload and Personnel

How many librarians know, for example, how much of their cataloging backlog is primarily clerical, how much subprofessional, how much professional and how many man-hours are needed to wipe it out? Can they report readily: "We have this many items of uncataloged materials; it takes so many man-hours to catalog an item; we need so much in the way of additional resources to become current in our work." Can they show proven figures to their budget officials to supplement their statements?

Do they know in their own library whether work in certain sections is slacking off—if, perhaps, they could use that personnel in some other activity? Even in a small library, it shouldn't be necessary for an administrator or supervisor to go around from desk to desk to see how work is moving. By reviewing properly-developed work reports, a supervisor should be able to tell at a glance what is happening to the workload in any part of the library. This is one of the functions of work measurement.

These are only two examples of one use of work measurement—balancing workload and personnel. It is a primary use, however, that of achieving a fair relationship between workload and the personnel and resources available to take care of it.

## Improving Operating Efficiency

A second area of usefulness which might be emphasized is the improvement of operating efficiency. Suppose a librarian notes from work reports that the unit cost in a certain activity is constantly going up. Wouldn't he be inclined to wonder why and to try to determine the reason and a remedy? Perhaps there is a need for more training, closer supervision or perhaps a check on leave might be in order. He might find that the situation was completely unavoidable, as in the case of an employee receiving longevity pay whose production is decreasing. There are innumerable situations about which nothing can be done, but it is the responsibility of an administrator to know why they exist and to make what adjustments he can.

To cite another example of improving operating efficiency, how many librarians know how much of their money goes into service activities and how much into support? "Service" in this context means reference, circulation and research; "support" means acquisitions, cataloging and other technical processes. This is the type of information which is readily available, particularly from annual summary reports. One may determine that the allocation of funds is what is wanted and necessary for effective operation. In a new library, it is perfectly understandable that a good deal of the budget should go into supporting activities, such as acquiring new materials and cataloging them, but once a library is established, the primary purpose is service and there should be a shift in the allocation of resources.

## Evaluating Performance

The third use made of work measurement is a very important one to the librarian and his staff from the human relations point of view. Does the librarian know exactly what is a reasonable standard of performance for processing books, filing and shelving? Can he back up his standard with figures? A performance rating discussion between supervisor and employee becomes much more objective when it is based on standards understood and acceptable to both. It is one area in which it is most important for figures to be really valid and based on average performances over a reasonable period of time. Then a supervisor can say to the employee, "This is a fair and honest standard; what can we do to help you meet it? Is there a reason why it cannot be met?"

How can work measurement affect performance rating? It actually prevents a supervisor from being personal and arbitrary in judgments, for he can't base a rating on extraneous factors when there is a standard which is factual and correct. Work measurement requires a supervisor to judge performance, not an individual.

Work measurement also helps a librarian compare his own operations with those of similar organizations. Every librarian is interested in knowing how much it costs him to catalog a book, as compared with costs in some library doing a similar type of cataloging. But before making comparisons, it must be determined that procedures, materials and working conditions are fairly similar. Comparisons can't be drawn, for example, between costs in large public and specialized governmental libraries.

### Budget Formulation

The last use which may be made of work measurement is in budget formulation. If an organization uses a performance budget, it is essential that it also have a method of measuring production. It is necessary for a library to know, for instance, how much it costs to put a book on the shelf, how many reference questions may be expected the following year and how much it will cost to answer each, how many man-years are needed for research and whether costs may be expected to rise or decline.

If, on the basis of previous performance, a library expects to acquire 5,000 volumes for cataloging next year, and the unit cost per item averages $.75, then it will be necessary to allocate $3750 of the library budget for cataloging purposes. This is something in black and white which can be presented to management people and it also shows that since costs cannot be further reduced nor the function eliminated, a backlog will result unless adequate funds are made available.

To summarize, then, these are some of the reasons why work measurement is a useful tool in managing a library:

1. It helps to balance available resources with estimated workload.

2. It helps in managing a library more efficiently and more economically.

3. It helps in objective evaluation of individual employee performance and in evaluation of the organization as a whole.

4. It helps the librarian plan for financial and personnel needs on the basis of sound production figures.

### Setting Up A Work Measurement Program

Here are some general rules or suggestions for setting up a work measurement program for a library.

First, and most important of all, secure employee participation at all stages of the program—in the planning, in the determination of functions to be measured and in the definition of terms. An unfortunate morale effect is almost inevitable if no preliminary explanations are made as to what is being done and why. On the other hand, if employees are told why the figures are needed, how they will be used and how the library will benefit, and if the functions and definitions are worked out cooperatively with them, they will be interested in knowing just how much each activity does cost per unit and how much their time is worth. The general climate of the organization and the effectiveness of communication generally between management and employees will largely determine how a work measurement system will be received.

The second rule — have reporting done by each individual employee, not by a supervisor. Only the employee can give an accurate account of his time, and in most cases, of his production; there is no point in setting up a system and developing figures unless they are accurate and honest.

Third, plan to review and revise the system continually. Because functions and definitions have been written down or because a form has been drawn up, should not mean that the system is complete and cannot be changed. It should be tried out for a period to see how effective it really is, whether it is bringing out the facts needed. It may be necessary to redefine certain activities or transfer certain elements or make other changes after a trial period. It may take a long time before a plan can be standardized.

Fourth, work report figures should not be used as final and definitive until they have been developed for a long enough period to produce a true average. In some cases, it may take several years to obtain a valid range of material and production for such an average.

## Listing Activities

The first step in developing a work report program is to list all activities performed. These are not at all uniform; even in government libraries, duties, activities and terminology vary greatly. One way of drawing up an activities list which has worked out satisfactorily is to ask all employees to list everything they do for a week or so. This shows the entire range of usual library activities. While there is a great deal of duplication in the original list, it is one way of getting complete coverage. In making up a composite list, there should be no overlapping and each activity should be clearly defined.

If it is desirable for a library to show unit time and cost for particular types of materials or particular phases of an operation, such as the cost of acquiring books, the cost of acquiring periodicals or the cost of acquiring free materials, this can be done by providing the necessary subdivisions under a major activity. Similarly, the cost of typing catalog cards, filing cards and doing descriptive cataloging can be separately listed under cataloging. Unless there is a specific need for this type of detailed information, however, it is better not to add a great many subdivisions to activities because they make the work report more time-consuming to maintain and compute. The work report should fit your needs and work for you,

Allow for operating and non-operating time and cost. Operating activities have an end result or an end product. Non-operating activities include administration, supervision and leave, generally. However, time spent supervising a particular type of activity, such as circulation, should be charged to the activity supervised. Only general supervisory and administrative activity is considered non-operating.

## Defining Activities

The second step in developing a work report system is to define activities—clearly, concisely, cooperatively. This is a most important step if the figures are really to mean something. For example, in the acquisitions count, are all items retained in the library counted, whether they are recorded or not? Are items counted which may later be discarded or given away?

In circulation, are items given away counted, or only those charged? Where will the others be recorded? Is a total series counted as one volume, or is each individual part counted? Is going to a file or a shelf, pulling out a report and sending it out, recorded as reference or circulation? Decide into which category certain auxiliary activities fall, for example, indexing Congressional hearings or annual reports.

Again, decide when an item is to be counted. Is an item counted as cataloged when the cards are all typed, when a book is ready for the shelf or when the cataloger has finished with it? There must be a specific point at which work is counted as completed, so there won't be a duplication of items counted or any items that are not counted.

Much of this can be best worked out in staff conferences where a librarian may say: "Here is an activity—what should be included in it? What should be eliminated?" Disposition should be made of each item coming into the library.

## Setting Up The Form

The third step is to set up a tentative form. It should be kept as simple, as clear and as self-explanatory as is humanly possible.

On the Civil Service Commission's Employee Work Report Form, MB 190a (see sample), the functions run down the lefthand side of the sheet and the days are filled in across the top. This format is designed for an agency-wide system with machine tabulating facilities. This is the reason for the various symbols and code numbers on the form. The Commission has a four-week reporting period, which corresponds to the pay period. Each work-report period covers 160 hours if the employee has been on the job the entire time and has not taken any leave without pay. Each Work Report Form has space for only a two-week period, at which time the figures are added and carried over to the four-week report.

It is necessary to allow space for the number processed in each daily block. In the Commission's form this is indicated by "N" above (for number) and "T" below (for time). The form must also include space for operating and non-operating time and for the status of the workload.

## Recording Time and Production

The last step in the work report program is to record time spent and number processed. This reporting should be done at the close of each working day without fail. Often staff members may have to be away unexpectedly, and the previous day's or week's work is unknown to anyone else. Filling in the form at the close of the day becomes a habit very quickly.

Every employee should have clear-cut directions for filling in the form, preferably in writing, so that he can refer to them when he is in doubt as to

| PERIOD ENDING (cc 5-10) | UNITED STATES CIVIL SERVICE COMMISSION | | NAME OF EMPLOYEE |
|---|---|---|---|
| ORGANIZATION CODE (cc 11-15) 190.00 | THE LIBRARY | | ANNUAL SALARY (cc 23-27) |
| HOURLY COST RATE (cc 16-18) | EMPLOYEE WORK REPORT | | |

| CHECK ONE: | ☐ 1. REGULAR (cc 19) | ☐ 2. SUPPLEMENTAL (cc 19) | ☐ 3. OVERTIME (cc 19) | ☐ _____ (cc 19) Other (Specify) |
|---|---|---|---|---|

| CODE NUMBER (cc 28-34) | FUNCTIONS | TOTAL TIME (cc 41-44) | TWO-WEEK TOTALS | | | DAILY COLUMNS FOR REPORTING TIME AND PRODUCTION |
|---|---|---|---|---|---|---|
| | | | LAST PERIOD | THIS PERIOD | | |
| 8.51.00.00 | ACQUISITION | I | | | N T | |
| 8.52.00.00 | CATALOGING | I | | | N T | |
| 8.53.00.00 | BINDING AND LABELING | I | | | N T | |
| 8.55.00.00 | CIRCULATION | I | | | N T | |
| 8.56.00.00 | REFERENCE | I | | | N T | |
| 8.57.00.00 | RESEARCH | I | I | I | T | |
| 9.01.10.00 | Administration, Supervision and Other Non-Operating Time | I | I | I | T | |
| 9.01.20.00 | Paid Leave | I | I | I | T | |
| 9.03.10.00 | Reimbursable Services Performed Under Commission Supervision | I | I | I | T | |
| 9.03.20.00 | Reimbursable Services Performed Under Other Agency Supervision | I | I | I | T | |
| 9.09.10.00 | Leave Without Pay and Compensatory Time Off | I | I | I | T | |
| 9.09.20.00 | Time Reported on Supplemental Reports | I | I | I | T | |
| | TOTAL TIME | I | I | I | T | |

A U. S. Civil Service Commission Employee Work Report Form. Library activities are listed on the left side and columns for marking daily time spent and number of items processed for each function run down the right side.

where he should list certain types of activities. It is very helpful if each employee has a "Definitions of Functions" guide available.

The administrator should occasionally review reporting to check that time is being properly charged to the correct activity and that the entire staff is consistent in its allocation of time on infrequent and unusual activities.

Compiling The Summary Work Report

The final stage in the work measurement program is to collect and summarize at regular intervals the daily time records which employees have kept, compute costs corresponding to the time amounts shown for each activity and relate the time and cost to the work performed during the report period. When summarization is completed, the report will show hours and cost for each operating activity, the number of items processed and hours and cost for administration and leave.

In the Commission's work measurement system, non-operating time (administration and leave) is prorated to each operating activity. Many libraries have systems of work measurement, but they do not charge their operating activities for the cost of administration and leave, and therefore their unit cost is, of course, lower. This is one example of the fallacies involved in making comparisons between libraries unless all operations and procedures are similar.

The details of the computations involved in prorating non-operating to operating costs are not within the scope of this article but they are not difficult to evolve if an over-all, as well as an operating, time and cost figure is desired.[2]

The last step in computing work report figures is showing the status of the workload. This is included on the back of the Commission's Summary Work Report Form and is computed by subtracting the number processed from the sum of the number on hand and the number received. The number remaining carries over from one month to the next. This is a figure to which management people are most apt to pay attention and it is an important figure for the library administrator to watch.

Conclusion

Some administrators may begrudge the time spent in developing, setting up and maintaining a work reporting system. However, it is only through actual experience with a system that one can determine whether the operating information it develops contributes enough to improved efficiency to make worthwhile the relatively small amount of time spent in maintaining it. For after all, a work measurement system is only as valuable as the use made of it.

CITATIONS

1. U. S. Bureau of the Budget. *Work Measurement in Performance Budgeting and Management Improvement*. Washington, D.C.: 1950, p. 4.

S<small>TANDARD</small> T<small>IMES FOR</small>
C<small>ERTAIN</small> C<small>LERICAL</small> A<small>CTIVITIES</small>
<small>IN</small> T<small>ECHNICAL</small> P<small>ROCESSING</small>

*Henry Voos*

SOURCE: From Henry Voos, "Standard Times for Certain Clerical Activities in Technical Processing," *Library Resources and Technical Services* 10 (Spring, 1966) pp. 223-227. Reprinted by permission of the author.

THE USE OF STANDARD TIMES has made possible great advances in office management and has expanded greatly over the last generation. Once we know how long it takes to type a character, we can estimate how long it will take to type a catalog card with 200 characters for a scholarly library or one with 100 characters for an elementary school library. The growth of the use of standard time units has been rapid. It is rare, at the present time, for the management study in office or factory work to begin with stop watch measurement of operations. Instead it starts with the handbook of standard times and the list of operations performed.

We have little information on standard time units for repetitive library operations. If the technique and data developed in the course of

* This article is a summary of a doctoral dissertation accepted by the Faculty of the Graduate School of Library Service, Rutgers University, in January 1965. Copies of the dissertation are available from University Microfilms, Ann Arbor, Mich.

the research here reported lead to more widespread availability of such units for library routines, they will result in the development of a powerful new tool for library planning, management, budgeting, and supervision.

Past library studies have been more concerned with costs of broad operations such as cataloging and circulation. However, costs vary depending on the economic area where the library is located, the average hourly wage of the particular employees engaged in the operation, and the time frame within which the operation is being reported. Cost is more subject to change than time is. Broad operations consisting of many components also tend to create more variances in comparative times than do the particular components of these operations. For this reason, the method used in the study was a form of the micromotion technique. This technique measured the smallest parts of the work being done. Among these were the typing time per key stroke, the lettering time per letter for marking books, and the property marking time per rubber stamping or embossing. The micromotion technique takes advantage of the fact that within the great variances in procedure, in library layout, in types of library personnel, and in supervision there are certain elemental parts of the total library technical processes which are similar. To further eliminate any variances such as those cited above, only actual work time was measured, with no allowance for idle time, transport time, or delay time in the processing operations.

The mean was used as the measure of central tendency. It was felt that it is the one valid measure which most librarians can apply with little or no statistical training.

A stop-watch was used for timing the operations measured. In addition, certain typing times were measured with an instrument designed by the author using a Veeder-Root Counter attached to a typewriter. This instrument counted the number of strokes and carriage returns and measured the time these took.

To be certain that the data were valid, the sample size of the work measured at the four libraries observed was determined in two ways: one was to use a nomograph, the second was to take subsets of the entire sample size and test the means of the subsets against the universal mean. The data evolved for certain tasks which are components of the technical processing operations are given in Table I. However, this data should not be applied to any library's operation without reading the detailed text in the dissertation to discover its component parts and the equipment and machinery used in the operation. Every effort was made to measure as many of the various ways of accomplishing a given task as possible.

In addition to the data presented in Table I, certain more general conclusions could be drawn from the study. Among these are:

1. The time required to perform clerical routines used in technical processing can be predicted under many conditions and for a wide variety of machines and devices. It is also obvious that many of these routines

TABLE I

Summary Table of Data

| Function | Mean Time (Sec.)[a] |
|---|---|
| Pasting book pockets | |
|     Manually | 8.75 |
|     Machine | 3.12 |
| Pasting pre-gummed date-due slips | 3.4 |
| Pasting pre-gummed book plates | 9.8 |
| Removing dust jackets | 2.5, 1.95 |
| Excising holes in dust jackets | 21.2, 9.6 |
| Replacing dust jackets | 6.6 |
| Measuring and noting book size | 6.6 |
| Removing plastic covers from storage racks singly | 4.7 |
| Folding and placing dust jacket into plastic cover | 27.1, 36.3 |
| Folding once, pasting and placing dust jacket into plastic cover | 10.23, 10.73 |
| Replacing covers and jackets on books | 5.2 |
| Scotch taping plastic jacket onto book, per edge | 4.07 |
| Pasting plastic jacket onto book, per edge | 3.0 |
| Erasing (Electrically) | |
|     Call number and subject | 89.8 |
|     Call number and imprint | 319.6 |
|     Subject | 29.9 |
|     Collation | 81.1 |
|     Imprint | 104.5 |
|     Call number | 30.9 |
| Embossing | 1.59 |
| Finding "Secret" page | 3.59, 7.68 |
| Rubber Stamping | |
|     Outside, per stamp | .81, .73 |
|     Outside and inside, per stamp | 3.72, 3.66 |
|     Accession number, per stamp | 3.31, .89 |
| Graphotyping, per letter | .64 |
| Running addressograph, per impression | 2.2 |
| Tying pamphlets, per bundle | 24.64 |
| Typing | |
|     All, merged, per stroke | .40 |
|     Manual, Royal, per stroke | .52 |
|     IBM Executive, per stroke | .30 |
|     Royal Standard Electric, per stroke | .39 |
|     Smith Corona, Manual, special, per stroke | .38 |
| Lettering | |
|     Hot or cold tape, per letter | 2.94 |
|     India ink, per letter | 1.31 |
|     White ink, per letter | 2.3 |
|     Type, per letter | .42 |
| Hot-type printing (Altair machine) | 25.4 |

[a] When two figures are provided, they are the results of different techniques for accomplishing the same function.

are common to a majority of libraries, whether they are public, college, university, school, or special libraries. This data does not depend on the number of people staffing the library. The data gathered can be used for standardizing statistical reporting, for performance measurement, for cost comparison, and for standardization of processes.

2. The use of standard time data will permit library administrators to know what deviation from standard procedures costs them and permit them to evaluate work simplification measures.

3. The knowledge of standard time data permits library administrators to evaluate the quality of supervisory practices in libraries. By simply deducting the predictable standard time, the percentages for personal time, fringe benefits, delay times, and transportation times from a total operational time, they are left with a difference that is largely attributable to supervision.

4. The data can be used to set up work standards on a broad operational basis by preparing a flow-process chart for each operation for which a standard is to be prepared. The standard time which is appropriate is used where applicable, and those for which no time has been yet established can be taken from available industrial engineering sources or by timing it oneself. A summation of the component parts will provide a rough operational standard. Once this standard has been established, performance should be measured against it during a six- to twelve-month period. If the efficiency rate deviates more than between 80 to 120%, the standard can then be adjusted to 100%.

5. A formula has been developed which will permit comparison of the most simple systems. The formula provides a means for graphing and a determination of a point of economic equality between systems. Indications are provided on how to develop the formula to encompass more complicated operations, such as computer versus manual tasks.

In addition to the foregoing general conclusions, some particular conclusions can be drawn on specific operations:

1. Despite maintenance and initial cost, a mechanical pasting device is more efficient and economical than a manual pasting operation.

2. Except for minor exceptions, the data presented in the literature on clerical library operations is relatively valueless for comparison purposes, because, by and large, no detailed breakdown of operation makeup or statistical techniques is presented.

3. If plastic covers are placed on books, it is more efficient to presegregate the volumes by size than to handle them individually. It is also better to affix these jackets with two gluings than by taping them individually to the books.

4. For recataloging operations it is cheaper to retype the entire card if the items to be erased take longer than 110 seconds. Furthermore, if esthetics is not the prime consideration, crossing out the old data, insofar as it concerns imprint, collation, series notes, and then typing the revised data above, below or next to it, is definitely cheaper.

5. Clerical routines such as typing can be done more effectively by

trained personnel who spend full-time on them, rather than by fragmenting the duties of library personnel. It is realized that in smaller libraries such specialization is not always possible. However, it then becomes necessary to plan and schedule one's work so that enough time can be spent on each unit task to overcome initial inertia.

6. Until we know the effectiveness of our property-marking techniques, it would be advisable to hold these to a minimum.

7. If LC card numbers are not readily available, it is cheaper to alphabetize the LC card orders than to look up these numbers before transmitting the order.

The study makes recommendations based upon the data and conclusions:

1. Additional time studies should be performed to cover as complete a range of library operations as possible.

2. Time standards, work measurement, and methods of evaluation should be taught in our library schools to help improve the quality of library supervision.

3. Book pockets with gummed or pressure sensitive backs should be tested.

4. Each library operation should be under constant surveillance to determine whether it is necessary and whether it accomplishes the task it was originally set up to perform.

5. Editors of library journals should make it their business to ensure that authors inform the reader exactly what the time and cost constituents of a described operation are when they publish descriptions of or comparisons of library operations.

6. The value and effectiveness of property markings should be studied.

# Conducting
# User Requirement Studies
# in Special Libraries

*Carole E. Bare*

SOURCE: Reprinted with permission from Bare, C. E., Conducting User Requirement Studies in Special Libraries, *Special Libraries* 57 (No. 2) p. 103-106 (Feb., 1966). Copyright © 1966 by Special Libraries Association.

Periodically it is necessary to canvass the user population of a special library to determine whether the objectives of the library are being met. Some sort of feedback system must be set up; a study of the requirements of the library's users can be made. Three methods of conducting user requirement studies are presented here.

The prime purpose of a user requirement study is to provide information on how accurately the librarians and administrators have interpreted various users' needs. Also, it shows users that it is their privilege as well as their responsibility to help the library meet their own informational requirements. Through periodic surveys, communication channels between users and the library may sufficiently open up so that users who don't know precisely what they want from a library become more aware of and more able to express their needs.

### First Steps in Developing a Survey

The question of who should be involved in the survey and what areas should be covered must be discussed, first of all. Librarians and administrative personnel can be direct investigators or can participate indirectly. There are advantages and disadvantages to either procedure. Users may express attitudes more openly if a neutral team administers the study. However, administrative and library personnel must be intimately involved in tailoring the study and interpreting the findings to the user group and make their own decisions on how the services should be modified.

The following areas should be covered in a library user requirement survey: content, communication channels, delays, special features, and centralization vs. decentralization.

To evaluate users' satisfaction with the content of the library these library functions should be covered: anticipation of future needs, scope, and up-to-dateness of acquisitions and announcements. To evaluate the personal communication channels between library personnel and the user, the survey may have to elicit responses about how well the following services are performed: clarification of requests, recommendations of specific written materials or sources in response to an inquiry (this includes bibliographies that provide suggestions for alternatives if the most pertinent references cannot be obtained immediately), referral to resource persons, and notification of whether or not material requested can be obtained or how well delays are estimated and communicated to the user. Special features such as abstracts, catalog format, location of the library, and microfilm utilization may also have to be evaluated.

If the centralization-decentralization issue applies, the following questions should be answered: where should the major library be located; what are the physical facilities with respect to noise, lighting, and browsing; should smaller special information centers be provided or encouraged; and what should their relationship be to the central library service?

The first step in conducting a survey is to enlist administrative support in the form of a cover letter indicating that the administration not only welcomes opinions but also urges users to participate.

Three major methods have been used to assess user requirements in scientific and technical libraries—the questionnaire, diary, and interview. The advantages and the disadvantages of each method point to the conclusion that the three methods are best used in combination. The optimal combination will depend on the nature of the particular organization.

Two types of pretesting the three techniques are possible. The opinions of administrative and library personnel and of several users on the content and format of the questions to be asked is most valuable. Therefore, the over-all plan, the specific items, and the format should be submitted to librarians and administrators for criticisms. The second pretesting method is to try the techniques out on a small group of users.

## Questionnaire Method

It is important to keep questionnaires short so that present and potential users can give their opinions with a minimum of time and effort. The following questionnaire characteristics help to insure widespread participation:

1. Clear instructions and an explanation of purpose. The majority of questions should only require a "yes" or "no" response or the use of a rating scale containing an even number of scale intervals or points. Thus the rater must take a position for or against an issue and cannot choose the mid-point of the scale and remain "undecided."

2. A few open-ended questions requiring brief narrative statements should be included and/or a few lines for comments can be provided for the yes/no or scaled questions.

3. A few questions can be included that ask for estimates of average frequency or a percentage of occurrence.

The example below combines all three formats:

Are you now using, formally or informally, your library (or information center)?

Yes-1 _____

No -2 _____                        _____

If so, to what extent has the information center been useful?

| 1 | 2 |
| --- | --- |
| not useful | somewhat useful |

| 3 | 4 |
| --- | --- |
| quite useful | extremely useful |

If useful, elaborate briefly on reasons:

_____

_____

_____

It is desirable to include the total present and potential user population in a questionnaire survey because frequently clerical personnel, who may not use the library services directly but who facilitate use for scientific and technical personnel, know a great deal about the services. Their opinions should be included, but their status and function should be coded as one of the variables for the analysis of the data. Other demographic data should be collected on the respondents: type of scientific and technical work, type of responsibility (e.g., supervisory, teaching, learning), amount of experience, age, sex, length of time at present job, academic background, and so on.

## Diary Method

A daily record sheet on which individual information retrieval transactions are to be tallied can be given to any type of library user. The content of the questions on which daily records are kept can include a range of items covering an evaluation of each contact with the library to the charting of how a user's information need was fulfilled, e.g., "Found material in own office," "Sent to another library," or "Colleague supplied information."

It is important that the instructions be clear and that participants are visited or monitored periodically, so that questions on how to fill out the daily record sheets can be answered, and uniformity of recording can be achieved. The majority of questions should be answerable by a check mark, but a few open-ended questions or space for comments should be provided, for example:

How soon is information needed?

| 1 | 2 | |
|---|---|---|
| a At once | Up to 2 days | |
| 3 | 4 | |
| 3-12 days | More than 13 days | _____ |

| 1 | 2 | |
|---|---|---|
| b Received when expected | Later than expected but still useful | |
| 3 | 4 | |
| Not soon enough to be very useful | Much too late | _____ |

Actions taken:

| | | | |
|---|---|---|---|
| a | Found material in my own office | Yes-1 No-2 | ____ |
| b | Colleague supplied relevant information | Yes-1 No-2 | ____ |
| c | Requested assistance from library | Yes-1 No-2 | ____ |
| d | Query answered by library | Yes-1 No-2 | ____ |
| e | Query answered by outside source | Yes-1 No-2 | ____ |

### Interview Method

This technique provides the most intensive look at user opinions, since it is possible to ask the reasons why opinions are held or why a certain process is preferred. Critical incidents and suggestions for correcting shortcomings can be elicited and explained fully by an interview.

The interview should be scheduled in advance and should probably not last over 45 minutes. At the time the appointment is scheduled and again at the beginning of the interview, the purpose of the interview should be stated as well as the interviewer's willingness to limit himself to the time the interviewee feels he can spare. Good use can be made of an interview schedule, i.e., an outline of topics to be covered in addition to issues and topics the interviewee wishes to bring up. The topics are the same as those mentioned for coverage in a questionnaire. However, it may be best to start the interview in an unstructured manner and then proceed to the data on the schedule.

After a brief statement of the purpose of the interview, i.e., that information is being sought on the information needs of scientific and technical personnel and on their requirements for library services, the interviewee should be encouraged to comment on anything in these two areas that comes to mind. If, during this open-ended phase, the interviewee provides information on any topics on the interview schedule, these topics can then be omitted from the second more structured phase during which specific questions are asked about library services and information needs. A mock-up of different indexes, such as KWIC, subject, title, author, or document series indexes, can be presented to elicit preferences for the library catalog organization.

### Evaluation of the Three Methods

An evaluation of the relative effectiveness of these three techniques depends partly on the specific objectives of the user requirement study. If a large organization is being served and the results of the study are to provide clear guidelines for specific items in a large budget, it is advisable to use all three methods so that questionnaire findings are interpreted and supported by the more specific findings of the diaries and interviews.

The advantage of a questionnaire is its more complete coverage of users. Most surveys have a return rate of about 60 per cent, and the opinions expressed are therefore more representative of the present and potential user group. Questionnaire findings can be analyzed for particular needs of special work groups and for differences of opinions between special user groups. The disadvantages of a questionnaire are the advantages of the two other techniques—the diary and the interview—for the responses to questionnaire items represent opinions in considerable less depth and therefore do not form as specific a base for the creation of services that could satisfy a majority of user requirements.

The advantages of depth and specificity afforded by interview and diary methods are somewhat offset by several disadvantages. The diary is the greatest burden for a user. While he observes his information-gathering habits closely, he may alter his usual pattern or he may recall his procedures only very selectively. The diary method, when employed for all kinds of surveys, has frequently yielded poor returns. However, the diary can be kept extremely simple and clerical help can be provided; it represents the most specific datum that can be obtained.

To fulfill most user requirements this degree of specificity may not be as important as the intensity and scope of information obtained by an interview. The weakness of the interview method is the cost of time of well-trained interviewers.

Depending, of course, on the particular purposes of the special library, the most successful pattern for a user requirement survey is to cover the total user population by questionnaire, to give out and monitor a few diaries, and to conduct 15 to 35 interviews. It may be feasible to start with the interviews and construct the questionnaire from the interview findings.

## Analysis of Data

Classification schemes must be constructed for categorizing and quantifying the responses to the open-ended questions. The responses to questions should be rated by at least two independent raters in terms of their appropriate categorization, and if discrepancies occur, a third rater can provide resolution of the differences.

Once all the data have been converted to scores, computer analysis can be of considerable help. Frequency distributions, correlations, analyses of variance, and other statistical methods are best if the number of respondents is large. Frequently diary and interview returns can be summarized in simple tables without the aid of a computer since, especially if the number of respondents is small, the data only serve to supplement the questionnaire survey findings.

It is essential that the analysis and interpretation of the data be communicated to participants in the study as soon as possible to insure their continued interest and to promote the users' feeling of a joint responsibility for the success of the library's services. Through such studies and the feedback of results the user will think more clearly about his needs and will become more sophisticated in his responses. Therefore the study should be repeated periodically. User requirements studies can represent an accurate and open communication channel between users and the library and can contribute to the mutual understanding of the problems of both groups.

# LOGICAL FLOW CHARTS AND OTHER NEW TECHNIQUES FOR THE ADMINISTRATION OF LIBRARIES AND INFORMATION CENTERS

*C. D. Gull*

SOURCE: From C. D. Gull, "Logical Flow Charts and Other New Techniques for the Administration of Libraries and Information Centers," *Library Resources and Technical Services* 12 (Winter, 1968) pp. 47-66. Reprinted by permission of the author.

THE WIDESPREAD INTRODUCTION of electronic digital computer systems for information processing has produced significant advances in management theory and practice in recent years. For example, two management devices, PERT and CPM,[1] undeveloped and impractical before computers, have been basic to the success of our outer space program. It is perhaps overlate in library development, but appropriate in this memorial to Miss Esther J. Piercy, to explore the application of new management knowledge and practice to the administration of libraries and information centers, and to sketch some directions in which research could be undertaken to benefit the management of information.

Librarians and information center managers are employing several devices or techniques to assist them in the operation and administration of libraries. Some of these techniques are relatively old, such as

(1) organization charts
(2) operating manuals (also known as staff manuals, divisional manuals, sectional manuals, etc.)

(3) position descriptions
(4) personnel or job administration procedures
(5) standardized forms;

others are relatively new and unusual in libraries, such as:

(6) functional block diagrams
(7) logical flow charts, or decision flow charts
(8) decision tables, and
(9) flow process charts.

A brief survey of the literature indicates that these techniques and devices have never been studied together to learn how they can be interrelated, nor whether they can be synthesized to provide an improved management package.

The following figures are introduced to show how the nine techniques could be integrated. The process of order-searching is used as the central example in the figures.

Fig. 1, the functional block diagram, is designed to show the relationships of the library, its operations, and its users. The process of order-searching occurs within the selection and acquisition box in this block diagram.

Fig. 2, is a hypothetical organization chart for a university library. The position of order-searcher is in the Order Division of this chart.

Fig. 3, is a logical flow chart for the process of searching monograph requests in any university library. The order-searcher should perform certain of the work on this chart, as determined by the many factors shown here and on the other figures.

Fig. 4, is a flow process chart. This type of chart is used in industrial engineering. It should be studied in conjunction with logical flow charts to learn which is more useful, whether both are needed, or if a combination is best.

Examples of text from operating manuals, job descriptions, and administrative procedures are generally available and unnecessary here.

Many of the problems of organization and of job performance are easily seen on the logical flow chart. If we assume that it is appropriate for the order-searcher to perform the activities shown in boxes 11, 22, 32, 42, and 54, and to make the decisions required in boxes 12, 23, 33, 43, and 44, who then should have the authority to approve the requests as asked in box 63? Should this approval be assigned to the order-searcher, a bibliographer, to an order-librarian, an assistant-order librarian, or the chief order librarian?

If logical flow charts are properly written, they will show all necessary actions and fundamental decisions. If these actions or decisions have been omitted from the operating manuals or the job descriptions, the logical flow chart will bring out the omissions. If the organizational positions of the actions and decisions are not clear, the logical chart will assist the librarian in choosing their placement and in selecting persons to be responsible for their performance.

Logical flow charts can be made extremely useful by adding information such as:

(1) The name of the position corresponding to every action box and question box.

(2) The status of each item on a flow line (for example, $A\overline{RL}$ on line between boxes 63 and 73 stands for an approved request for a monograph not already in the library collection).

(3) The number of items passing from one box to another box in a certain period (for example, RL 37/1965 between box 12 and box 2 means that 37 requests were returned in 1965 to requesters because the books requested were already in the library).

(4) The average time per item required to complete an action or make a decision (14 min./request in box 22).

(5) The equipment required. (A typewriter for box 91).

Some of the problems of organization, responsibility and job performance are not so easily seen on a logical flow chart, especially rearrangements designed to achieve optimal performance of library operations and services. The objective of optimal design can be expected to require applied research efforts, and gaps in the knowledge acquired through applied research can be used to suggest basic research programs.

The objective of this research is to learn if the techniques and kinds of data previously listed can be integrated and manipulated by manual or electronic methods to provide a framework against which the problems of attaining optimum organization of individual libraries can be solved.

While the actual steps in research must be developed during each research effort, the type of effort can be suggested here. For example, humanly produced flow charts are susceptible to errors which may not be found until experience shows up the errors during actual operations. An attempt can be made to develop a technique for checking the accuracy of humanly produced flow charts by computer methods. Let us assume that such checking can be done successfully with a computer, since programs already exist for the production of flow charts from input data.[2]

Humans are readily able to vary the sequence of operations and the wording of actions and questions in constructing flow charts. They are, however, too easily lost in perceiving the effects of introducing this variety, and they tire easily if required to write down several different flows for comparison. It should be possible to program a computer to accept one or more changes, from humans or automatically, and print out the results as new flow charts or tables.

By introducing real or supposed quantities into the computer formulas and outputs, and solving for successive answers, it should be possible to obtain a computer produced flow chart which will provide reasonably optimum operations for a given library and its individual characteristics. This result is the objective of the research.

Even if the first research investigation is limited to a few departments, such as technical processes, reference services, and circulation, there is such a variety of operations and such an extraordinary diversity of materials, records, and user requirements flowing through them, that it will be all too easy to create a model which will tax the capacities of very large computing systems. Considerable care must be taken to confine the model to significant decisions and operations.

At some point in conducting the research, it should be possible to achieve a reasonably complete one-to-one relationship among the organization charts, functional block diagrams, logical flow charts, operating manuals, job descriptions and job administration procedures; and then to report on the work. It should be possible to write the report to be applicable to both manual and electronic situations in libraries. Even limited research would encompass a sufficient number of functions, individual operations, and individual jobs to illustrate the applications in the principal areas of libraries.

Although examples of block diagrams and logical flow charts are extremely numerous in management literature and computer programming documentation, and are growing more common in the literature of librarianship and documentation, there are almost no examples of good directions for making block diagrams and logical flow charts. The following directions, used in my classes recently, are offered to fill this lack.

### The Preparation of Block Diagrams

A block diagram is a graphic representation of groups of operations, processes, personnel, equipment, products, etc., which collectively are a system. It provides an overview of the system at the gross or macro level. It should be prepared in a clear uncomplicated format to show the whole system at a glance.

The primary objective of a block diagram is to indicate the paths along which information, control signals, materials, etc., flow among the parts of the system, although the exact relationships and specific flow directions need not be shown in full.

The preparation of a block diagram is an art, not a technology. The preparer selects the important operations, processes, personnel, equipment, information, control paths, products, etc., to put in the diagram. He selects the enclosure symbols or boxes from available templates, and artfully writes the labels for the boxes. Pictures, drawings, and cartoons can be substituted for the boxes. Connecting lines and arrows can be added to reflect the relationships among boxes and any of these can be widened, blackened or colored to emphasize relative importance.

The preparer can construct a block diagram from a written description of a system or from data gathered in systems analysis work, or he can create one from systems design work or from information in his head.

A handy technique for constructing a block diagram from a written text is to number every page and every line of every page. Read the text rapidly and write out appropriate boxes, etc., showing the page and line numbers for each box. Use one small slip of paper for one box. Select only the most general and most important things in this reading.

Arrange the loose slips on a large surface to obtain a preliminary picture of the block diagram. Reflect on the wording, arrangement and relationships. Combine boxes where possible. Refer again to the written text to clarify understanding and to fill in significant gaps.

Affix the slips temporarily to a sheet of paper with rubber cement. Draw lines and arrows to show flows and relationships. Describe this draft to yourself or another person, and improve it further. Copy the improved draft, and add a descriptive caption, your name and date.

*The Preparation of Logical Flow Charts*

A logical flow chart, also called a decision flow chart, is a graphic representation of the flow of information, control signals and materials within a system. Such a chart is considerably more detailed than a block diagram, but less detailed than a computer program. Logical flow charts are based on two-value (or yes-no) logic which is basic to digital computer systems and to the representation of decisions in computer systems. Exact relationships and specific flow directions must be shown. These charts are particularly useful guides for writing computer programs, which must include instructions for literally every operation performed by the computer.

The preparation of a logical flow diagram is an art, not a technology, although there are a few rules to be followed. The preparer selects the important operations, processes, personnel, equipment, information, control signals, products, etc., to put in the diagram. He selects the enclosure symbols or boxes from available templates and writes the questions and actions within them. He adds connecting flow lines and arrows to show flows and relationships, and widens, blackens, or colors them to emphasize relative importance. The preparer selects the order of operations and organizes the placement of boxes to facilitate the users' understanding of the chart. The clarity of the chart is in direct proportion to the artistry of the preparer.

Flow charting symbols and meanings have been standardized since 1965 through the work of committee X3.5 of the United States Standards Institute. These standards have been published and most of the symbols are available in templates from the major computer manufacturers. The symbols used in this article are derived from these standards, with some modifications and additions.

Logical flow charting is an extremely effective graphical method to provide clear descriptions of information systems already in operation. This method is also very effective as a problem-solving technique in the design of new or improved information handling systems.

Logical flow charts reveal what decisions must be made, where or when they must be made, and what effects result from decisions.

Logical flow charts have this cyclic pattern: STATUS—DECISION—ACTION —STATUS—DECISION—ACTION . . . etc.

*Logical Flow Chart Symbols*

Decision points are expressed as *question* or decision boxes in flow charts. While the diamond shape is standard, the large modified oval provides more space for text:

Once a decision is made, an action must be accomplished to implement the decision. Rectangles are used for *action* or process boxes in flow charts.

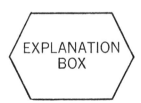

Explanations may be required to state what material is being processed, the condition of the material, etc. *Explanations* may be enclosed in irregular hexagonal boxes.

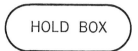

Since it is not possible to maintain a continuous flow of materials through an information system, delays can be represented within a *hold* box in a small modified oval shape:

This symbol can be used also to represent other conditions, such as

Lines are used to show flow. Flow is in one direction only, as shown by arrows.

FLOW LINE

Two small circles can be used as onpage connectors for a flow line broken by the format of a chart.

ONPAGE CONNECTORS

Since it is not possible or convenient to place all boxes for one information system on a single sheet, offpage reference symbols for IN and OUT are required.

OFFPAGE CONNECTORS

The basic symbol for input-output is the parallelogram.

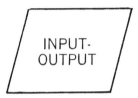

Specialized input-output symbols are used for punched cards, magnetic tape, punched tape (normally paper tape), and written or printed documents.

SPECIALIZED INPUT-OUTPUT SYMBOLS

*Rules for Making Logical Flow Charts*

A. Question Boxes

A question box may contain only one question, and only two answers are permitted, Yes and No. Although several inbound lines are permitted, the preparer is restricted to using two outbound lines from one question box.

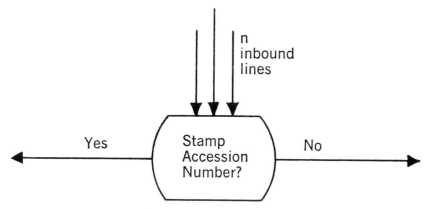

restricted to 2 outbound lines

If three answers are required, two question boxes must be used:

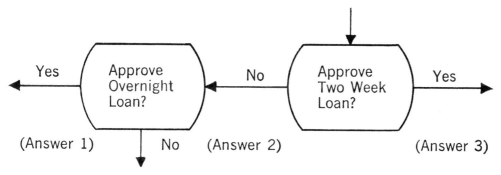

Always label the outbound lines with Yes or No.

All outbound lines must be connected to an action box; i.e. no loose ends are permitted on flow lines.

It is desirable to write a question in the form: Verb—Object—Question Mark. Modifiers can be inserted if needed. The subject noun is usually implied by the kind of chart, e.g. **CLERK** Stamp Accession Number?

B. Action Boxes.

Write directions in action boxes in this form: Imperative verb—Object. Modifiers can be added if needed.

> Stamp Accession Number on Page 1.

Several actions can be placed in one box if they all result from the decision taken in the previous question box, and if they do not require decisions of their own.

C. Identification of Boxes.

Identify all boxes. While numbers are simplest, a coordinate grid system is useful for large charts; i.e., letters on the horizontal axis and numbers on the vertical axis, with each box identified by the intersection of two axes.

D. Reference Boxes.

For onpage connectors, the OUT box contains the number of the box referred to, and the IN box contains the number from which reference was made.

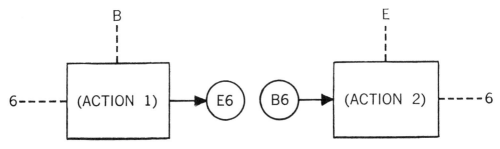

The OUT box must contain the number of the box to which its line is to be connected:

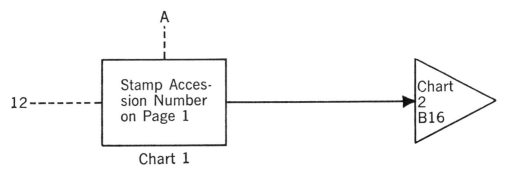

Chart 1

The IN box must contain the number of the box from which the inbound line comes:

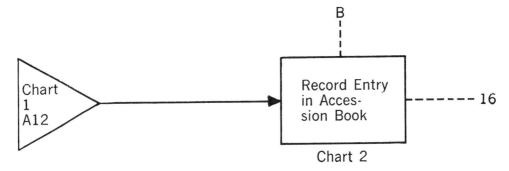

Chart 2

E. Direction of Flow.

Flow is only in one direction, and must be shown by arrows. While the customary direction on the page is left to right and top to bottom, other directions may be used for enhancing clarity.    ⟶

F. Iterative Loops.

Loops are frequently required to show processes which are repetitive in nature:

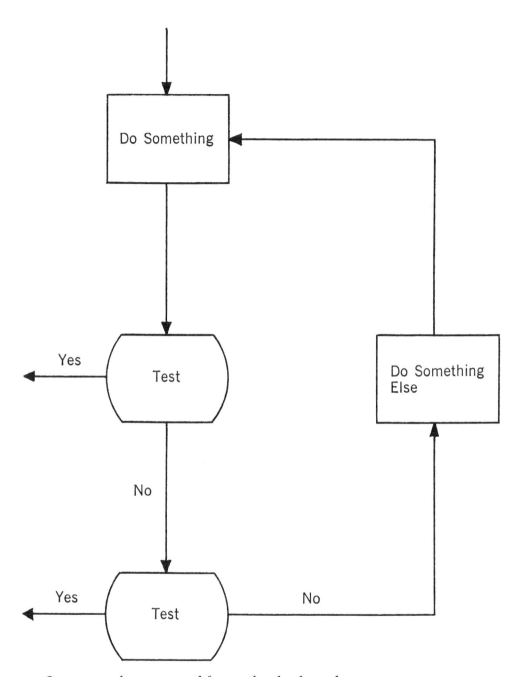

Loops may be connected forward or backward.

Beware of creating a loop in which the material never satisfies a question. The Yes answer must be possible to permit the material to leave the loop!

## G. Identification of Materials

Identify the materials which proceed along flow lines with letters and other symbols, e.g., C for Catalog Cards, S for Slips. Show this identification for everything on every flow line. A negative status is shown by placing a bar over the symbol, e.g., RĀ can mean a purchase recommendation not approved by the order librarian.

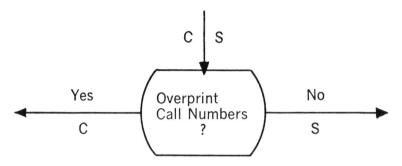

Show the meaning of the letters in a legend box.

Flow charts can be made more informative by adding quantities per unit of time, time to accomplish an operation, etc., along flow lines and in the various boxes.

*Hints on Making Flow Charts.*

It is often desirable to construct flow charts by writing each individual box, its text and connecting lines, on a 3 x 5″ or 2 x 3″ slip in whatever outline form is required and with whatever text is necessary. The slips can be fastened to a window, sheet of paper, the wall, etc., by means of Scotch masking tape or plastic cements which will peel off easily. This removability is particularly desirable for preliminary work in which errors are frequent, because the good slips can be removed and replaced in another position with the lines being connected up anew to show the changes, while the incorrect slips are destroyed and replaced by new, correct slips.

There is another technique for improving the arrangement of an unsatisfactory flow chart, or portion of a flow chart, on which the boxes are already numbered or identified on a grid. Establish the number of vertical columns and horizontal rows you have available in the new space for the work you are undertaking. Then label the columns with letters and the rows with numbers. On the new space work out a satisfactory arrangement by placing only the old box numbers in the new positions, as required. For example, temporarily put box no. 34 in column B and row 2. Put the corresponding grid number on the box of the unsatisfactory chart. Show the connecting lines on the new chart. Each box in the old and new charts then has a double number, here 34 and B2, and the transfer of slips can be made from old to new when the new format is satisfactory.

It is often useful to photocopy a chart in process, with date and number showing its sequence in development. If two photocopies are made, one copy can be cut up for transfer to improved drafts.

*Problems Encountered in Preparing Logical Flow Charts*

Since logical flow charts are often prepared from block diagrams, from descriptive text, or from systems analysis data, there are often failures to transfer the operations from the sources to the draft flow chart. Such failures can be reduced by adopting a method of checking off the information on the sources to make certain each item is transferred.

The use of imprecise, misleading and confusing terminology in the boxes is another type of problem, perhaps indicating that the charter does not understand the operations, and at least meaning that the reader of the chart is hindered in comprehending the operations. The asking of the wrong kind of question, or the poor definition of the contents of a file are examples here.

The symbols used to identify materials and to show status are often omitted from flow lines. These omissions can result in serious and ludicrous errors. Without these symbols, it is all too easy to combine several classes of materials on one line and to process them alike, when at least one class is to be treated differently from the others; or to direct an action upon a class which never reaches the place where the action should occur.

It is easy to leave a flow line unfinished. While open-ended lines can be mistakenly omitted anywhere on a chart, they are most commonly found after a question box, where they are very serious, often leading to the omission of significant suboperations.

It is easy to direct an action that requires a significant decision and fail to put the question box ahead of the action. This situation is equivalent to leaving one flow line open after a question box.

Any failure to make outbound and inbound connections with proper grid symbols, and to check to see that both parts of a pair of boxes are filled out properly, results in an open flow line.

The loops required to perform necessary operations are sometimes omitted or they are connected improperly, forward or backward, with the result that the material no longer fits the chart as it is moved from box to box.

The following references are helpful:

Bolles, Shirley W. "The Use of Flow Charts in the Analysis of Library Operations." *Special Libraries,* 58:95-98. Feb., 1967.

Carr, John W., III. "Educating the Computer—and the Man Who Guides It." *In* American Management Association. *Pioneering in Electronic Data Processing.* New York, AMA, 1956. pp. 65-80.

Clark, Gerald R. *Introduction to Flow Charting.* Washington, American University. 1961. 8 p.

Colman, Harry L. *Computer Language; An Autoinstructional Introduction to FORTRAN.* New York, McGraw-Hill, 1962. pp. 6-16.

Dennis, Sally F., "The Use of Electronic Computing Machines for Literature Searching." Chap. 19. pp. 581-591, and Appendix, pp. 592-600, of *Tools for Machine Literature Searching,* by J. W. Perry and Allen Kent. New York, Interscience, 1958.

DeVos, Henry. "The Techniques of Flow Charting." *Journal of Accountancy,* 118:84-87. October, 1964.

Dougherty, Richard M. and Heinritz, Fred J. *Scientific Management of Library Operations.* New York, Scarecrow Press, 1966. pp. 52-65.

Hattery, Lowell H. and McCormick, Edward H. *Information Retrieval Management.* Detroit, American Data Processing, 1962. p. 106.

Honeywell, Inc. Electronic Data Processing Division. "Standard Honeywell Templates for Flow Chart Symbols." pp. 79-85 in its *Glossary of Data Processing and Communication Terms.* 3d ed. Wellesley Hills, Mass., Honeywell Inc., 1965. The X3.6 Subcommittee symbols are published here.

International Business Machines Corporation. *IBM Reference Manual; Flow Charting and Block Diagramming Techniques.* White Plains, N. Y., IBM, 1962 [?]. 27 p. C20-8008-0.

International Business Machines Corporation. Data Processing Division. *Data Processing Techniques; Flowcharting Techniques.* White Plains, N. Y., 1963 [?]. 33 p. C20-8152.

Melin, John S. *Libraries and Data Processing—Where Do We Stand?* Urbana, University of Illinois Graduate School of Library Science, 1964. 44 p. (*Occasional Papers,* no. 72)

Sacks, Edward I. "Picking the Best Design with Flowcharts." *Data Processing Magazine,* 8:22-26. December, 1966.

Schultheiss, Louis A., *et al. Advanced Data Processing in the University Library.* New York, Scarecrow Press, 1962. pp. 79-84.

Schultheiss, Louis A. "Techniques of Flow-Charting." *In* Clinic on Library Applications of Data Processing. *Proceedings.* 1963. Ed. by Herbert Goldhor. Urbana, University of Illinois, Graduate School of Library Science, 1964. pp. 62-78.

Sippl, Charles J. *Computer Dictionary,* Appendix N: "Flowcharting (Logic)." Indianapolis, Sams, 1966. pp. 300-306.

Stych, F. S. "Teaching Reference Work; The Flow Chart Method." *RQ,* [Reference Services Division of ALA], 5:14-17. Summer, 1966.

Swenson, Sally. "Flow Chart on Library Searching Techniques." *Special Libraries,* 56:239-42. April, 1965.

Touloukian, Y. S., *et al.* "Systems and Procedures Developed for the Search, Coding and Mechanized Processing of Bibliographic Information on Thermophysical Properties." pp. 78-91 *in* Symposium on Thermal Properties, Purdue University, 1959. *Thermodynamic and Transport Properties of Gases, Liquids, and Solids; Papers.* New York, American Society of Mechanical Engineers, McGraw-Hill, 1959. [Published earlier, 27 p., mimeo., at Lafayette, Indiana, Thermophysical Properties Research Center, Purdue University, July 1, 1958.]

United States of America Standards Institute. *American Standard Flowchart Symbols for Information Processing.* USA Standard X3.5-1966. New York, The Institute, 1966. 11 p. Includes bibliography different from above list.

### REFERENCES

1. Archibald, Russell D., and Villoria, Richard L. *Network-based Management Systems (PERT/CPM).* New York, Wiley, 1967.

   Levin, Richard I., and Kirkpatrick, Charles A. *Planning and Control with PERT/CPM.* New York, McGraw-Hill, 1966. [PERT = Program Evaluation and Review Technique. CPM = Critical Path Method.]

2. Scott, A. E. Automatic Preparation of Flow Chart Listings." Association for Computing Machinery, *Journal,* 5:57-66. 1958.

   *Documentation Aids System,* IBM Program no. 1401-SE-12X and Reference Manual H20-0177.

   (The Auto Flow Computer Documentation System was introduced in 1966 by Applied Data Research, Inc., of Princeton, N. J.)

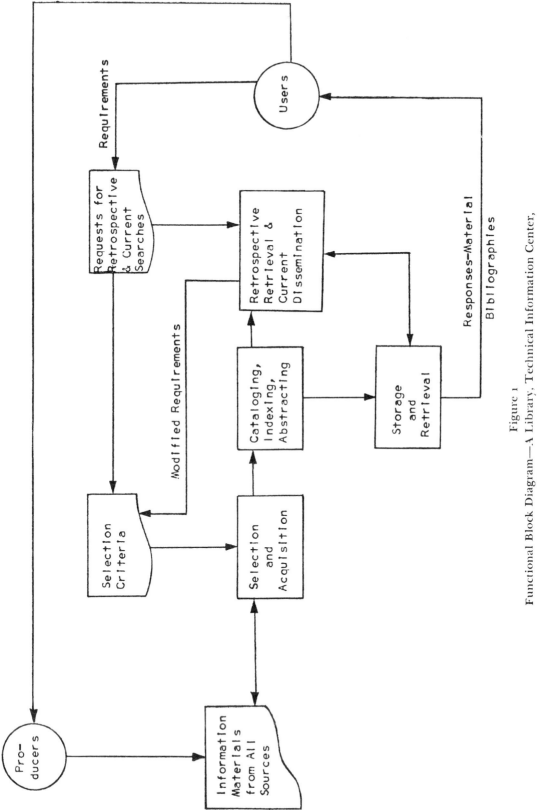

Figure 1

Functional Block Diagram—A Library, Technical Information Center, or Storage and Retrieval System

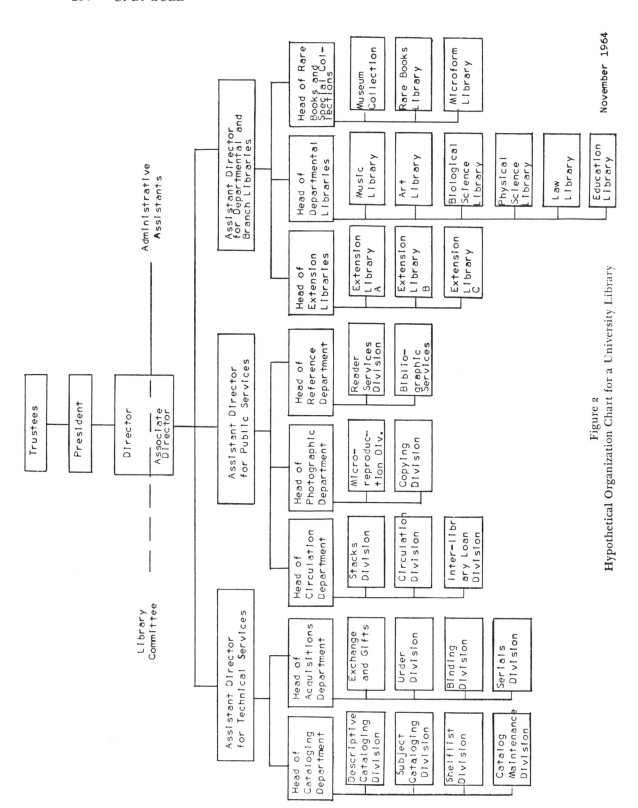

Figure 2
Hypothetical Organization Chart for a University Library

### Figure 3
Logical Flow Chart
Searching Monograph Requests in any University Library

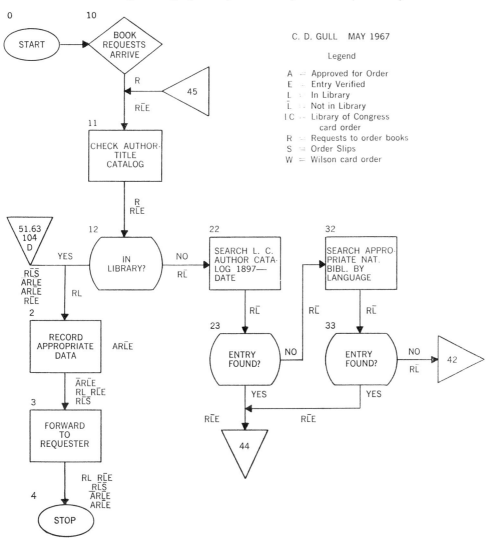

Figure 3 (Continued)

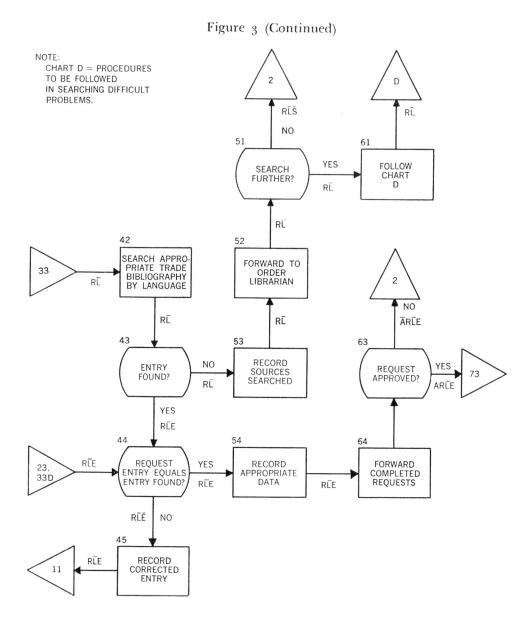

NOTE:
 CHART D = PROCEDURES
 TO BE FOLLOWED
 IN SEARCHING DIFFICULT
 PROBLEMS.

Figure 3 (Continued)

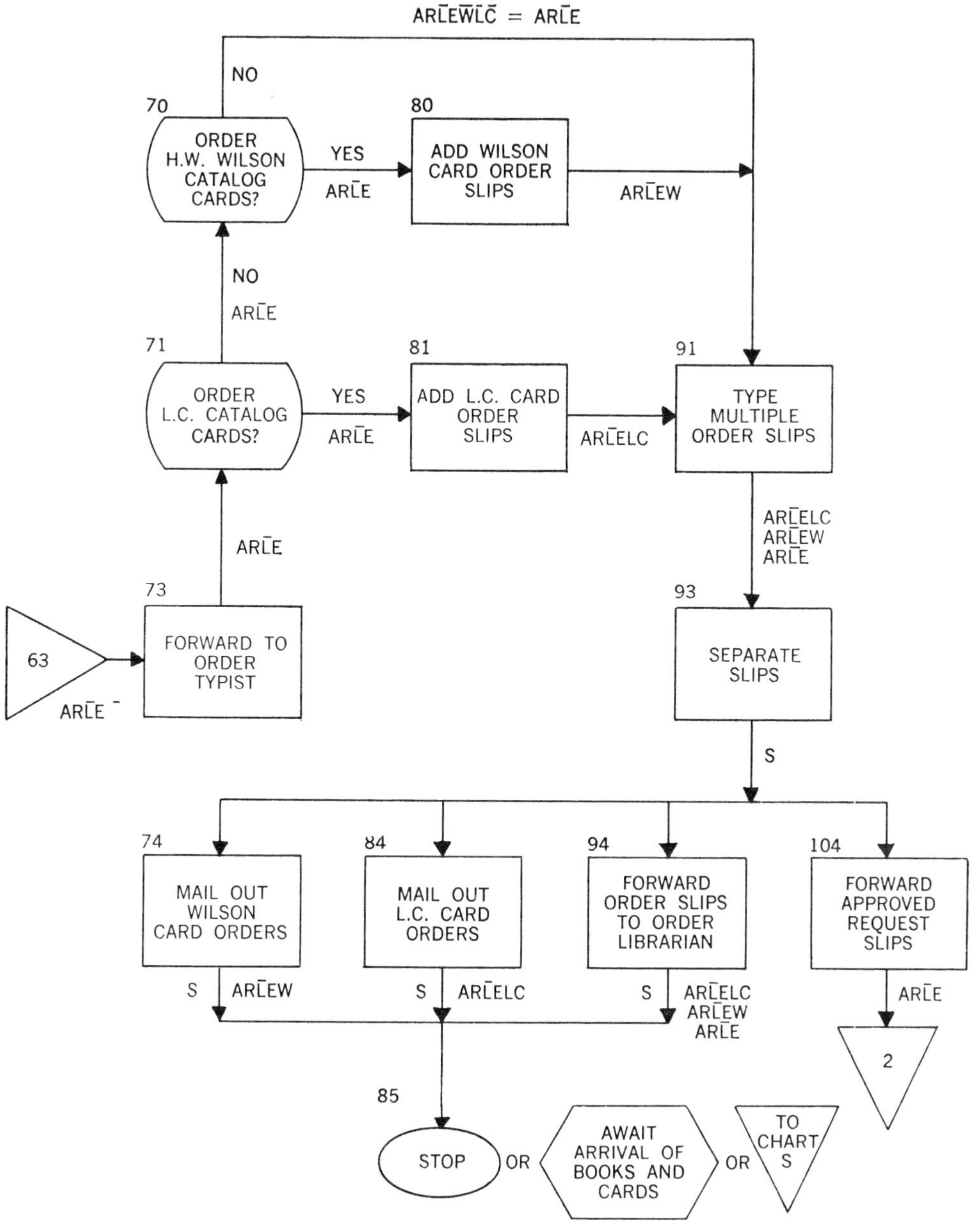

$\overline{ARLE}\overline{WLC} = \overline{ARLE}$

# By asking key questions about each step...

**WORK SIMPLIFICATION PROGRAM**

## PROCESS CHART

PROCESS CHARTED __Certification__    UNIT ____Certification____

____Procedure____    DIVISION, BRANCH, ETC. ____Field____

DATE ____October 18, 1963____

| O IN FEET | TIME IN MIN | OPERATION | TRANSPORT | STORAGE | INSPECT. | STEP NO | DESCRIPTION OF EACH STEP (SHOW WHAT IS DONE—WHO DOES IT) |
|---|---|---|---|---|---|---|---|
| | | ○ | ○ | ▲ | □ | 1 | Incoming case at master register desk |
| | | ● | ○ | △ | □ | 2 | Case entered in master register |
| | 10 | ○ | ○ | ▲ | □ | 3 | In outgoing basket |
| 120 | | ○ | ● | △ | □ | 4 | To file clerk |
| | 30 | ○ | ○ | ▲ | □ | 5 | In incoming basket |
| | | ● | ○ | △ | □ | 6 | Case entered in file room register |
| | | ● | ○ | △ | □ | 7 | File searched for previous action |
| | | ● | ○ | △ | □ | 8 | File pulled and attached to case |
| | | ● | ○ | △ | □ | 9 | Charge-out slip made on file |
| | 40 | ○ | ○ | ▲ | □ | 10 | In outgoing basket |
| 180 | | ○ | ● | △ | □ | 11 | To clerk A |
| | 120 | ○ | ○ | ▲ | □ | 12 | In incoming basket |
| | | ○ | ○ | △ | ■ | 13 | Case checked against file for change of address |
| 120 | | ○ | ● | △ | □ | 14 | To correspondence clerk |
| | 35 | ○ | ○ | ▲ | □ | 24 | Case to analyst |
| | | ● | ○ | △ | □ | 25 | In incoming basket |
| | | ○ | ● | △ | □ | 26 | Statement of recommendations prepared |
| 10 | | ● | ○ | △ | □ | 27 | To typist |
| 50 | | ○ | ● | △ | □ | 28 | Statement typed |
| | 60 | ○ | ○ | △ | □ | 29 | To case editor |
| | | ○ | ○ | △ | ■ | 30 | In incoming basket |
| 130 | | ○ | ● | △ | □ | 31 | Statement checked for form |
| | 120 | ○ | ○ | ▲ | □ | 32 | To section chief |
| | | ● | ○ | △ | □ | 33 | In incoming basket |
| 130 | | ○ | ○ | △ | □ | 34 | Draft revised for conformance to policy |
| | 120 | ○ | ○ | △ | □ | 35 | To case editor |
| | | ○ | ○ | △ | ● | 36 | In incoming basket |
| 50 | | ○ | ○ | △ | □ | 37 | Statement checked |
| 60 | | ● | ○ | △ | □ | 38 | To typist |
| | 30 | ○ | ○ | △ | □ | 39 | Final draft of statement typed |
| | | ○ | ● | △ | □ | 40 | To analyst |
| | | ○ | ○ | ▲ | □ | 41 | In incoming basket |
| 40 | | ○ | ● | △ | ● | 42 | Reads |
| 90 | | ○ | ○ | ▲ | □ | 43 | In outgoing basket |
| | 240 | ○ | ● | △ | □ | 44 | To section chief |
| | | ○ | ○ | ▲ | □ | 45 | In incoming basket |
| | | ● | ○ | △ | □ | 46 | Reviewed and signed |
| **1100** | **965** | **13** | **14** | **14** | **5** | TOTAL | |

**BEFORE ➤**

**46**

Figure 4. Flow Process Chart

# ...these results were achieved:

## WORK SIMPLIFICATION PROGRAM
### PROCESS CHART

PROCESS CHARTED __Certification__     UNIT __Certification__
__Procedure__     DIVISION, BRANCH, ETC. __Field__
DATE __October 19, 1963__

| IN FEET | TIME IN MIN | OPERATION/TRANSPORT/STORAGE/INSPECT | STEP NO | DESCRIPTION OF EACH STEP (SHOW WHAT IS DONE—WHO DOES IT) |
|---|---|---|---|---|
| | | | 1 | Incoming case at master register desk |
| | | | 2 | Case entered in master register |
| | 10 | | 3 | In outgoing basket |
| 120 | | | 4 | To file clerk |
| | 5 | | 5 | In incoming basket |
| | | | 6 | File searched for previous action |
| | | | 7 | File pulled and attached to case |
| | 35 | | 16 | In incoming basket |
| | | | 17 | Statement of recommendations prepared |
| 10 | | | 18 | To typist |
| | | | 19 | Statement typed |
| 80 | | | 20 | To section chief |
| | 120 | | 21 | In incoming basket |
| | | | 22 | Draft revised for conformance to policy |
| 130 | | | 23 | To case editor |
| | 60 | | 24 | Incoming basket |
| | | | 25 | Statement checked for form |
| 50 | | | 26 | To typist |
| | | | 27 | Final draft typed |
| 40 | | | 28 | To analyst |
| | 20 | | 29 | In incoming basket |
| | | | 30 | Reviewed and signed |
| 500 | 270 | 10  9  9  2 | | TOTAL |

◀ AFTER

30

Figure 4. Flow Process Chart (Continued)

INSTITUTIONAL IMPLICATIONS

OF AN

AUTOMATED CIRCULATION STUDY

*Floyd Cammack*
*and*
*Donald Mann*

SOURCE: From Floyd Cammack and Donald Mann, "Institutional Implications of an Automated Circulation Study," *College and Research Libraries*, 28 (March, 1967) pp. 129-132. Reprinted by permission of Floyd Cammack.

THIS PAPER is a report, based on initial, limited data, of a few of the questions and answers made economically possible through the combination of machine-readable circulation and student records at Oakland University. The specific answers to the questions asked in this study are of secondary importance, except as their relative status may suggest general trends. The questions themselves, together with similar, more refined types of inquiries, are ones which become readily and regularly answerable with any circulation system that combines call number information with borrower identification in an easily tabulatable form.

In the case of the Oakland system, this transaction record is a punched card produced automatically from a machine-readable book card, a machine-readable borrower's card, and a time clock. Its primary function is input to the main circulation records system. By accumulating these cards after their information has been processed through the computer, there develops a store of book-and-borrower data which in turn can be run against student records containing current information on grades, curriculum, and class.

The analyses performed thus far are basically counting mechanisms. The approach to each analysis was to determine, within the charge data, the possible span of the quality- or identifier-field to be analyzed, such as the day of charge, class of book charged, etc. A counter for each unit of the span was set up in computer memory, the file of charge cards was passed through the computer, which incremented the appropriate counter for each charge, and the contents of the counters were then printed.

No forecasting or statistical analysis was done by computer at this time. The study merely lends insight into the relative volumes of the various characteristics of the data. By establishing the counters in memory prior to giving the data to the computer on each run, the necessity of large-scale card sorting was all but eliminated. All programs were written in the FORTRAN programing language and were run on the university IBM 1620 computer which has card, disk, and printer capacity.

The charge cards were first separated from the discharge cards. Three analyses were then made on the charges: Use by Day (run time: 50 min.), Use by LC Class (85 min.), and Use by Time of Day (55 min.). The charges for students were then split out and sorted into student number sequence. These cards were then run against the student cumulative performance file which is maintained by the university, and student summary cards were punched. These cards contain student number, number of charges, cumulative Grade Point Average, curriculum, and class. The remaining four analyses were then run: Use by GPA (20 min.), GPA by Use (20 min.), Use by Class (20 min.), and Use by Curriculum (20 min.). Copies of detailed data not presented here can be made available on request.

The initial tabulations, after four months of operation, provided answers to the following inquiries.

*Question:* Could personnel scheduling at public service desks be more efficiently or economically handled?

*Answer:* Yes. A tabulation of dates of charges showed a striking consistency by which charging activity tended to follow a cyclical pattern, consistently giving certain days of the week by far the heaviest workloads. Desk staffing patterns and shelving labor could be brought into better relationship with an unexpectedly predictable activity cycle. In answer to the same question, the number of changes per hour showed four regular peaks with definite implications for desk staffing and improved closing procedures. Unexpectedly, and in contradiction to the staff's impressions, charging activity did not tend to cluster around class intervals, but spread quite evenly throughout each hour of the service day.

*Question:* Which portions of the library's collections are used most heavily, and which are used least?

*Answer:* A breakdown by LC classification letters showed English literature,

history, philosophy, education, and economics topping the list in that order. The lowest significant use areas involved mathematics and the physical sciences, the totality of which did not equal use in English literature alone.

Such an answer might well raise a number of subsidiary questions concerning adequacy of the collection for student use, library orientation for science students, comparative figures for similar institutions, the numerical relation of science majors to other majors, and perhaps the science faculty's awareness of the library's resources together with the whole question of departmental goals and book budget allocations.

*Question:* Students majoring in which subjects tend to be the heaviest library users, and does this correlate as expected with the answer to the immediately preceding question?

*Answer:* Secondary education majors in modern foreign languages and English head the list for average number of charges per person, followed closely by liberal arts majors in the same fields. Similarly high averages support history and philosophy use patterns mentioned above. Majors in physical and biological sciences average below the mean, but again with their secondary education counterparts somewhat more active than students in the College of Arts and Sciences.

The suggested implications of these tabulations would appear to be that students tend to read largely within their own fields of study (although a further breakdown by individual students' charge records would be required to confirm the conclusion) and that, at Oakland at any rate, students planning teaching careers use the library more than arts and sciences students with the same subject specialization.

*Question:* Do upperclassmen tend to use the library more than lowerclassmen?

*Answer:* Decidedly yes. Of those who used the library during the list period, juniors and seniors averaged almost twice the number of charges for freshmen and sophomores. Somewhat more enlightening are the implications of the class breakdown shown in Figure 1 below. The freshman year would appear to be a "non-library" year, in at least a portion of which only 10 per cent of the class found it necessary to check out a library book. By the sophomore year, the number has increased sharply and by the junior year, apparently almost all students are library borrowers. With the implementation in 1966 of a new lower division curriculum at Oakland, it would seem worthwhile to establish a detailed analysis of freshman library activity (or inactivity) in cooperation with the administrative officers responsible for the curriculum. The inclusion of librarians within the freshman advising program, and the presentation of library orientation programs, apparently require examination and increased emphasis.

With an over-all library non-use figure of 42 per cent during the test period, it appears likely that freshmen are the odd-men-out. Unless their "in-house" use of library facilities is considerably higher than other indications would imply, major attention should be focused on the library's relations with lower classmen.

*Question:* What is the relationship between library use and academic achievement?

*Answer:* Extremely close. As illustrated in Figure 2, even a limited body of

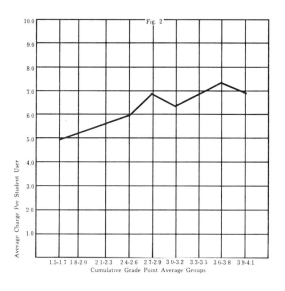

Fig. 2

data shows a direct, positive correlation between the borrowing of library materials and cumulative grade point average. Using the test period as a sample, the cumulative GPA of students who borrowed at least one book during the time was 2.73, while that for students who borrowed no books during the period was 2.54; implying again that regular

Fig. 1

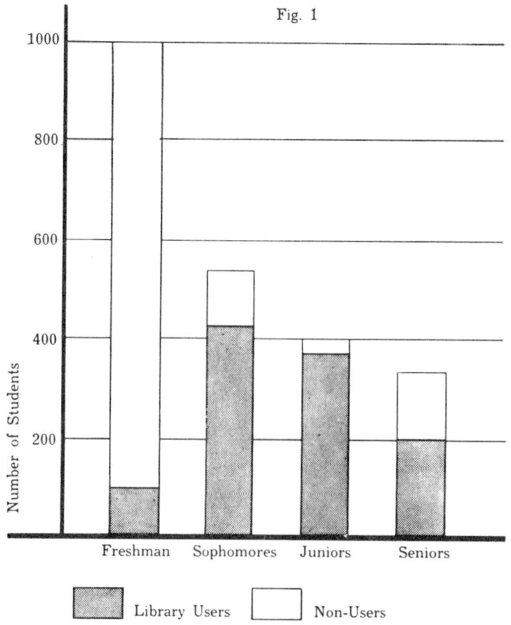

library borrowers are likely to make better grades than do nonborrowers. It may also be that entrance examination scores are higher for the former group. A semester-by-semester tabulation of these relationships could begin to establish profiles of student achievement by class-year and subject. Further sorting of data could provide individual "reading profiles" for individual students and groups of students, answering in part the old question "What does a 'good' student do that a 'bad' student doesn't?"—and vice versa.

Sidestepping for the moment any discussion of causal relationships between academic achievement and library use, the library's position (if not function) within the undergraduate educational process would appear, simply on the basis of circulation figures, to be demonstrably significant. When this correlation is viewed in connection with freshman and sophomore use patterns, perhaps academic librarians (at least at Oakland) should set out to "recruit" users in their first two years of college work. It might well be that college librarians should take another look at the public library's operation and staffing of young adult collections.

As the store of transaction data continues to increase as a by-product of the circulation system, it will soon reach the point at which reliable answers can be found to a number of additional questions. The ability to identify nonborrowers quickly and accurately should provide useful information for academic advisers. The borrowing records for commuters, dormitory students, students in specific courses, part-time students, and faculty members are within easy access, allowing runs of the total accumulation of data or of any desired sample. While most manual circulation systems generate the same type of data, machine-readable data allow easy and rapid tabulation at a fraction (estimated at 1/5) of the cost of the same information developed manually. Initial faculty and administrative reaction to the availability of this type of information has very quickly answered the question, "Who cares?" Requests for special runs have ranged from the professor who asked for periodic "traffic reports" on his assigned reading lists to a department chairman who expressed a near-unethical degree of interest in the reading habits of his faculty members in connection with promotion decisions.

With the system designed to go online as soon as practicable, the provision of remote terminals could put the answers to these and many other questions literally at the fingertips of librarians, administrators, and faculty members, with data always up-to-date and instantly available.    ■ ■

# IV

# APPLYING SYSTEMS
# ANALYSIS CONCEPTS

**Evaluation of Existing Systems
Design of and Decisions
for New Systems
Implementing the New System
Following-up on the Installation
of a System
Example of Cost Analysis
in Systems Design
Case Studies of Systems Analysis
Applied to Library Problems**

## INTRODUCTION

The concept of systems study or analysis consists of three interdependent phases*:

1. Analysis, which is the accurate delineation of the requirements placed on a system; the current procedures by which the requirements are met; the outputs of the system in satisfaction of the system's requirements; and the inputs used to generate the outputs. The four items under analysis represent concurrent identification of the areas of inquiry, coupled with the charting of all operations, functions, decisions, and actions, the gathering of data produced and forms used, the listing and evaluation of available personnel and equipment, all synthesized into a report of existing conditions.

2. Evaluation, which is the detailed examination of current procedures with respect to their adequacy to implement the mission of the system.

---

*As given in, E. A. Chapman, *et al. Library Systems Analysis Guidelines*. New York: Wiley, 1970. p. 19.

3. Design, which is the action taken by validation of the existing system, by modification of it, or by substitution of a newly designed system to satisfy the demands being placed on the system. (The problems to be solved and the techniques employed in this design phase are well stated in, R. L. Ackoff. *A Concept of Corporate Planning,* New York: Wiley-Interscience, 1970, pp. 87–111.)

Systems study and planning for improved operations are inextricably intermeshed activities. Problem recognition is a prelude to systems analysis and the development of plans for desired ends. Systems analysis is indeed the basis of effective planning to reach operational goals. Such analysis supplies the means of validating the efficacy of plans—do they work as envisioned or projected.

This section on applying systems analysis concepts opens with discussion of the second phase of a systems study or analysis—the evaluation of existing library operations. Chapman *et al* supply guidelines for determining how well current operations meet the demands placed upon them, bring attention to the useful scientific management technique of "management by exception" and emphasize the criticality of a well prepared report of findings and recommendations. The elements of good report writing are treated extensively. A caveat is given the analyst: however well prepared his report of findings is, management for a variety of reasons may have to restate the objectives originally given the analyst. Thus, he may be required to repeat his evaluation instead of proceeding to the design phase of the systems study. As is true of the complete process of systems analysis, its components can become complicated and confusing unless a standardized pattern of investigation is firmly followed. Determinations to be made in the evaluation phase are sequentially indicated with supporting examples where needed.

Greenwood, in clearly understood and encapsulated form, supplies guidelines for planning, organizing and conducting a systems analysis of data processing operations from a business systems management view. Applicability to library systems management's use seems evident. All the basic principles of systems analysis and how to implement them are given: study staff organization and orientation; planning of the study outlined in six steps from definition of the problem through the time schedule within which the analysis is to be completed; all culminating in a cost estimate of the study and formal announcement by management of what is to go on and why. Such announcement is designed to gain the cooperation of the regular staff of the system by showing management's direct interest in the project and the benefits in the working conditions of the staff. The content of the systems study announcement is suggested.

Kilgour, well recognized in the library profession as a leading proponent of computer-based information control and supply, presents systems analysis and design concepts geared to "information processing machines." Emphasis is placed upon the study and identification of library user requirements and the need of the computer in meeting those requirements effectively and with timeliness. Systems should be designed with the user as an integral part of them. Concern is with the principles governing the analysis and design of computer-based information systems rather than on the techniques of systems analysis and design based upon the stated principles. As pointed out, the techniques for planning and designing a computer-assisted system differ very little from those used in the analysis and design of any new library system. The difference lies in the absolute need for imaginative and logically detailed design required for successful computer manipulation of data. Proposed procedures' design must be rigorously tested and measured against several factors mentioned here before coming to a final decision.

Installation of recommended procedures is usually a more time-consuming and difficult task than all other phases of the systems study together. Neuschel briefly defines installation as getting from where you are to where you want to be, and, not withstanding the best of plans, that faulty installation inevitably leads to failure in implementing recommended procedures. The author with extensive experience in business systems analysis, supplies practical instruction in installation methods, the problems to be solved and the mistakes to be avoided. An hypothetical installation plan is furnished, illustrating the extent of detailing required for useful guidance. He further discusses the installation responsibilities of the analyst and of the staff which will implement the new procedures. Personnel training for this is also covered. The importance of testing procedures before their introduction into the system is emphasized; so are follow-up and review for a time, to see how well the system is working in serving the operational objectives sought.

The important subject of installing recommended procedures resulting from a systems study or analysis is further treated in the extract from the Association for Systems Management monograph, *Business Systems.* The subject of follow-up on the application of new procedures receives extensive attention. Again, practical and precise instruction is supplied for installation planning and methods of installing new procedures. The merits of three methods are evaluated. The controls needed during installation to ensure the required outputs of data in a smooth and timely fashion, are designated. The scheduling of the steps of installation is set in relation to available personnel, delivery of new equipment and forms, development of job instructions, training of personnel and so forth. The roles of the analyst, supervisors and systems staff in installing the new procedures are delineated.

Although the paper by Klintøe is an analysis of the operations of the specialized Danish Technical Information Service for industry throughout Denmark, it is notable for the policies and procedures applied in identifying user requirements and in evaluating and costing the services given. The methods of measuring the inputs and outputs of the system in terms of what is accomplished together with analysis of costs, are adaptable to a systems analysis and evaluation of a library's information or reference services. Since exact cost/benefit measurement of library information services is very difficult if not impossible, justification of such services is reached in a trade off of operating expense evaluated against increasing used demand and acceptance.

Erisman's selection is a condensation of his careful and thorough systems study report on the processes necessary to return a book to its proper location on the shelf after "home-use." The processes studied were (1) Removing books from the book drop, (2) Converting "Numerics" into books adapted for the new automated circulation system, (3) Discharging books returned from home use, (4) Sorting books into call number order, and (5) Shelving the books in the stacks. Information was gathered on the characteristics of the system: objectives and requirements; personal; facilities, equipment, and conditions; and forms, records, and reports. Time studies were conducted and the results analyzed. Some recommendations are made but no major conclusions are reached. The report is primarily a documentation of the present system.

Morris describes the design of a reclassification project begun with investigation of the elements of several other similar projects elsewhere. As the result of this study the project was designed satisfying most of the objectives of systems analysis:

efficiency, economy, productivity, timeliness and accuracy. The project's design is well depicted by flow charts already discussed in this book as effective worker training devices and valuable tools for transmitting logical work flow information about operations and understanding the procedures involved.

# EVALUATION OF THE CURRENT OPERATING SYSTEM AND REPORT OF FINDINGS

*Edward A. Chapman, Paul L. St. Pierre*
*and*
*John Lubans, Jr.*

SOURCE: From Edward A. Chapman *et al*, *Library Systems Analysis Guidelines*. New York, Wiley, 1970. Reprinted by permission of the publisher. Copyright © 1970 by John Wiley & Sons.

APPLICATION OF MANAGEMENT CONCEPTS

With completion of the analysis phase the analyst is prepared to proceed with the second phase of the systems study—evaluation of how well the current operating system fulfills the demands placed on it. Based on a thorough understanding of present methods and procedures used in the current system, supported by the survey worksheets of requirements, personnel, equipment, the outputs and inputs of the system, together with completed sample copies of all forms used and records produced, the analyst is prepared to evaluate the system to determine how effectively, efficiently, timely, how accurately, and at what cost, the present system meets the requirements of each operation.

It is in this phase of the systems study that flow charting seriously comes into play.

Other techniques of scientific management applicable in library systems study are discussed in this chapter—the "management by exception" concept and the preparation of reports and recommendations in a persuasive manner and cogent form for consideration by management.

Statistical sampling and job analysis for developing work standards were briefly discussed in Chapter 4 in connection with the analysis and understanding of current procedures.

The evaluation phase culminates in a report of findings including recommendations for maintaining the status quo, eliminating or transferring certain operations to another system, reorganizing given operations, combining operations, increasing or decreasing staff, recasting the functions and responsibilities of professional personnel, or designing a completely new system.

The analyst should be aware of the options open to management on its receipt of the evaluation report. He should anticipate such contingencies as that management will give full acceptance to his recommendations, requiring the analyst to be prepared to proceed with the final phase of designing a new system or modifying the present one; that management may give initial disapproval of the recommendations, asking for further substantiation of them; or that management may place certain restrictions on the operations of the system, calling for review of the original recommendations for possible accommodation to such restrictions. These restrictions may arise from a need to economize, a change in management's goals, the elimination of one or more previously stated requirements of the system, or the abrogation of decision on the type of system (computer based or noncomputer based[1]) originally opted for by management.

In any event the adequacy of present procedures cannot be evaluated without thorough knowledge of what management expects the system to do, such as attaining maximum efficiency, highest productivity, maximum accuracy, and least operating cost. Because the confluence of these four factors, however acceptable and laudable as goals, is unlikely if not impossible, management is compelled to specify the primary goal. For example, if it is highest productivity management desires, it is likely that lower operating costs and maximum accuracy cannot be anticipated. Conversely, least cost is not entirely compatible with the goals of maximum efficiency, productivity, and maximum accuracy.

In the evaluation of current methods and procedures the analyst's report of findings and recommendations for revision, elimination, or maintenance of existing methods and procedures is based on specific determinants.

---

[1] The noncomputer-based system is defined as one employing conventional manual methods and/or any and all types of electromechanical equipment that may be activated to produce independently of a computer, required informational output by keyboard, punched card or paper tape devices.

1. There are requirements placed on the syst m that are germane to the successful operation and management of the system and of the total system of the library including those requirements logically contributing to the meeting of the requirements of related systems. To cite an example of the interrelationship of requirements among systems, a requirement of the preorder search subsystem of the acquisitions system is the supply of bibliographic information needed for ordering a title. In satisfying this requirement more frequently than not the searcher uses a source containing additional bibliographic data also needed in the cataloging system. The preorder search requirement should include the capturing of this data needed in the cataloging system in order to eliminate the costly, time-consuming repetition of the search.

2. There are inputs/outputs and controls of the system that adequately fulfill its requirements within the total system. Evaluation of the system's inputs/outputs is (a) to determine that information is received when needed for timely decision and action and (b) to eliminate the production and supply of information serving little or no purpose in meeting the demands placed on the system. The matter of balances in book fund accounts is taken as an example. Without accurate and timely reports of the status of accounts, control of expenditures in an orderly and equable manner throughout the budget year cannot be maintained. Similarly overexpenditure in given accounts can be discovered too late for any corrective action, leading to disruption of the system's functions and effective administration and planning. Obviously in the latter case the report management needs is one that indicates what and when individual accounts reach predetermined minimums for exercise of decisional control at the proper time. In further illustration book purchase requests submitted in variant form unrelated to the logical arrangement of bibliographic data best suited to searching and ordering procedures can only result in slowed and inefficient manipulation. What is needed here is submission of the book request in standard format.

3. Further determinants are those current methods and procedures judged adequate for processing the work loads. Again using the acquisitions system as an example management's goal is to attain a processing rate of 200 orders per week, which is at present not being achieved by current procedures and staff. The existing procedures must be evaluated with respect to their applicability and need in meeting the stated goal. On the one hand certain methods and procedures found unnecessary can be eliminated, thereby allowing the current staff to satisfy the desired processing rate; on the other hand, with present methods and procedures judged to be valid, additional personnel is indicated in order to reach management's objective.

4. The staff must be capable of fulfilling the system's requirements under the currently applied methods and procedures. Involved here are size of the staff and the capabilities, capacities, and skills or special training of individual workers. Depending on the results of staff evaluation additional workers with the necessary skills may be needed or perhaps workers should be reassigned to jobs fitting their levels of competency.

5. The equipment available must allow application of efficient methods. Changing, replacing, or adding a piece of equipment frequently leads to greater efficiency and improved productivity of a system. The application of efficient methods to required cataloging procedures by the addition of a simple piece of equipment is illustrated. Instead of manually copying cataloging data from the *National Union Catalogue* to produce needed cards for the library's catalog, a quick-printing enlarging camera can be procured to produce a near-standard image of the *National Union Catalogue* entry which in turn can be used as master copy to produce the number of copies of the card needed in cataloging by copying equipment. The addition of clerical help may be avoided by the increased productivity of the current staff through more efficient methods made possible by introducing one piece of relatively inexpensive equipment.

In arriving at the foregoing determinants the analyst uses the following "tools" obtained in the planning stage and analysis phase of the systems study: (a) managerial statement of goals enunciated in the planning stage of the study; (b) existing procedural manuals including job analyses and work samplings; (c) the worksheet prepared on each requirement; (d) preliminary survey worksheets; (e) worksheet prepared for each output; (f) summary worksheet prepared for all requirements and their associated outputs; (g) the worksheet prepared for each input; and (h) the summary worksheet summarizing all inputs.

When completely executed, the worksheets contain all the data the analyst needs for the evaluation of current procedures. It remains for him, prior to submitting his report of evaluation, to summarize systematically this mass of data supported by flow charts of existing operations.

Having come to some conclusion about the adequacy of the present system, the analyst should (a) compute the cost of processing a unit of material through the system; (b) measure the productivity of the system; (c) determine whether the system provides data promptly for timely action; and (d) evaluate the accuracy of information supplied.

During evaluation the analyst reviews the answers he has received in the analysis phase of the study and considers additional questions arising in his mind in the course of arriving at his initial conclusions. In testing his findings and conclusions the analyst may use the flow chart as a simulator of the system and its components. By following the flow of work through its associated functions, decisions, and actions he verifies or disproves the results that can be obtained under various operating conditions. This process, with examples, is now given.

1. Each component, as well as the overall system, should be appraised under increasingly stepped-up workloads to determine the maximum capacity of the system at which it can continue to fulfill its requirements. Does the present workload and accompanying cost meet management's goals? In the acquisitions system three factors patently affect the workload—the number of requests received, the number of orders processed, and the number of books received. The magnitude of workloads throughout the system depends on the volume of book purchase requests. By varying the number of requests entered into the system the ability of the subsystems (preorder searching, ordering, receiving, accounting, and reporting) can be evaluated and their maximum capacities estimated.

In the circulation system two factors affect the workload here—the number of books charged out and the number of books returned. The magnitude of the workload throughout the system depends essentially on the volume of books loaned. By increasing this number and noting its corresponding effect on the operations of the circulation system, such as the sending of overdues or shelving of books, the ability of the subsystems to meet the increased volume can be evaluated.

2. Current reporting requirements should be appraised to determine whether they stipulate statistical or other informational detail pertinent to the needs of management at various levels. Does management need all such reports or can the principle of management by exception be applied so that it receives only the information required for decision making or taking justified action? For example, in the cataloging system the supervising librarian need not be given a report on the size of the backlog of books to be cataloged so long as the size of the backlog stays beneath a predetermined level. As soon as the backlog has reached the problem zone management is notified.

3. Each input/output should be appraised for sufficiency in meeting the internal and external requirements placed on it. Do the outputs of a comptroller's office furnish sufficient financial data to satisfy the requirements of the acquisitions system? If not, what recommendations should be made to correct this? If a total system is one of management's goals, do the present outputs of the preorder search subsystem furnish the necessary bibliographic data required by the cataloging system and are the data furnished in ready-to-use form? If not, what modifications can be made in the form of the outputs of the preorder search subsystem; what effect would such modifications have on efficiency, capacity, and cost of the preorder search subsystem's operations; what effect would such

modifications have on the number and kind of staff needed in the subsystem?

## MANAGEMENT BY EXCEPTION

Management by exception[2] is a technique of scientific management broadly applied in industry and business but generally not so consistently or consciously applied in the management of libraries. The technique is applicable at all levels of responsible action within the library's organization. It is the principle whereby management receives only that information on which action is indicated. To apply this principle it is necessary for management to be able to specify the information required for taking action and to set the point at which such action may be taken.

The timely supply of information about what *has not* happened under stated conditions rather than information on all that *has* happened necessitates a sifting out of the "exceptional" data from the mass of data not requiring action. Only that information signaling the need for corrective action, adjustment, and decision is supplied. Before management by exception can be applied, the conditions or parameters of control actually required for the exercise of assigned responsibilities and decision making should be precisely delineated.

The technique can find only limited application in manual procedures because of the inordinate consumption of time in manually reviewing data files item by item. However, the extent to which this technique can be used in computer-based systems is limited only by ability and imagination in defining the parameters of operational control needed. Among the following examples the objectives are not new but most of them suggest action not economically feasible by manual methods. The process of claiming issues of periodicals not received is an exercise in management by exception. Serials librarians have done this from time immemorial but the missing factor has been timeliness of action to assure reasonable receipt of the claimed issue due to laborious hand methods. If computer-based serials control is applied, a computer report may notify the serials librarian of all issues of foreign journals not received in a predetermined interval and produce the claim forms for these issues. As another example, the director may be troubled about a seeming overload in the acquisitions system. Trying to pinpoint the reason for an increasing backlog of book

[2] Aspects of management reporting by exception and management's informational needs are discussed in many sources including C. L. Littlefield and Frank Rachel, *Office and Administrative Management,* 2nd ed., Prentice-Hall, Englewood Cliffs, N. J., 1964, pp. 83-104.

orders, he does not ask for a record of all requests searched but just for those representing books already in the collection. Thus he and the acquisitions librarian by applying the principle of management by exception find that one third of the searches are for books already in hand or on order, indicating a significant reason for the overload in acquisitions. Again, the director of the library does not need to know the current balances of all book accounts each month but he does need to know in time for taking action that a given account has fallen to a predetermined minimum; or he may wish to see monthly reports on the balances of all book accounts beginning the seventh month of the fiscal year in order to decide whether the same rate of expenditures can be continued for the remainder of the year.

The concluding examples of the exercise of management by exception indicate computer rather than manual handling. The acquisitions librarian does not need a weekly report of all books ordered but rather a continuous report of items on order for a predetermined period and not yet received at the time expected. For keeping a measure of the extent of the cataloging backlog and for taking action on books for which Library of Congress catalog cards have not been received within a predetermined period, the catalog librarian needs a report of the number of books in process and a list of the titles to be originally cataloged in the absence of Library of Congress cards rather than a nonselective listing of all books in process.

In evaluating the system the analyst should test the reports received at each level of management to see if the principle of management by exception can be applied under current procedures and if applicable, what the effect is on the efficiency, capacity, and operating cost of the system. The design of a system should eliminate unnecessary reports and substitute reports of information actually needed for specified conditions requiring surveillance for action.

## REPORT OF FINDINGS AND RECOMMENDATIONS

Finally, the study staff should submit to management a written report precisely stating its findings, conclusions, and recommendations. The study staff should provide enough detail to enable management to make the correct decision in relation to overall organizational goals. Depending on these recommendations and the decisions of management the study staff may or may not proceed to the culminating step of a systems study— that of design.

It is customary to think of reports as formal documents with such attending characteristics as formal language, exceedingly impersonal tone, great care in the making of claims and conclusions, inappropriate jargon, and inflated writing using redundancy and repetition. These characteristics can be obviated by using simple, direct language to state what needs to be stated, and no more.

Ideally the analyst in the actual writing of the report should have already, in a sense, written most of it because he has collected and organized all of the data needed to prepare the report of findings. He can follow his own investigation, can put together those parts of the report that contain information best known to him, and can finish the report by filling in any gaps from the information he has already recorded. The analyst should be able to state in specific terms in writing what the objectives of the study were and what was accomplished by the study. The accomplishments of any study should be statable in specific terms within the compass of a short paragraph of two or three sentences. This paragraph will become, then, the topic paragraph of the conclusion section of the report and as such is the most important statement in the entire report.

The task of filling in the gaps of secondary information will take a good deal of the time required for writing the draft of the report. This part of report writing can be difficult if the analyst does not have sufficient orientation to the organization of the report.

## ORGANIZING THE REPORT OF FINDINGS

Some authorities seem to think it is detrimental to list the generalized outline of a report but if the writer will realize that the outline is generalized and that it should be used judiciously and changed when he believes it is necessary to change, a good outline is very useful. The outline follows:

Front matter
    Cover with title and identification
    Title page
    Tables of contents and figures
    Foreword, preface, and acknowledgements
Body
    Introductory summary
    Statement of problem
        Objectives of study

      Scope of study
      Possible outcome
    Method(s) of study
      Planning and schedules
      Personnel
      Equipment and materials
      Procedures
    Results
      Collected data
      Interpretation of data (reduction)
      Method(s) of data analysis
    Conclusion
      Conclusions
      Recommendations
      Implications
  Back matter
    Appendixes
    Detailed data sheets
    Computations
    Large figures and drawings
    Appended documents and correspondence
    Bibliographic matter
    Index (only if report is voluminous)
  Back cover

## SEQUENCE OF WRITING

One of the hazards resulting from an outline such as the one above is the inference many people draw that the actual writing of the report should be done in the order indicated. A more appropriate order is suggested here.

The first piece to be written could be that labeled "Statement of Problem," and in particular the subsection "Objectives." The next section should then be Conclusions, followed by Recommendations. Then the Results section can be sketched out fairly fully and Method(s) come next. From here on the order of the writing does not make as much difference as in the earlier parts, with two exceptions. The Introductory summary should be very nearly the last section to be actually written. This section, frequently called "Abstract" or simply "Summary," is the most-read part of the report and should therefore be the best. The last thing, usually, is putting together the table of contents and other similar tables. The table of contents should be started with an outline such as the one here, with appropriate changes made to that outline as the report is developed, and with the notion that the outline will indeed become the table of contents by the addition of page numbers.

## POINT OF VIEW

The report writer needs to know what point of view he should use in the writing. It is most likely that he will need a dualistic point of view: first, he is the expert communicating to peers; second, his writing should be organized from general to particular; it should have the inverted pyramid style of the news reporter. The writer of the report has done the investigation that he is reporting; he should know more about it than anyone else. On this basis alone he is the expert. The general-to-particular organization of the writing itself is called for because it is true that more than one person will make real use of the report and that these uses themselves will vary and that any one reader will not 'have to or want to read all the various sections of the report. If every section and subsection is so organized, all readers will be able to grasp the total import of the report without having to struggle through every paragraph.

The best device to indicate the orientation of the report to the reader is the liberal use of headings. The organization and selection of these begin with the early outlining of the report. Just as the outline for the report becomes the table of contents, it also serves as the basis of the list of headings. The writer should take advantage of this in checking several times during the actual writing to see that he is using the outline to indicate headings and to check and change his outline as he includes additional headings. An excellent discussion of the formating of headings is included in Ulman and Gould.[4]

## FUNCTIONS OF PARTS OF THE REPORT

Following are brief statements of functions of the individual parts. The order here is that of the outline listed earlier.

*Cover.* Two functions are performed by the cover. It physically holds the report together and provides sufficient identification for the

[3] J. N. Ulman, Jr., and J. R. Gould. *Technical Reporting,* Rev. ed., Holt, New York, 1959, pp. 17-19.

locating and handling of the report as a document; the cover should provide for easy access into the report itself through convenient binding in terms of the thickness of the report.

*Title Page.* The title page contains the informative title and complete identification as to author's name, position, organization; date and place of publication; place of the report in any series, including serial enumeration, if any. It is usually desirable that all the information on the title page be included on the cover.

*Table of Contents, Illustrations, Figures.* Informative, complete, and highly utilitarian guides to the reader as to location, parts, size, and interrelationships are supplied by the front material tables. Such tables should be made as legible as possible and should occupy as much space as is needed to assure completeness and legibility. They should not be crowded.

*Foreword, Preface, Acknowledgements.* These sections are considered optional; they should be used to indicate something about the justification for the writing of the report, any special circumstances that tend to make the report different from what the reader might expect, and any unusual help given the writer/analyst. If these sections are used they should be kept brief.

*Introductory Summary.* Most readers appreciate a summary that is not a linear abstract. The summary should give short versions of the important parts of the report, particularly the objectives of the study, the method of study (this should be very brief), the specific results, the conclusions, and the recommendations. If the writer has followed carefully the general-to-specific order of writing, the beginning sentence or two from each of these sections will serve as the essence of the summary. Many writers indicate that the summary can best be written by simply repeating such sentences verbatim and providing adequate transitions. The introductory summary should be considered by the writer as a document that may have to stand in stead of the entire report; it is therefore a very important part of the report and should be worked on with great care and a good deal of energy and thought.

*Statement of Problem.* This section defines the study (it should not define the report except as an incidental matter). It is probably the second most important part of the report.

*Method(s) of Study.* In many reports this is an invaluable section but if the method is routine and well known, the section may be quite brief and even nonexistent.

*Results.* Although many report writers consider this section as very important, often it is not and frequently it is not even read thoroughly, if at all. One of the problems here is that "Results" is confused with

"Conclusions." The only practical function of this section is to state in abbreviated form the actual data collected, so that the reader, if and as he wants, may verify the conclusions from the actual observations.

*Conclusions.* The conclusion functions as the answer to the question, "Precisely what was accomplished by this study?" If this question is not well answered, the reader is likely to gain the impression that not much, if anything, was accomplished by the study.

*Back Matter.* Customarily the back of the report serves as the repository for all the other material that did not seem to fit into the body of the report. This is too often the case. Instead, the back of the report should contain items of information that may be of good use to some readers. The analyst should attempt to make a reasonable judgment as to the potential utility of the material and if in doubt should not include it.

## USE OF ILLUSTRATIONS AND GRAPHICS

If the information to be presented concerns primarily a description of a space, such as a reading room or a set of offices, and if the information is primarily dimensional, a dimensioned drawing such as a floor plan is more communicative than words. The argument can be carried on to other characteristics besides dimensions the conclusion being that other forms of illustrations (sketches, perspective drawings, flow charts,) and so forth may prove superior in their communicability to words. When this is the case, the "nonverbal" may be used first and supported by verbal interpretations.

A table containing specific numerical values is easier to use than a graph from which certain values must be derived. If the specific numerical values are not the major concern but trends and comparisons are, bar charts, column charts, and even pie charts are much more communicative than tabular material.

Most illustrations and graphics should be introduced with text material and it is probably best that the writer follow the nonverbal with a verbal interpretation. This means that the nonverbal material will be worked into the text of the report linearly—this is, so that the reader is almost forced to study the nonverbal as he reads the verbal material. Illustrations and graphics should not be located in other parts of the report, even if only a page away. A discontinuity occurs when this happens and when the reader gets to the page where the nonverbal material is, he tends to look immediately at it rather than at the continuation of the verbal material he is reading.

Finally, nonverbal material should be presented in good style. The art work should be neat and professional. Legends should be plentiful, accurate, and informative. In regard to the use of mathematical expressions over English, Kapp states ". . . every writer, before embarking on mathematical forms of presentation, should ask himself the question: is your mathematics really necessary?"[4] Keeping in mind who the readers of the report may be, the use of mathematical formulas may not be the best means of communication. Rather their use may complicate decision making by management to implement the report's recommendations.

Douglass,[5] Schmid,[6] Sigband,[7] Tichy,[8] Ulman[9] and Wyld[10] are considered excellent treatments of report writing and contain good bibliographic material.

[4] R. O. Kapp, The First Draft, in *Computer Peripherals and Typesetting,* by Arthur H. Phillips, HMSO, London, 1968, App. 3, p. 611.

[5] Paul Douglass, *Communication Through Reports,* Prentice-Hall, Englewood Cliffs, N. J., 1957.

[6] Calvin F. Schmid, *Handbook of Graphic Presentation,* Ronald Press, New York, 1954.

[7] Norman B. Sigband, *Effective Report Writing for Business, Industry, and Government,* Harper and Row, New York, 1960.

[8] Henrietta J. Tichy, *Effective Writing for Engineers, Managers, Scientists,* John Wiley, New York, 1966.

[9] J. N. Ulman, Jr., and J. R. Gould, *op. cit.*

[10] Lionel D. Wyld, *Preparing Effective Reports,* Odyssey Press, New York, 1967.

# Applying Systems Analysis Concepts
## Design of and Decisions for New Systems

ORGANIZATION AND PLANNING;

AND, CONDUCT OF

THE SYSTEMS ANALYSES

*James W. Greenwood, Jr.*

SOURCE: From James W. Greenwood, Jr., *EDP: The Feasibility Study; Analysis and Improvement of Data Processing* (Systems Education Monograph No. 4) Washington, D.C., Systems and Procedures Association, 1962. Reprinted by permission of publisher. Copyright © 1962 by Systems and Procedures Association. (now Association for Systems Management).

<u>Chapter 1</u>

ORGANIZATION AND PLANNING

A.  ORGANIZING THE STAFF

Data processing activities usually are not restricted to individual organizational

units but tend to cross organizational lines.  Further, the products of data processing

are used on an organization-wide basis and have broad management implications.  For

these reasons and others, the head of each organization needs to take a direct interest in

the analysis and improvement of data processing.  He should give his full support to the

study and should ensure the full support of his top staff.

For the actual conduct of the analysis he should designate a responsible official

within the organization and assign an adequate staff to assist that official.  The staff mem-

bers assigned to the analysis should be organized into a Survey Staff and an appropriate

number of Task Forces according to the amount of work involved.

Both the project leader and the members of his survey staff and the task forces should

be thoroughly familiar with organization objectives, policies, activities, procedures and

management problems.  Most of the members of the survey staff and the task forces should

have a knowledge of, and skill in the use of, management analysis techniques.  At least some

of the members of the survey staff should also have some understanding of the general capabili-

ties and limitations of the various types of office machines and data processing equipment.

a. <u>Survey Staff</u>

A survey staff, under the leadership of a responsible official reporting directly to the

head of the organization and consisting of top level management or program analysts or

their equivalents, should serve in a continuing capacity as the planning and coordinating group for the conduct of the systems analysis and improvement.

b. Task Forces

A number of task forces conducting specific studies under the technical direction of the survey staff may also be assigned. Each should have a specific responsibility for the conduct of the systems analysis and improvement in a defined organizational or functional area.

## B.  ORIENTATION AND TRAINING

All members of the survey staff and task forces should secure a basic orientation in systems analysis, data processing techniques, and machine capabilities prior to starting the survey. Regular courses in these subjects are offered by many universities.

Selected members of the survey staff should also secure training in programming of electronic accounting machines and electronic computers.

## C.  PLANNING

In order to ensure the most effective use of available resources and to provide a basis for a logical approach, the organization should have a detailed plan for the conduct of the systems analysis and improvement.

After appropriate orientation of its own members, the survey staff should prepare such a detailed plan.

By way of preparation for this step, the survey staff will need to collect and organize in summary form some information of a general nature about the organization, the activities, the operations, the data needs, and the current costs of data processing. This information will serve as a basis for drawing up the plan for the systems analysis and improvement. The detailed plan will include the following steps:

1. Define the Problem(s).

2. State the Objectives.

3. Define the Scope.

   (a) Identify specific activities.

   (b) Identify organizational units.

   (c) Identify specific data processing operations.

4. Determine the Methods to be Used.

5. Plan the Organization.

6. Establish a Time Schedule.

Each of these items is discussed in more detail below:


a. Define the Problem(s)

The survey staff, in drawing up a detailed plan, should first identify and describe the specific problem (or problems) to be studied. The problem definition should be in sufficient detail to serve as a guide to the members of the task forces and to inform other members of the organization who will be involved in, or concerned with, the activities to be studied.

The problems may be simple or complex. For example, one problem might involve a simple process of gathering and compiling statistics on production, or type of accident, or sales. Another problem might involve a complex process of integrating the flow of large volumes of data in several separate operations concerned with different but related activities.

Whatever its nature, each problem should be specifically defined so as to focus

the attention of the task force and to facilitate the study.


b.  State the Objectives

The basic objective of the analysis and improvement of data processing is set

forth in the Introduction.

In addition to this basic objective, each individual study of a problem will have a

number of specific objectives.  The survey staff should spell out these specific objectives

in terms of the problem to be studied.  These objectives will vary considerably from

study to study.  Some specific examples might be:  increased speed of data compilation,

analysis and publication; reduced volume of files; improved quality of reports; prevention

of further increase in clerical and overhead costs; introduction of common language in

related but separate data processing operations; integration of separate systems, in-

creased mechanization of routine decision-making.

Each objective should be clearly and explicitly stated; and the listing of objectives

should be as comprehensive as possible so that the task force may have clearly in mind

just what it is expected to accomplish.


c.  Define the Scope

A definition of the scope of the survey will identify the particular data processing

operations to be studied, the organizational units in which the operations are performed,

and the activities involved.

The areas of study should be as sharply defined as possible so that the survey staff

may properly focus its attention and so that the task forces may be given non-overlapping

assignments.

In addition to identifying the areas to be studied, the survey staff should list the studies in the order in which they will be made. This will establish the priority so that the most important or most urgent studies will be given earlier attention. In assigning priorities to the several areas to be studied, the survey staff will consider among the more important factors the relative costs of the present systems of data processing, and the potentiality of effecting savings or improvements.

Of course, both in the identification of the data processing operations to be studied and in the assignment of priorities, the survey staff will need to provide for some measure of flexibility. This is essential because of the inter-relationships among activities and data processing operations and because of the continually changing nature of the organization, activities, and operations.

d.  Determine the Methods to be Used

In order to ensure a logical, systematic analysis, as well as consistency of results and comparability of the findings of the several task forces, the survey staff should determine and specify the methods and techniques to be used in the study and the skills and facilities which will be required. The survey staff should include in the assignment of each task force:

(1) A detailed outline of the specific types of information to be secured.

(2) A list of the sources of such information, including:

   (a)  The names and titles of persons to be interviewed.

   (b)  The published materials to be reviewed.

   (c)  The source records to be examined.

   (d)  The equipment installations to be observed and inventoried.

(3) Specific instructions as to the manner of presentation and format of reports, charts, tables, graphs, and other materials.

e.  Organizing the Work

The foregoing steps will enable the survey staff to estimate the amount of work involved and the type and amount of resources needed to do that work.  Determining how to most effectively utilize those resources is the next step.  These are the activities involved in organizing the work.  More specifically, the survey staff should:

1.  Estimate the amount of work involved in each task force assingment in terms of man-days or man-months, by the various types of skills required — e.g., management analysts, clerks, typists, statisticians.

2.  Determine the resource needs in terms of:

(a)  Number of personnel by classification, with an indication of the estimated duration of each assignment.

(b)  Types of skills needed from other sources on part-time or consulting basis.

(c)  Types and amount of special equipment, supplies, space, if any.

3.  Determine the number, size, composition and duties of the necessary task forces.

4.  Determine the respective roles of the members of the survey staff.

5.  Develop plans and procedures for:

(a)  Making assignments to the task forces.

(b)  Coordinating the work of the task forces.

(c)  Preparation and review of task force reports.

(d)  Preparation and presentation of survey staff reports.

f.  Establish a Time Schedule

The final step in the planning stage is the establishment of a time schedule.  The survey staff should establish for each individual study or task force assignment, and for the work of the survey staff as a whole, a list of specific, identifiable stages with a target date for the completion of each stage.

g.  <u>Prepare the Announcement, Estimate Costs</u>

When the survey staff has completed its planning, it should prepare for the head of the organization a draft of a formal announcement of the proposed analysis.  The contents of such an announcement are suggested in the next chapter.

At the same time the survey staff should present to the head of the organization an estimate of the costs of the proposed analysis.

Chapter 2

CONDUCT OF THE SYSTEMS ANALYSIS

A.  EXECUTIVE ACTION

The head of the organization should review the proposed plan and, when it meets

with his approval, make a formal announcement of the study as indicated below.  After

the formal announcement, the survey staff should assign the task forces to begin the

systems analysis.

a.  Approve and Announce Plan

The head of the organization should carefully review the proposed plan with the survey

staff and such other officials as he may designate.  When he has approved the plan, he

should formally announce the study to all members of the organization.  The announcement

should:

1.  Indicate the reasons for the study.

2.  Set forth the principal objectives of the study.

3.  Identify the expected benefits of the study.

4.  Solicit the cooperation of all the members of the organization.

5.  Give appropriate assurance to all members of the organization on the subject

    of job security.

b.  Coordinate Activities

The head of the organization should make provisions to insure that:

1.  The conduct of the analysis will be coordinated with activities of a related nature

and will be in conformity with established policies.

2.  Reports will be made to him at regular intervals on the progress of the study.

3.  Appropriate consultation and cooperation will be had at all levels in the organi-
    zation throughout the course of the study.

## B.  SURVEY STAFF RESPONSIBILITY

a.  Inventory Current Systems

The aim of this and the following steps is to secure the facts about the data processing
operations currently being performed.

The primary purpose of the analysis is to provide a sound basis for developing improvements in the data processing system.  In order to provide such a basis, the task forces
should:  (1) secure sufficient information and so organize it as to ensure a clear understanding of the current data processing system; (2) isolate and define the strengths and
weaknesses of the system and any specific problems; and (3) evaluate the system in terms
of sound management principles.  The analysis will include:  inventorying current data
processing operations and equipment; recording the current practices and procedures; and
evaluating these practices and procedures.  The detailed steps are set forth below:

1.  Identify and list for each activity or organizational unit the current data
    requirements:

    ( a )  What information is needed?

    ( b )  In what form?

    ( c )  At what times?

    ( d )  For what purposes?

2.  Identify and list for each activity or organizational unit the current data processing
    operations by major processes (e.g., payroll, production, sales, safety).

b.  Record Current Procedures

The task forces should record all of the pertinent facts which will be needed for
the analysis of the current practices and procedures utilized in each of the data processing operations listed in the above-mentioned inventory.  The types of information needed
and the sources of information are indicated below.

c.  Information needed

For an adequate analysis of current data processing practices and procedures, the
task force should record the following types of information:

1.  The specific details of the data requirements listed in the above-mentioned
    inventory.

2.  The general functional and structural organization of each organizational unit
    included in the study.

3.  Detailed functional and structural organization of each organizational unit devoted
    to data processing, with an indication of relationships to other personnel and positions
    involved in data processing.

4.  Staffing of data processing units.

5.  Step by step practices and procedures showing the operations as actually performed
    and the actual flow of data.

6.  Sample copies of all forms and reports.

7.  Equipment utilization.

8.  Personnel utilization.

9.  Office layout.

10.  Workload; and workload distribution.

11.  Costs of operation.

12. Inter-relationships of current data sources, data processing operations, reports and other forms of data output.

d. Sources of Information

1. Organization charts.

2. Manuals.

3. Other instructions: procedures, bulletins, circulars, etc.

4. Accounting system.

5. Project files.

6. Minutes of meetings.

7. Reports—regular and special.

8. Correspondence files.

9. Forms files.

10. Internal audit reports and files.

11. Direct observation of operations.

12. Interviews with operating personnel, supervisors, data users.

# Applying Systems Analysis Concepts
## Design of and Decisions for New Systems

SYSTEMS CONCEPTS

AND

LIBRARIES

*Frederick G. Kilgour*

SOURCE: From Frederick G. Kilgour, "Systems Concepts and Libraries," *College & Research Libraries* 28 (1967) pp. 167–170. Reprinted by permission of the author and of the American Library Association, the publisher.

Unhappily, the word "system" has acquired so many meanings that it has become a coat which fits nearly every wearer. The classic library system consisting of a group of libraries or a major process such as a circulation system is not the system of the modern systems analyst. A system is an on-going process that produces some wanted operation and is thought of as a whole rather than as an assemblage of pieces and procedures. A system is a dynamic event. The telephone system of the United States and Canada is an oft and appropriately cited example of a huge, complex system designed of many subsystems to function as a totality. Any one of its more than ninety million customers can communicate with any other no matter with which company he is a subscriber or whether he has an ancient magneto telephone or a modern pushbutton instrument.

Modern systems, in addition to being all-encompassing, on-going processes, are information based. Many systems have information in their processing phase, such as a telephone message or a book in a library system. The important aspect of the information base is not the information in process in the system, however, but rather information about what is in process. It is the use of this information, about the form of information in process, to control the system, that distinguishes the modern system from a collection of procedures.

The computer is the instrument of choice in processing information and is obviously the device to employ in monitoring and controlling an information-based system. The computer culls information from the processing and calculates action to be taken, or not to be taken. Although modern systems thinking began to arise about half a decade before the computer was generally available, the rapid extension of systems thinking had been due mostly to the use of computers.

Techniques for designing new library systems do not differ much from older techniques. What is different are the questions that are asked, and the subsequent answers obtained. In particular, it is necessary (1) to formulate the problem, clearly, (2) to choose appropriate objectives, (3) to define factors relevant in the environment, and (4) to employ ingenuity in inventing new systems and segments thereof. It is also important to judge accurately the reliability of cost and other data, but this judgment was equally important in setting up efficient procedures.

Tests to be employed in trade-off studies among proposed procedures design include (1) quality, such as the quality of files, (2) performance, such as need and reliability, (3) compatability, as with other systems and input data, (4) flexibility, to achieve future expansion, (5) simplicity, particularly in regard to operation, (6) time required to install the system, and (7) costs, including initial costs and repetitive costs.[1]

Systems are often spoken of as being "total," but it is not always easy in practice to achieve totality. Nevertheless, a system should be thought of as being comprehensive. Examples of such comprehensiveness are the telephone system of the United States and Canada to which reference has already been made, and the power grids which also include both countries.

Management-information components should be designed into systems to facilitate direction of the system. Here the information processor produces information reports to management, so that management has an accurate picture of events in the system. Management in turn may change parameters in the information processor to alter control of the system. Similarly, management may shift human components in the manual processing organization, or even change that organization.[2]

Feedback is present in such systems since the information-processor, or computer, controls the flow of materials in the system on the basis of information derived about that flow. The process of the use of management information described in the previous paragraph is a similar example of feedback in the system.

A library, a group of libraries, or a national library network can be viewed as an information transfer system. A single library has at least four major subsystems: (1) an information store; (2) communications; (3) real property; and (4) users. The information store consists of two types of information: (1) books, pamphlets, serials, manuscripts, reports, films, records, and pictures; (2) card catalogs, index and abstract journals, printed catalogs, and bibliographies. The communication subsystem could contain telephones and automobiles; real property, buildings and grounds. The user is the customer who generates information transfer with his needs for specific information.

From the systems point of view, these four major subsystems meld into a whole but can also be further subdivided into lesser subsystems. It it important to observe that in the library building design process, all four subsystems are treated as an integrated whole, but that the design of information transfer, the design of the catalog and book arrangement do not specifically integrate with the modern user subsystem. This dislocation exists because there has been no major redesign of cataloging and classification processes during the past half century. Rather, cataloging and subject classification are currently used on a conglomerate of a priori bibliographic, classificatory, and economic principles largely developed in the four decades preceding the first World War. Librarians strive to give the best possible service with such organizations of catalogs and books to each user when he appears, but the time has come to analyze acquisition, cataloging, and classifying library information in terms of users' needs—as integrated with the user subsystem.

The modern library user has impressively different needs from those who

[1] A. D. Hall, *A Methodology for Systems Engineering* (Princeton: Van Nostrand, 1962), pp. 105-107.

[2] A. M. McDonough and L. J. Garrett, *Management Systems; Working Concepts and Practices,* (Homewood, Ill.: R. D. Irwin, 1965).

used libraries at the turn of the century. Information needed by modern users occupies a new position that is central in our society and has become a major national resource—a position which it has come to enjoy only since World War II. From a systems viewpoint, it appears that the library information store and its processing have not kept pace with the changing status of information and information users. Or to put it another way, during the last half century, library science—relative to library use—has become increasingly a priori in principle and increasingly self-existent in practice.

The advent of that remarkable information-processing machine, the electronic digital computer, forces an examination of the library as a system in the modern sense of that word. If it is correct that library information processing has not kept pace in principle with library information usage, computerization of present-day library processes will achieve no more than would replacing a cart and horse with an automobile. Analysis of the library into four major subsystems suggest that three (information store, communications, and users) should be computerized. But to achieve dramatically improved systems performance, it will be necessary to treat users as an integral part of the library and to do extensive investigations on the characteristics of users, on information needs of users, and on their uses of library information.

Investigation of use of library materials, of users' needs, and of characteristics of users will involve answering a host of questions. Once the user is considered part of a library system, it is then necessary to know something of his productivity and to evaluate that productivity—an evaluation which only a few have attempted.[3] It will also be necessary to

know users' costs. At the present time, cost effectiveness in libraries is judged only on the amount of money spent in processing one item in a given procedure. American libraries have increasingly tended to use cataloging, classification information, and subject-heading work of the Library of Congress because it is cheaper. The appropriate classification for the Library of Congress might be entirely inappropriate for a local library, however. Moreover, the amount of subject analysis done by the Library of Congress, and most other libraries, for that matter, has been reduced to less than an acceptable minimum because of the engulfing growth of card catalogs. This reduction in subject accesses to information has lessened the cost of library procedures, but it most certainly has increased user costs. Indeed, it is probable that this diminution of subject analysis has been one of the major factors in generating the current library crisis.

Little is known of the information needs of users and there are relatively few studies available from which to learn something of those needs. Recently, the Systems Development Corporation reviewed some four hundred and fifty such studies and concluded that there were only fifty-eight that had objective and reliable information.[4] Moreover, most such studies treat the user as not being a part of a library system, and therefore neglect the costing aspects. This neglect has led to the suggestion of systems taking a week or more to produce information, but it is clear without doing a study that if users' costs are to be kept at a minimum, this access to information should be in terms of minutes and not days.

There is also confusion as to what con-

[3] See, for example, Ben-Ami Lipetz, *The Measurement of Efficiency of Scientific Research* (Carlisle, Mass.: Intermedia, Inc., 1965).

[4] U.S. Federal Council for Science and Technology. Committee on Scientific and Technical Information, *Recommendations for National Document Handling Systems in Science and Technology*, Appendix A, (Springfield, Va.: Clearing House for Federal, Scientific, and Technical Information, 1965) II, 2-8.

stitutes "new" with respect to information supplied to users. Altogether too often, "new" is taken to mean "recent." Indeed, some studies of users' needs have concentrated on recent information as though it constituted all information that was new. New information should be interpreted as information new to the user; indeed, it may have been published decades ago but still be valuable and new. The confusion of "new" and "recent" has unnecessarily narrowed user-needs studies, so that older material has been unwarrantedly neglected.

Finally, design of modern library systems will profit from studies yielding knowledge as to how library materials are used. It is important to know which volumes are used and to what use they are put. Citation studies do not reflect use of information and these studies have tended to show that books were little-used. Studies done at the Yale medical library and elsewhere[5] have revealed, however, that scientists use books in significant numbers. It was also found at Yale that use of books by researchers was nine-tenths research related.

[5] F. G. Kilgour, "Recorded Use of Books in the Yale Medical Library," *American Documentation,* XII (October 1961), 266-69; Helen Kovacs, "Analysis of One Year's Circulation at the Downstate Medical Center Library," *Bulletin of the Medical Library Association,* LIV (January 1966), 42-47; L. M. Raisig, Meredith Smith, Renata Cuff, and F. G. Kilgour, "How Biomedical Investigators Use Library Books at the Yale Medical Library," *Bulletin of the Medical Library Association,* LIV (April 1966), 104-107.

Acceptance of the premise that the user is an integral part of a library system leads to extensive reorientation of design of library systems. Certainly, it is a goal of the new systems to supply the user with information when he needs it. If he needs it in his laboratory or his study, it should be available to him there. It may well be that librarians will find themselves engaged in an atavistic activity leading to systems in which the user will hardly, if ever, enter a library. It seems quite probable that before the end of the century, information servicing to faculty and students on new campuses of multi-universities will not involve a huge library building. Rather, there will be an information store available by consoles in laboratories, studies, and dormitories where faculty and students can have rapid access to information required.

Of course, universities already fortunate in the possession of huge numbers of library materials will continue to possess these holdings and increase them. These great holdings will be the source of the automated systems of the future.

In summary, the most novel aspects of new library systems will be their character of being information based and controlled by information processing machines. Also, they will include the user as part of the system, and the systems will be specifically designed for rapidly and specifically fulfilling users' needs. ■ ■

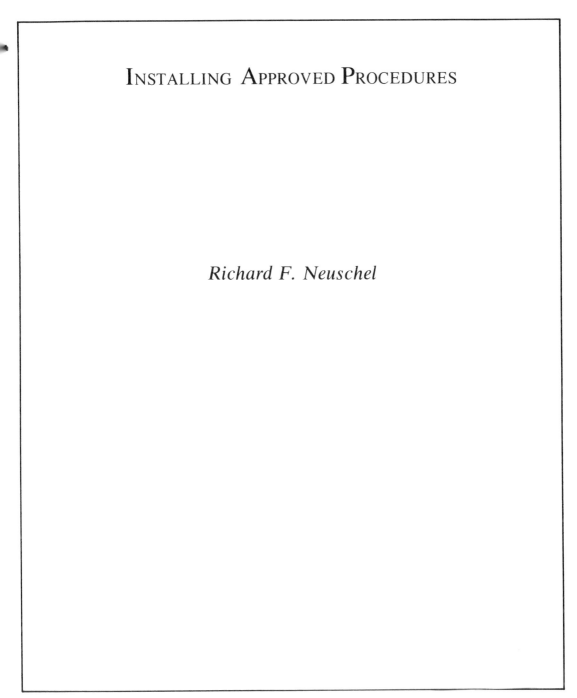

INSTALLING APPROVED PROCEDURES

*Richard F. Neuschel*

No other phase of procedures work is so consistently haphazard and imperfect as that of putting approved plans into operation. It avails little, of course, that the plans are workable if they are not made to work. And yet, in the author's experience, more procedure failures result from muddled, slipshod installations than from any fundamental defects in the procedure itself.

There are two principal causes of this condition:

1. Procedures analysts tend to feel that they have accomplished their mission when the line organization has accepted their recommendations. By the time this point is reached, they have been through the often long and difficult processes of research, analysis, development, and persuasion. Once the victory is won, the tedium of implementation seems anticlimactic to them. They are more eager to meet the challenge of new problems than to buckle down to the prosaic task of threshing out operating details. And since procedures installation is more closely related to the operating function than to the planning or staff function, analysts are inclined to believe that responsibility for making the change-over cannot be separated from responsibility for operating the new procedure once it is in effect.

Line supervisors, on the other hand, are concerned chiefly with day-to-day operations. They do not usually have the time, or often the background, to carry out the approved program by themselves. The analyst forgets that the plan which he understands thoroughly may still be only a skeleton outline to the operator. And because of these limitations, an earnest beginning is too often followed by halfway measures, imprudent compromises, loss of enthusiasm, and finally a bogged-down program.

2. Soundly conceived procedures may be doomed to failure be-

cause the time and preparation required to install them properly are greatly underestimated. It is a safe generalization to say that the installation of most procedure changes takes more time and hard work than all the previous phases of the project combined. On a major study where the survey, analysis, development, and presentation work took two months, the installation might take six. It matters little that the new procedure, forms, records, and reports have been developed in complete detail, for these are still only a blueprint—a picture of your ultimate destination. As the previous chapter suggested, installation means getting from where you are to where you want to be. And the job of bridging that gap is often much greater than is generally realized. Forms and equipment must be ordered and their availability coordinated; operating instructions must be written and tested and the personnel indoctrinated in them; the new procedure must be refined and detailed points of operating mechanics settled; the cutover must be planned; records must be converted; and preparatory changes must be made in related procedures. These are typical of the steps that must be planned and carried out in advance if the actual change-over is to be made without confusion, disruption of service, or a waste of the effort that went into the whole procedures study.

## RESPONSIBILITY FOR THE INSTALLATION

It has already been implied that the analyst who developed the procedure should play a leading role in its installation. The nature of his participation, however, should change in midstream, that is, between the task of planning the installation and that of actually carrying it out. During the initial step of planning the change-over, he should retain the initiative. He should continue to spearhead the program, even though at this point the participation of line personnel will be greater than in any previous part of the study. In this way, installation planning not only benefits from his singular knowledge of the new procedure but it also serves as the means for transmitting his knowledge to the operating supervisor. It is the transition step between the analyst's leadership of the project and the assumption of that leadership by the line organization.

Responsibility for actually making the installation belongs, of course, to the line organization. During the change-over the analyst may continue to keep a close watch over progress and results and to confer regularly with operating supervisors. But he has no direct authority over performance of the work except at the specific request of a supervisor who may consider it expedient to delegate such authority temporarily. Under this circumstance the analyst is temporarily accountable to his delegant, and his authority over the installation extends no further than the delegant's area of responsibility.

## PLANNING THE INSTALLATION

The steps involved in the planning of a major procedures installation do not differ fundamentally from those outlined in Chapter 8 for the initial planning of the study. That is, the total job to be done must be broken down into its components, and each must be assigned and scheduled.

**Timing the change-over.** As a beginning point, decisions should be reached on three related questions concerning the time at which the cutover to the new procedure will be made.

1. Must the new procedure be inaugurated at a particular time? This requirement arises frequently when the changes to be made affect the basic character of the accounting system or any other activity that should be kept consistent throughout the calendar or fiscal year. Where an inflexible deadline must be met for these or other reasons, the dangers of confusion, breakdown, or excessive installation expense are greatest. Because the available time is invariably all too short, the preparatory work must be planned and scheduled with extreme care and its progress followed just as closely.

2. Should the entire procedure be installed at the same time or should it be installed in parts? This question is closely related to the first, since changes that must meet a specific deadline must also be installed wholly at that time. Where this requirement does not exist, however, the analyst usually has a choice between concurrent and successive installation.

If the changes are not of major importance or too great in number, they can usually be made simultaneously without danger of serious disruption. A complex series of changes, on the other hand, should ordinarily be handled progressively so that the installation is not jeopardized by too rapid a rate of change. Obviously if too many adjustments are undertaken simultaneously—either at one or several places throughout the organization—the supervisory force cannot coordinate them properly or the work force assimilate them smoothly.

One simple but effective way of spreading the installation load is to separate those changes involving the elimination or combination of work from those involving new methods of operation. Usually changes of the first type can be made quickly and at a time when the installation crew is ordering forms and equipment, writing job instructions, and taking other preparatory steps. In this way much of the underbrush is cleared away and part of the operating personnel is freed to help with the changes still to be made.

Where the installation is undertaken one step at a time, the analyst must be careful that it does not spread over so long a period that momentum is lost or that the disruption or hardships accompanying the transition are unnecessarily prolonged. Another danger to be equally guarded against is the temporary loss of parallelism between related activities or transactions. In his article on the installation of a punched-card system, Chapin states: [1]

[One argument for] making a clean break from the old system [is the] close interrelation of various records such as cash and accounts payable, accounts payable and distribution of expenditures, pay roll accrual and pay roll distribution. In other words, if certain data on punched cards represent a debit entry, it may be helpful to have the offsetting credit entry on punched cards also.

These dangers are real but they can be avoided. They do not argue against progressive installation so much as they emphasize the need for skillful installation planning.

3. The third question to be answered is at what point in the installation of the new procedure should the existing procedure be discontinued? As a protection against unanticipated problems or hidden weaknesses in the new procedure, should both the new and old be operated on a parallel basis for a given period? If not, should some parts of the old procedure be temporarily retained? Decisions on these points should be made early in the planning process, for they will materially affect the nature of the preparatory work, particularly in the selection and training of personnel.

In general, parallel installations should be avoided wherever possible, for they always mean added expense and often create delays

---

[1] Troy A. Chapin, Jr., "Some Problems of Installing a Punched-card System," *N.A.C.A. Bulletin,* Vol. XXVII, no. 3, 1945, p. 112.

in the processing of source documents. Again, skilled planning should greatly reduce the hazards of installation and the need for clinging to the old procedure. Trial installations are often advisable, but they can be confined to a small sample of the work so that complete duplication of procedure is avoided. As Chapin points out: [1]

> Should the test runs not have been conclusive, there is more reason to duplicate the work by continuing the old system for the first month or two, but under such conditions there may be a question as to whether the conversion should be made at that time or delayed until plans have been more completely worked out.

**Planning the preparatory work.** When these questions of timing the change-over have been decided, the next step is to determine what preparatory work needs to be done before the new procedure can be put into effect. This is essentially a job of listing all the measures that must be taken to equip and train the operating personnel for carrying out the new routine. One reason why so little time is usually allowed for performance of this work is that analysts neglect to think it through fully when the installation plans are laid. They either overlook important parts of the job or fail to break it down into sufficient detail to make realistic scheduling possible.

The inexperienced analyst can minimize these dangers by reviewing his approved procedure plan step by step and by asking the following questions about *each* of its *new* operations:

1. Is this operation described fully enough to cover every contingency, or does it need further testing, refinement, or amplification?

2. What new equipment, forms, supplies, work aids, or other clerical tools are needed to carry it out?

3. How much indoctrination will the worker need to perform the operation satisfactorily?

**Fixing responsibility and scheduling the work.** Having determined specifically *what* work needs to be done, the persons responsible for the installation should complete their plans by deciding *who* will carry out each step and *when* it should be started and finished. The analyst must let the operating supervisors make most of these decisions, for the installation timetable will be difficult enough to meet without risking the danger that they feel no responsibility for having set it up. Nevertheless, the analyst can be of real help to the

[1] *Ibid.,* p. 114.

supervisory force by urging that the following requirements be met when responsibilities are assigned and schedules prepared:

1. The key individuals on whom the success of the installation will depend should be relieved of enough other responsibilities to ensure adequate manning of the job.

2. Where no inflexible deadline must be met, the actual cutover should be scheduled to take place when the minimum of confusion will result. However well operations are planned and controlled, many departments have slack and peak periods during the month. These variables should be carefully studied when the change-over date is being set.

3. The timetable must be *realistic*. It must make allowances for obstacles, problems, delays, and unanticipated preparatory work. Above all, it must recognize the need for "staying in business" during the transition period.

Once the schedule has been drafted, the analyst should make sure that every person participating in the installation receives a copy. In addition, he will do well to make the key executives who are interested in the program aware of the estimated time required to complete the change-over. Persons who have never participated in a major procedures installation seldom realize the magnitude of the preparatory task. Unless they are told in advance what to expect, they are likely to become impatient over lack of results and to regard subsequent explanations as impromptu excuses.

**Revising the schedules.** As the preparatory work progresses, the need for additional preliminary steps often becomes evident. These should be incorporated in the formal installation schedule immediately so that they will not be forgotten.

If the preparatory work has not been completed when the installation date approaches, and if there is no deadline requirement, the change-over should be postponed and the remaining work rescheduled. Starting such installations under a handicap usually results in prolonged change-over periods, extra expense, excessive overtime, and undue strain.

One word of caution, however: every effort should be made to avoid repeated postponements. At best, they cause loss of enthusiasm and delay in realizing the benefits of the study. At worst, they raise doubts that the procedure is practicable or suspicion that it was poorly planned. Clearly the best protection against these hazards lies in developing the installation plan with great care, in following its progress closely, and in taking whatever measures are necessary to keep it on schedule.

**Hypothetical installation plan.** The following hypothetical case suggests a format for recording the installation plan and illustrates the amount of detail required if the plan is to serve as a useful guide.

## Plan for Installing New Order Processing Procedures

| Action step | Responsibility of | Scheduled dates |
|---|---|---|
| 1. Hold a series of meetings with the headquarters personnel concerned to outline the new procedures and the steps by which they will be put into effect. | Holmes<br>Bates | 5/2–5/4<br>5/2–5/3 |
| 2. Prepare and send to regional managers a brief outline of the approved procedures changes affecting the branch and regional offices. | Bates | 5/4 |
| 3. Eliminate immediately the following records, files, and other clerical activities of the order department:<br>  *a.* Customer sales record card.<br>  *b.* Commission record.<br>  *c.* Duplicate file of open orders maintained by order schedulers.<br>  *d.* Manifest of orders released to plants.<br>  *e.* Duplicate editing of orders received from branch offices.<br>  *f.* Checking of unit prices on invoices to customers and maintenance of an invoice file.<br>  *g.* Preparation of daily summary of shipments by product class and sales region. | Holmes<br>Gardiner | 5/5<br>5/5–5/13 |
| 4. Determine the number of clerks who will be made available by the work eliminations listed in step 3 and arrange for their transfer. | Holmes<br>Gardiner | 5/6<br>5/6–5/13 |
| 5. Contact various suppliers of duplicating equipment, witness demonstrations, and visit companies using the equipment to determine the type and model that will best meet our requirements; issue the necessary purchase requisitions for the equipment and operating supplies. | Bates | 5/5–5/20 |
| 6. Procure order schedule boards and sorting equipment for the order department. | Bates | 5/9–5/13 |

## Plan for Installing New Order Processing Procedures (cont.)

| Action step | Responsibility of | Scheduled dates |
|---|---|---|
| 7. Place order for visible reference panels and strips. | Bates | 5/9–5/13 |
| 8. Discuss tentative designs of following new or revised forms with various forms printers, clear any changes with the operating supervisors concerned, and release the final designs to the purchasing department for ordering:<br><br>   *a.* Order-invoice duplicating master for factory shipments.<br>   *b.* Order-invoice snap-out set for shipments from branch warehouses.<br>   *c.* Customer master information card.<br>   *d.* Customer sales-statistical card. | Bates | 5/16–5/27 |
| 9. Pending availability of new order-invoice duplicating master for factory shipments, have a supply of facsimile forms printed for use by the branch offices in place of the present order forms. | Bates | 5/20 |
| 10. Order window envelopes and binders for new invoice form. | Bates | 5/27 |
| 11. To ensure uniformity of order preparation, make a survey of the sizes of type face and the condition of typewriters in the branch offices, order and billing departments, and plant shipping departments. Plan the necessary shifts of equipment and initiate the purchase of any new equipment required. | Bates | 5/31–6/10 |
| 12. Work with equipment suppliers to develop details of procedure for transmitting urgent orders to the factories by wire. Place orders for equipment and forms and arrange for training of personnel to operate the equipment. | Bates | 5/31–6/10 |
| 13. Develop order department personnel program.<br><br>   *a.* On the basis of the revised plan of organization and division of work, prepare for the salary committee a definition and evaluation of each new or revised job. | Holmes<br>Gardiner | 5/9–5/11<br>5/9–5/20 |

## Plan for Installing New Order Processing Procedures (cont.)

| Action step | Responsibility of | Scheduled dates |
|---|---|---|
| *b*. Determine the number of persons required in each job. | | |
| *c*. Reappraise the qualifications of the present personnel in light of the increased responsibilities of the job. | | |
| *d*. Arrange any necessary personnel transfers. | | |
| *e*. Select and train any new personnel required. | | |
| 14. Draft an organization announcement listing the new responsibilities of the order department. | Holmes | 5/11 |
| 15. Instruct mail-room personnel to route to the order department instead of to product managers all purchase orders, branch sales orders, and correspondence relating to specific orders or shipments. | Holmes | 5/12 |
| 16. Notify branch managers that their phone or wire inquiries about specific orders or shipments should be made to the order department instead of to product managers. | Holmes and Halverson | 5/12 |
| 17. Discontinue the product managers' maintenance of the following records:<br>*a*. Open- and closed-order files.<br>*b*. Order-backlog record.<br>*c*. Daily and cumulative record of plant production. | Holmes and Halverson | 5/13 |
| 18. Develop policy covering the assignment of shipping priorities. | Holmes | 5/16 |
| 19. Prepare instructions governing determination of the plant from which orders should be shipped. | Holmes | 5/17 |
| 20. Write detailed job instructions for each of the following positions in the new order department:<br>*a*. Order scheduler.<br>*b*. Order clerk.<br>*c*. Statistical clerk. | Holmes | 5/18–5/20 |
| 21. Review these instructions with order department personnel, make any necessary modifications or additions to the instructions, and issue them in final form. | Holmes and Gardiner | 5/23–5/25 |

## Plan for Installing New Order Processing Procedures (cont.)

| Action step | Responsibility of | Scheduled dates |
|---|---|---|
| 22. Make the approved layout changes in the order department and move the credit and traffic departments into the adjacent area. | Holmes Gardiner Wade Callahan | 5/26–5/27 5/26–6/4 5/30–6/4 5/30–6/4 |
|    *a.* Prepare final layout drawings. | | |
|    *b.* Reach agreement with the building superintendent on the tagging of furniture and equipment, the scheduling of partition and telephone changes. | | |
|    *c.* Initiate moving and rearrangement authorization. | | |
| 23. Work out the details of tying brokers' orders into the new procedure. | Holmes | 5/31–6/3 |
| 24. Develop and carry out plan for transfer of customers' special packing, shipping, and invoicing instructions from headquarters to branch offices. | Holmes Gardiner | 5/31–6/3 6/3–6/17 |
| 25. Expand the number of classifications into which order data are summarized for production planning and shipment-scheduling purposes. | Holmes Ryerson | 6/6–6/10 6/10 |
| 26. Obtain from the credit manager a list of preferred credit risks and type this information on Linedex panel strips for use by order department clerks in checking orders. | Holmes Wade Gardiner | 6/6–6/10 6/8–6/17 6/13–6/24 |
| 27. Work with credit manager on development of standard collection form letters and a standard mailing schedule. | Holmes Wade | 6/6–6/10 6/6–6/17 |
| 28. Draft standard practice instructions covering the following procedures and clear them with the operating supervisors concerned: | Holmes Bates | 6/13–6/14 6/13–6/24 |
|    *a.* Processing of orders and invoices covering factory shipments. This should include: | | |
|      (1) Distribution of order-invoice master. | | |
|      (2) Information to be added at each point. | | |
|      (3) Copies to be duplicated at each point and their distribution. | | |
|    *b.* Processing of order-change notices. | | |
| 29. Investigate any suggested revisions, points of disagreement, or areas of possible diffi- | Holmes Bates | 6/23–6/24 6/20–6/24 |

## Plan for Installing New Order Processing Procedures (cont.)

| Action step | Responsibility of | Scheduled dates |
|---|---|---|
| culty brought out by the comments received on the draft of standard practice instructions in step 28. Revise the instructions, if necessary, and issue them in final form. | | |
| 30. Make a study of the elapsed mailing time via air mail and first-class mail between each of the company's plants and each branch office. Issue instructions to each plant specifying which service to use in mailing shipping notices to each branch. | Holmes | 6/15–6/24 |
| 31. Set up for billing department pricing clerks precalculated charts showing invoice amounts for varying quantities of each product. | Holmes Mitchell | 6/15–6/17 6/15–6/24 |
| 32. Draft a procedure instructions manual covering all operations of branch sales offices. | Holmes | 6/20–6/22 6/27–6/28 |
| 33. Analyze the tabulating department's report schedule and determine the steps necessary to speed up monthly reports of shipments by branch office, product class, and customer. | Bates Holmes | 6/27–7/15 6/29–7/6 |
| 34. Develop plans and make arrangements for installing the new procedures on a trial basis at the Philadelphia branch office. | Holmes Halverson | 7/7–7/8 7/8 |
| 35. Develop standard unit time allowances for each clerical operation in the branch sales offices. | Holmes and Bates | 7/11–7/15 |
| 36. Make pilot installation at Philadelphia branch according to the following sequence of steps: | Holmes Bates | 7/18–7/22 7/18–7/29 |
|    *a.* Explain new procedures to office personnel. | | |
|    *b.* Eliminate activities and operations not required under the new procedures. | | |
|    *c.* Convert existing records to their new forms. | | |
|    *d.* Make test runs of the new order-processing procedure using facsimiles of the factory and warehouse order-invoice forms. | | |
| 37. On the basis of this pilot installation: | Holmes | 8/1–8/5 |

# Plan for Installing New Order Processing Procedures (cont.)

| Action step | Responsibility of | Scheduled dates |
|---|---|---|
| *a.* Revise the branch office procedures instructions manual. | Bates | 8/1–8/12 |
| *b.* Refine the standard time allowances for branch office clerical operations. | | |
| *c.* Determine the required clerical complement for each branch under the new procedure by applying the standard time allowances to the branch's known work volume. | | |
| 38. Develop installation program for other branch offices. This will include: | Holmes | 8/8–8/12 |
| *a.* A detailed list of all the steps that have to be taken to change over to the new procedures. | | |
| *b.* The timing of each step. | | |
| 39. Make arrangements with representative of the duplicating equipment supplier to train machine operators for the order and billing departments. | Bates | 8/15 |
| 40. Train a representative of each regional sales manager to assist the branches in his territory in installing the new procedures. | Bates | 8/15–8/19 |
| 41. Send to each branch manager a copy of the standard practice instructions manual and the program for installation of the new procedures. | Holmes and Halverson | 8/17 |
| 42. Issue memorandum of instructions to outside warehouses covering details of the new order-invoice procedure. | Holmes | 8/15 |
| 43. Give to each order scheduler a copy of the order-preparation instructions issued to the branch offices. | Holmes | 8/16 |
| 44. One week before adoption of the new order-invoice forms, issue instructions on the following points to the order, credit, traffic, plant shipping, and billing departments: | Holmes | 8/18–8/22 |
| *a.* Cutover dates. | | |
| *b.* How to handle in-process orders which were originated under the old procedure. | | |
| *c.* Potential problem areas requiring special supervisory attention during the transition period. | | |

## Plan for Installing New Order Processing Procedures (cont.)

| Action step | Responsibility of | Scheduled dates |
|---|---|---|
| 45. Visit each of the branch offices to review progress of the installation, to assist with any problems that may have arisen, and to reach agreement with the branch manager on his required complement of clerical personnel. | Representatives of regional managers | 8/22–9/16 |
| 46. During the first month that the new forms are in effect, make a 100 per cent audit of (*a*) invoices covering warehouse shipments and (*b*) order-invoice masters covering factory shipments. | Bates, Mitchell, and Gardiner | 9/1–9/30 |

## DOING THE PREPARATORY WORK

The nature of the preparatory work required on major procedures installations has already been indicated. The following paragraphs outline more fully the various types of work to be done and the objectives to be achieved in carrying out this part of the job.

**Preparing personnel for the change.** The first move in setting the stage for the change-over should be to recognize the personnel problems involved and to deal with them fairly, openly, and unhesitatingly.

This means, first, that decisions must be reached on what employee changes are to be made and how they will be handled. Here are some of the typical questions to be answered:

1. How many workers will be displaced by the new procedure?

2. Who specifically will they be?

3. Can some of them be transferred to other jobs without jeopardizing, in the long run, the benefits of the study? If so, which workers should be transferred and to what jobs?

4. How many must be released? Who will they be?

5. What will be done to minimize the hardships of their release? What will our policy be on termination pay? How can we help them find jobs elsewhere?

6. How many employees must be trained to perform work requiring new skills? Will the evaluation of their jobs be affected?

Second, every effort must be made to develop favorable employee attitudes toward the change before it is installed. Once decisions are reached on a major procedures problem, its human relations phases tend to be sidetracked by its technical aspects. But the analyst must remember that, even where a skillful job has been done in earlier parts of the study to promote understanding and interest, the employee's greatest concern is over the decisions finally reached and how they will affect him. If he must await the actual change-over before learning what these decisions are, the gains already made in winning his acceptance are likely to be unnecessarily lost. Even where some employees will be displaced by the change, there is less risk and greater profit in disclosing your plans to the work force promptly and openly than in seeking to "protect" its members until you are ready to move.

In the first place, rumor flourishes where knowledge and understanding are missing. And under these circumstances, the imagined consequences of a contemplated change are inevitably much worse than they actually turn out to be. Thus, needless uncertainty and unrest can be prevented by a frank and early discussion of the facts.

A second reason for outlining the proposed course of action in advance is that by doing so you give the employees an opportunity to think about it, discuss it, and become accustomed to it. Having made the necessary mental readjustment by the time the installation is begun, they are likely to take part in it with greater acceptance and enthusiasm.

Finally, nothwithstanding the effects of a change, when employees feel that nothing has been hidden from them, they are also more likely to feel that they have been dealt with fairly. Even where some releases are necessary, skillful handling of the situation can convert an apparent liability to employee morale into an asset. A forthright acknowledgment of the problems plus a clear explanation of what will be done to minimize the hardships of displacement can have an incalculably favorable effect on employee attitudes throughout the whole organization.

Most of what should be covered in this initial explanation to the work force has already been implied. It is not necessary to describe the new procedure in complete detail since that will be done later during the employee-training phase of the preparatory work. The objective at this point should be to outline:

1. The major features of the new procedure.
2. The reasons why the changes are being made.

3. How and over what period of time the installation will be made.
4. The effect of the new procedure on manpower requirements.
5. The steps the company has taken and will take in the interests of employees affected by the change.

**Testing the new procedure.** A second important preparatory requirement is to make sure that the new procedure meets the tests of actual operating conditions. This can be done in either of two ways: (1) by putting a sample of actual work through the planned operations; (2) by converting fully to the new procedure in one of several like units of an organization while the other units continue operating under the old procedure. The latter type of trial operation is usually referred to as a "pilot installation."

Preliminary testing by either of these methods has many values, the net effect of which is to take the panic out of a major installation. The most obvious advantage, of course, is that testing enables the analyst to settle many minor details previously overlooked and to make certain that the new procedure will meet every operating contingency. It also serves as a means of checking the completeness and intelligibility of new forms and written instructions. In addition, where extensive changes are to be made in many similar groups throughout the company, a pilot installation provides the best opportunity for training the installation crew. On changes limited to one area, the testing process can usually be combined with the worker training program either by using actual work or transcripts of it. A further advantage is that performance of the new routine under real or simulated conditions permits unit time allowances for each operation to be refined so that labor savings can be determined more precisely. It also frequently points the way to additional short cuts and improvements that can be incorporated in the procedure before its inauguration. Finally—and this is perhaps of greatest importance—a successful test run supplies the confidence needed to ensure a quick, smooth installation.

So much for the benefits of pretesting. Now let us look at some practical aspects of carrying it out. Here are three rules that will help the analyst get the maximum results from this part of the preparatory work:

1. Treat the preliminary tryout just as you would the actual installation. That is, conduct it as a tightly controlled experiment.

Be thoroughly prepared before starting. Make sure that everyone knows in advance what is expected of him. Assemble all the necessary working tools, supplies, instructions, and forms or duplicated fac-similes. Simulate actual working conditions as closely as possible. Unless these precautions are taken, many of the important points, which the test should bring to light, will be lost in the casual, slipshod method by which it is conducted.

2. If the test is being made by the sample method, be sure that the transactions represent a cross section of the varying conditions that will have to be met when the procedure is in actual operation. Include a number of the more difficult, nonstandard cases; for, in designing a new procedure, analysts tend to concentrate on the normal routine without giving too much attention to that small part of the work requiring special handling.

3. Give a great deal of thought to selecting the unit in which a pilot installation is to be made. Try to pick a group that has a reputation for open-mindedness and willingness to adopt new ideas. The other units will be watching the experiment closely, and their attitude toward the change will be very much influenced by the attitude of those who first install it.

**Designing and ordering forms.** As soon as the testing process is finished, any new or revised forms required under the contemplated procedure should be designed in final form and ordered from the printer. If, as a result of the trial run, the original designs were sub-stantially changed, it is wise to clear the final drawings with the operating supervisors concerned so that the cost and delay of further alterations in the printer's proofs can be avoided. This is also a logical time to reach decisions on the disposition to be made of superseded forms and to arrange for their recall.

**Obtaining equipment and rearranging facilities.** If changes or additions in office equipment are to be a part of the program, the project recommendations often will not have gone beyond specifying the type of equipment needed. Where this is true, an important part of the preparatory work will be to decide the exact make, model, capacity, and quantity of equipment to be obtained.

If more or less floor space will be required under the new pro-cedure, if the location of a department is to be changed, or if equipment and machines are to be rearranged, as many of these changes as possible should be made before the installation so that quarters and facilities are ready in advance. The preparation of physical facilities should also include the efficient arrangement of such apparently minor items as work tables, racks, cabinets, etc., for these items often play an important part in securing smooth installations.

**Preparing work aids and procedure instructions.** The various codes, tables, charts, indexes, wiring diagrams, and other work aids required under the new procedure should be prepared early enough to allow time for testing them before the change-over is made. Preparatory measures of this kind are often put off until they overlap the installation, thereby causing delay and confusion during its initial stages.

One of the most demanding problems in launching a new procedure is translating a general plan of action into a form and language that can be readily understood by the persons upon whom success of the plan depends. All too frequently analysts assume that this requirement is met simply by drafting standard practice instructions covering the new procedure. Although this is the first prerequisite, it is still only part of the job. It tells only what operations are to be performed *after* the installation is completed. It is like giving the workers a new and unfamiliar goal without telling them how to reach it.

On a major change-over, therefore, instructions on the steps to be followed *before* and *during* the installation are as important as those covering the ultimate procedure. Specifically, the following subjects should be covered:

1. The preparatory work to be done by the operating personnel.

2. The procedures to be followed during the transition period. If thoroughly prepared, these instructions are often more elaborate and detailed than those covering the new procedure itself. They should include such matters as the time at which to discontinue each part of the old procedure, the method of handling transactions that were started under the old procedure and are still in process when the cutover is begun, the provision of special checks and controls during the critical phase of the conversion, and so on.

3. The procedure and organization structure that will be in effect when the installation is finished.

4. The effect of the new procedure on departments that will not participate in it directly. For example, changes in a cost accounting or payroll function may be of concern to a foreman even though he plays no part in its performance.

In preparing such instructions, the analyst will do well to err on the side of imparting too much rather than too little information. He must remember that because of his long exposure to the problem, his mind supplies many of the minor details with which others are not familiar. A set of instructions that seems complete and understandable to him may have many gaps to the operating personnel.

In a procedure involving many departments, the usefulness of written instructions can also be increased by preparing the instructions in separable sections so that each group can be given only the material with which it is concerned. In some instances the analyst will want to go even beyond this point and prepare detailed breakdowns of individual jobs.

Finally, provision should be made for distributing copies of the instructions to everyone affected by them. One of the common deficiencies here is to prepare only enough copies for the department heads concerned. The result is that relatively little written information reaches the great group of clerical personnel upon which management must depend for efficient execution of its policies and plans.

Other suggestions for the preparation of written instructions and the development of procedures manuals are given in the next chapter.

**Training the personnel.** In most cases, the job of indoctrinating workers in the new procedure cannot be satisfactorily done just by the issuance of written instructions. Although such material forms the basis for indoctrination, it is still only a guide, a reference device, or a set of rules. If the instructions are to be properly executed, they must usually be supplemented by some sort of follow-up, encouragement, or training of supervisors and workers. This might take the form of group discussions, demonstrations, on-the-job coaching, or practice sessions during and after the testing of the procedure.

The amount of training required will be slight in some cases and great in others, but the need will be present in practically all. Where simple changes are to be made, the process of education will be simple. Under these circumstances a brief group meeting ordinarily will suffice. But where many or radical changes are to be made, a more elaborate and carefully planned training program of several weeks' duration may be needed. This is particularly true of projects involving the introduction of an electronic computer or of tabulating, bookkeeping, or other office machines that require specialized skills in their operation.

An important point to be borne in mind is that training should not be limited to the mechanics of the work, for enthusiastic and intelligent performance will depend just as much on an understanding of the broader aspects of the plan. Each employee should be helped and encouraged to understand the objectives and principles involved as well as the manner in which his job fits into the whole routine. A scheduling clerk, for example, should be told how the anticipated changes in a production-control system will help shorten the production cycle and reduce inventories. Employees who are being retrained to operate a new electronic computer or tabulating-machine installation should be taught the relationship of their work to the whole accounting system and the part that each basic type of machine plays in the whole process. It is easier for the trainee to learn the operation of a specific machine if he knows the fundamental principles of punched-card accounting. A further advantage of this broad approach is that it helps the worker understand the effect his errors or delays will have on subsequent parts of the procedure.

Against this background, each operator should then be fully instructed in the work he is to perform. Next should come a period of actual practice, accompanied by periods of discussion. The amount of time given to this part of the training will vary, of course, depending on the degree of change, the type of equipment to be used, etc.

A word is needed here about responsibility for the training program. In general, the workers should be trained by their own supervisor, who, in turn, has been indoctrinated in the new procedure by the analyst. This arrangement not only reflects supervision's intrinsic responsibility for worker education but also equips the supervisor to make better decisions during the installation period and to keep the routine adjusted to changing requirements thereafter. Exceptions to this rule may occur when a tight deadline must be met or when highly mechanized operations are to be adopted. Under these conditions, the supervisor would presumably delegate much of the training job to outsiders, including the analyst and a representative of the equipment manufacturer.

**Revising related procedures.** Any required revision in auxiliary routines, which feed into the main procedure, should be planned and made before the key changes are installed. This spreads the work and provides a chance to iron out minor difficulties before the larger job is taken.

The same principle can often be applied to activities that follow or depend on the main procedure. For example, the forms of various reports can sometimes be changed before their preparation under the new plan is actually begun. This may permit an allied procedure, which depends upon these reports, to be revised first.

**Making preparatory change-overs.** As a conclusion to this section it is safe to say that any work that can be done before the installation should be placed on the preparatory schedule and not left for the installation schedule. Study will show many items of this kind. Records may be transferred or converted in advance, partial routines may be established, control systems set up, files rearranged, etc.

## MAKING THE INSTALLATION

Assuming the intrinsic soundness of the new or revised procedure, the success of the project will be assured if the preparatory work has been done with skill, foresight, and thoroughness and if the installation is made under calm, capable supervision.

The following paragraphs contain a number of suggestions and cautions on this final step of putting the procedures plan into operation.

**Keeping on schedule.** Since unexpected problems and troubles frequently develop, the actual installation seldom proceeds as smoothly as planned. And when these difficulties arise, they tend to divert attention from the original plan and to provide excuses for delay. To prevent this occurrence, the progress of the work must be vigilantly watched, especially during its initial stages; for if any lagging behind schedule is permitted to go unremedied, its effects are likely to be cumulative and to increase the difficulties experienced at other stages. One qualification, however, is that you should not let your zeal for keeping on schedule force you into beginning a new step until any preceding steps on which it depends have been satisfactorily completed.

**Omitting planned features.** In the stress of installation there is a tendency to do without certain features of the original plan, to ignore parts of the instructions, to leave apparently minor things until later, or to get along without certain facilities. This danger should be guarded against, for in all likelihood these items were included to facilitate the whole program. Their omission only serves to increase the problems later on.

**Making emergency decisions.** One of the most common requirements during a large-scale installation is the issuance of supplementary orders when the regular instructions do not cover a particular situation. Obviously such decisions should not be made without thorough consideration of all phases of the problem, even though the work in question must be temporarily set aside.

In manning the installation organization, it is well to station qualified supervisors or advisers at strategic points so that these wrinkles can be quickly ironed out. Needless to say, the persons selected must have a full knowledge of the principles and objectives of the procedure and an ability to visualize the effects of their decisions on other parts of the plan.

As emergency decisions of this character are reached, they should be made known to everyone affected. In addition, the change should be recorded on a master copy of the procedure instructions so that the whole structure will be kept in balance.

**Making in-process changes.** Closely related to the need for issuing supplementary instructions is that of polishing or refining the original procedure as operating experience reveals opportunities for further improvement. These opportunities should be exploited immediately if doing so will not materially slow down the conversion. If it will, they should be set aside until the more important parts of the procedure are being performed satisfactorily.

Occasionally an installation will strike such a serious obstruction that the procedure will have to be basically changed. Should this happen, the installation work should not be stopped immediately, for it will often be found that the problem is not so grave as originally thought and that much of the work can be salvaged.

**Summary.** Here in summary form are 10 guideposts that will help to keep the installation team out of trouble:

1. Be ready before starting.
2. Keep up to schedule.
3. Avoid rash, emergency decisions.
4. Anticipate and eliminate crises.
5. Don't let minor kinks dampen your enthusiasm or your confidence in the plan.
6. Keep all phases of the change-over coordinated by informing

executives and supervisors promptly of any changes in the original procedure or the installation plan.

7. Prevent dissension among the personnel.

8. Don't require continuous or excessive overtime work of the installation crew. If the change-over is falling significantly behind schedule because of lack of personnel, get some extra temporary help.

9. Avoid disruption of service.

10. Don't sacrifice thoroughness for speed.

## INSTALLATION FOLLOW-UP

Even when the new procedure is in operation and the installation is considered complete, the job of the analyst is not over. One of his important continuing responsibilities is a systematic follow-up to ensure not only that the gains originally planned are realized but also that they are held or furthered.

This function has two parts: (1) a thorough checkup shortly after the installation to see that the change is stabilized; (2) a longer-range program of periodic audits to ensure that the procedure is properly maintained.

The initial follow-up step should be thought of as continuing until all phases of the new procedure are fully in effect and operating satisfactorily. Where the change-over is relatively simple, this review may be considered part of the installation work and made as the final step in that process. Where widespread changes have been made, however, a certain period of operation is required before the new procedure is shaken down and ready for review. Thereafter, continuing follow-up may be needed over a relatively long period of time before the desired results are achieved.

**Follow-up objectives.** The analyst should develop his plans for the installation follow-up in terms of these objectives:

1. To ensure that all parts of the new procedure have been put into operation and that the written instructions are being adhered to.

2. To make any further modification or refinements for which operating experience reveals the need.

3. To measure the results achieved against those originally predicted and to take whatever action is necessary to get the maximum benefits from the new procedure.

**Points to be checked.** Following is a check list of the major points to be covered in reviewing the newly installed procedure. The thoroughness of the check will vary from point to point in any particular review. In some areas, a quick visual inspection will suffice; in others —particularly where difficulty was experienced during the installation—a detailed audit of the work will pay dividends.

1. *Elimination of Old Procedures.* Make sure that no parts of the old procedure are being continued unintentionally. Where two procedures have purposely been operated in parallel, make whatever tests or comparisons are necessary to satisfy the executives and supervisors concerned that the new procedure can safely replace the old.

2. *Effectiveness of the Procedure.* Study the over-all operation of the procedure to see if the objectives for which it was designed are being achieved. For example, are reports being prepared in the prescribed manner and being submitted on time? Also, check to see whether or not all new services and reports are being fully utilized. If not, further educational work may be desirable to explain their use, form, or purpose, or certain parts may be eliminated.

3. *Operating Expense.* Wherever expense reduction is among the project objectives, check the savings actually realized against the estimate made as part of the project proposal. There is a tendency to delay in taking advantage of the expense reductions that a new procedure makes possible. This applies particularly to excess manpower. Where you find this condition exists, determine the reasons for the delay and work out with the appropriate executive or supervisor a specific timetable for realizing the balance of the savings.

4. *Procedure Instructions.* Check the written instructions on the procedure in sufficient detail to see whether or not they are being followed. If not, find out the real reason, whether it is indifference, misunderstanding, failure to receive copies of forms or instructions, insufficient follow-up or coaching by the supervisor, unworkability of the instructions, etc. Having determined the cause, you can then decide whether further training, better supervision, or other corrective measures are needed, or whether the variation is justifiable and should be incorporated in a revised set of instructions.

During this part of the review, ask those who are engaged in the work for their suggestions on further refinement or improvement.

5. *Clerical Performance.* Check the quantity and quality of work produced against the performance estimates made in the original proposal. If performance is not up to standard, work with the super-

visor to develop a plan for increasing production or realigning the work force. Where the written instructions call for the maintenance of clerical production records, check to see that these are being properly kept and utilized.

Also, as part of this step, check the presence and use of controls over accuracy of the work. These devices are of particular importance during and just following the installation, for they frequently provide the only reliable means for detecting errors in performance or failure to follow the procedure.

6. *Facilities.* Make sure that all the prescribed facilities, supplies, and work aids are available and that they are being used as planned. Check the layout of the office against the original diagrams and determine whether the anticipated reduction in movement or backtracking has been realized.

**Improvement programing.** Following each such review of the installation, the analyst should prepare jointly with the line supervisor a definite program of steps to be taken and results to be accomplished before the next review. This process should be continued until no change-over work remains and the procedure is considered satisfactory. Under this arrangement, the first step in each subsequent review is to see that everything on the previous improvement program has been done.

A written program of this type should also be prepared at the end of the formal installation work to serve as a guide during the period before the first scheduled review.

## SUBSEQUENT OPERATING AUDITS

The study of a given procedure should not be regarded as coming to an end when the changes have been fully installed, but rather as the beginning point of a development process that will continue indefinitely. Inefficient or unnecessary steps will creep back into the procedure. New personnel, not fully aware of the purpose behind the activity, will take over its performance. The basic requirements to be met will change. New equipment or techniques will open the door to further improvements. For all these reasons the postinstallation review should be followed by regularly scheduled audits to make certain that effective operation of the procedure can be and is being continued.

This continuing review should be a responsibility of the procedures staff, for in this way the staff is given an instrument to influence the final effectiveness of procedures work. Frequently, however, the staff will find that its efforts can be multiplied by calling on the internal auditing staff to make the actual review. Where this is done, both groups should work closely together in scheduling the reviews, in developing a detailed check list of points to be covered, and in reviewing the results of each audit.

## INSTALLATION AND FOLLOW-UP


*Systems and Procedures Association*

SOURCE: From *Business Systems* by the Systems and Procedures Association, Cleveland, Ohio, 1966. By permission of the publisher. Copyright © 1966. Systems and Procedures Association (now Association for Systems Management).

## INTRODUCTION

No single phase of a systems study can be selected as the most important step in the process, because they are all essential. But, when you find an efficiently operating system or procedure, you can be sure that there was a well-planned and executed installation phase.

However, sometimes the beneficial results of good systems design are never maximized because the installation task was not properly met by the systems analyst. A common tendency among systems men is to feel that the job is done when everyone is sold on the new system. At that point they lose interest, and are off to conquer new territories. This is a dangerous tendency, from the standpoint of personal reputation and the company's purse strings. No idea, no matter how simple or revolutionary, will live up to its promise unless it is successfully and completely installed.

Even after a system is installed and operating, the systems man hasn't finished the job. Checking or following-up the new processes should be accomplished; otherwise, the analyst cannot be sure the new procedure is working efficiently for the best interests of the company and its employees. Follow-up undoubtedly is the most neglected of the systems analysis steps, primarily because it is done *after* the system is installed. Following-up on the application is important, though, from several aspects.

1. It's the only factual feedback the analyst gets regarding the success or failure of his brainchild.
2. It enables the analyst to consolidate any gains, plug any loopholes, and apply the brakes to any backsliding.

3. It enables him to sensibly revise the system to fit previously un-known factors.

## INSTALLATION

For an installation to succeed, all the steps to be accomplished dur-ing the establishment of new procedures or the conversion of existing ones must be anticipated and documented in an installation plan. As is true of almost any long-range program, the more thorough the plan, the more successful its execution. Installing a new system cer-tainly falls into this generalization—sometimes to the extent that a poorly devised, incomplete plan may completely nullify any potential gains from the system.

Selling new ideas to management entails overcoming executive sales resistance. Implementation requires facing the same arguments (probably in ten times the volume) from clerks and operators who will actually do the work. This can be ruinous, unless they are sold on the new system and give it enthusiastic, or at least "cooperative" support. The best way to obtain this support is to go to people with a compre-hensive, well-documented, installation program and schedule. If you show how simple it is, and how much help will be available, even die-hard "old-timers" often can be won over. Occasionally, a system will have to be installed right through one of these stubborn ones, but it's a situation to avoid if possible.

### Planning

Installation planning should be part of the systems study. Investi-gation and analysis soon reveal the necessary or desired changes. It is wise, at this time, to begin noting the proper changes and conversion controls to insure a smooth changeover. Points to remember during in-stallation pre-planning are: plan early, select the type of conversion method, and determine the proper controls to be used during installa-tion.

*Plan Early*—The more complicated the system, the more vital it is to pre-plan the installation phase. An excellent approach is to outline the installation plan in conjunction with selling proposal to Manage-ment. This adds weight to your proposal by showing the system will work and can be installed. Lay the groundwork for the time and costs involved in the installation rather than have them as rather cruel and unexpected surprises!

Whether or not installation plans are included with the selling proposal, it is important to start recognizing installation problems as soon as the new procedure or idea is developed. This will often point up impracticabilities or inconsistencies in original ideas. It's far better to discover these points at this stage than during actual installation. Planning installation steps and check-points in advance also help overcome early arguments, which usually are manifestations of people's resistance to change.

*Select Type of Installation*—There are several types of installation plans—each suited to a particular problem or degree of complexity in the system.

a. "All-at-Once Type"

If the procedure to be installed is simple and doesn't involve heavy volume or too many departments, this plan is usually best. Formalized instructions are distributed to everyone involved which say, in effect, "We'll start this new way at 8:00 a.m. tomorrow." The "all-at-once" plan is a nice, clean way to install a program, but *not if any of the following factors are present:*

1. If the change is radical or involves new machines.
2. If the change involves several far-flung locations, and you don't have adequate installation help. (In other words, you can't expect to be in two places at once.)
3. If the personnel are not 100% sold on the proposal or new system.
4. If the volume of transactions is sizeable. (A good way to judge this is to consider the length of time it would take to recapture a day's transactions under the old system. Suppose the new system failed on the first day of installation. Would it take only an hour to process the work in the old manner, or is it going to take a dozen people all day?)

The "All-at-Once" method should be used with the utmost caution. The factors which could damage the effectiveness of the new processes should be few and easy to spot and correct.

b. Piecemeal Type

This is, by far, the easiest way of doing larger volume jobs. In fact, on very large, long-range systems, it is the only way to do it. The "piecemeal plan" means just that—take small chunks of the whole program and install them individually to satisfaction, then move on to the next chunk, etc. This is the relaxed way of doing the job—maybe it's too relaxed, because it will invariably take a lot longer than anticipated! But, this in itself is not bad, because it allows a gradual, controlled conversion.

Participants in the new system are "spoon-fed", and therefore have an opportunity to "live" with the changes and become accustomed to the new ways of performing their tasks.

A word of caution if the piecemeal plan is used. Be sure that every step is well documented and that due dates are met as anticipated! Don't let the interest in the new system die from inattention, or seeming inattention, because of the long period of time involved.

*Don't* use the piecemeal plan if:

1. Time is important.
2. If the system involves data used for either daily, weekly, or short term reports. Installing such a system, or portion of a system, a piece at a time can cause a periodic report to show data collected in the *new* way and some in the *old* way.
3. If the dollar savings are quite dramatic, don't prolong the installation any longer than absolutely necessary. Try to get the system into cost savings operation as soon as reasonably possible so the beneficial rewards can be enjoyed.

c. Parallel Type

For large installation projects and those involving intricate processing, the parallel plan is by far the best. This is similar to the "piecemeal" plan in that bits and pieces of the system are installed, but the old system continues to operate as well. Obviously, this is the safest way. Any emergency can be met without panic; everyone gets the "feel" of the new system before they're asked to produce under fire. But don't forget it's also the most expensive installation plan because theoretically, it's going to require twice the staff.

A major change to a payroll system would be a good example of when to use a parallel operation. The one thing a firm can least afford to miss is its payroll, so almost any cost that guarantees it is cheap insurance.

A parallel installation plan may be chosen, regardless of cost, in order to completely sell the system to participating employees. Often a radical change may be accepted by the operators because top management has endorsed it, but they may still be hesitant because they don't have enough faith (or knowledge) in the value of the proposal. By using parallel operations, the systems analyst offers participants an opportunity to build confidence in the new methods because they produced the same results in a cheaper or swifter manner.

*Determine Installation Controls Needed*—During the installation period, tight control must be maintained over every phase of the new system. To do this, set up a "warning system" within the system which forecasts potential bottlenecks and affords sufficient clues for correcting any problems, errors, or fall-downs. The key here is the control measures employed must be able to feed error conditions in minimum time. If errors are caught too late, they will damage the accuracy and effectiveness of those adjacent systems which rely on the new system for input. If a significant error condition in a large system is not captured in time, chaotic records result and a huge expense of time and money is required to rebuild accurate files. There are several control devices or techniques available.

a. "Program Check" Points

One way of establishing control is to develop check points through which every document or data flows. A periodic review of the processing to date at each check-point indicates when all isn't going smoothly. It's hard to describe how or where these check-points should be established because they vary from system to system and company to company. The idea is similar to "stops" in a computer program, where the computer physically stops processing and supplies totals to the operator. If everything looks okay, he starts again and goes to the next stop. When the program is debugged (or the installation complete), these stops are removed and work is uninterrupted.

b. Accounting Control Totals

If the new system involves accounting entries or ledgers, it will have control totals as a matter of course. This is standard accounting practice. But it may be beneficial to expand this thinking into other areas—at least during the installation period. The control total established need not be a significant number within itself; e.g., if the controller wants to be sure that all invoices were received and processed, you could accumulate invoice numbers as the first and last steps in the system and compare the totals. There are many combinations of numbers or items which can be accumulated in this way in almost any system, but the control totals established are meaningless and use-less( so called "garbage" totals) except as a check on the system.

c. Paper Flow Controls—Log Sheets, Numerical Controls, etc.

A simple system of logging input and output can serve quite effectively as a control during the installation. Perhaps sequentially numbering documents as they enter the system and checking the sequence as they leave would be adequate. These controls involve extra effort during the installation, and they will undoubtedly slow down the processing time. However, if a complex system is being installed, dollars and time can be saved. As the installation progresses towards completion, it may be possible to drop some controls as unnecessary steps, but don't relax too soon. It is far wiser to be over-controlled at the beginning than to be under-controlled.

d. Timing Controls

A new system will probably have arrival and departure of documents, materials and data built into it. Therefore, a factor of elapsed time should be incorporated into your document control measures. The systems analyst should know the proper time interval between certain activities within the system. Recording and appraising time span controls permit analysis of the operation and show any bottlenecks. As the system becomes commonplace to the participants, expect the time span to become shorter.

*Scheduling*

After developing the installation plans and choosing the method applicable to the new system, the analyst should consider when the system is to become operational. The scope of the new system and the degree to which it must be integrated with other procedures often determines the proper time to initiate changes. However, the analyst must insure a starting date in advance to allow enough time to complete all preparatory activity, but not a date so distant everyone forgets the matter in the interim. In scheduling the installation, four important points should be considered: time, personnel, forms and equipment, and job instructions.

*Timing*—The most important factors here are the accounting periods already pre-established. This is critical because the new system must relate to established accounting time conventions within the organization. To avoid confusion, start on an "even" date if possible, to skip any "stub periods" so periodic reports won't be excessively distorted; e.g., a payroll system should start on the first day of the pay period, or a weekly report should be installed on the first day of the week, etc.

*Personnel*—The number of people needed to initiate and sustain an orderly transition should be estimated during planning. When scheduling personnel needs, keep in mind the following:

a. Vacations and Holidays

Take these into account when planning the schedule. Excessive absence can throw a curve.

b. Availability of Extra People

In developing the installation schedule, you have to consider the availability of the extra help needed to get the thing off the ground. In planning a parallel installation, this can be the key point in your entire schedule.

*Forms and Equipment*—Most medium-to-large scale conversions require new or revised forms and new equipment. The installation plan should include realistic timing for the preparation of new forms and the acquisition of new equipment. Two factors to consider in this area are:

a. Delivery Time

Any new forms required should be on hand *before* installation. This can be a problem because some multiple copy or continuous forms take up to 90 days for delivery. Similarly, any new equipment needed should be ordered well in advance, so that delivery dates precede the actual installation date to permit operator training.

b. Possible Substitutes

If installation must move forward without delay, jury-rig some forms in order to get started. This takes time too, so here's another spot where good pre-planning will pay off. Substitutes for equipment are more difficult (sometimes impossible, if the new machine is the hub of the new system), but rentals, loaners or temporary modifications to existing equipment may suffice.

*Job Instructions*—Allow enough time in the schedule for the preparation, checking and dispersal of job instructions to be used both *during* and *after* installation. A well-documented program is reassuring to the participants during transition because it helps them to perform their tasks properly.

### Announcing the Plan

The success or failure of the installation may very well depend on the communications ability of the analyst. Although he successfully sells his proposal and receives approval to proceed with the transition, he still must effectively communicate the nature, scope and related benefits of the new system to all of the people involved in, or affected by, its operation. The best way to accomplish this is to document and publish the installation schedule. It should be distributed with the intention that comments and suggestions are desired.

*Let Everyone Know Program*—In a company-wide program, consider a series of training sessions or meetings. If it's a one-department affair, be sure everyone involved is personally briefed. The more they know, the smoother the transition. This dispels rumors and their associated poor relations.

Key individuals in the new system warrant special attention. Impress them with the importance of their processing step with regard to the entire program. Be sure they understand their new function, and generate (if possible) some enthusiasm toward their new or changed responsibility. Show them how much work they're saving, but calm their fears about being replaced by a transistor. This is a big order, but any success pays off in fewer installation problems.

Before starting an actual installation, be sure to cover the following points:

a. Installation Schedule

b. Necessary Job Instruction

Normally, the only detailed instructions published are those which will stay in effect throughout the life of the system. In extremely complex situations though, special, short-term, written instructions may be required on how to do the job during the installation steps.

c. Flow Charts (If Practical or Beneficial)

Flow charts are an excellent way of detailing the entire installation plan and those processing instructions needed during the installation period. Some applications don't lend themselves to flow charts, but if they can be utilized at all, they should be.

d. Obtain Management Endorsement in "Public" Announcement

If possible, get management to publicly state it approves the new system and expects concrete accomplishments as a result of it. This "public" approval shouldn't be couched in such a way that it says "do it or else" because this could kill the whole project. It should explain why the change will be beneficial to everyone and indicate that cooperation with, or recognition of, the systems staff's efforts during the installation period will be appreciated.

### Training Personnel

*In-Plant Training Sessions*—Individual training, or training sessions for small groups at a time, are both acceptable ways of teaching the individuals new techniques. These sessions can be very informal, on a person-to-person basis, or they can be scheduled sessions taught by someone from the training department and included in the regular company training program.

*Vendors' Schools*—If the new system involves machines, it's an excellent idea to send the operators (and maintenance people, if you have your own service department) to the machine companies' operator training schools. Every office machine company offers this training, so make use of it.

*Pilot Installation—On the Job Training*—If the new system is complicated or involved, test it before trying to install it on a universal basis. Pilot installations also provide an excellent opportunity to train the people involved right on the job.

For a pilot installation, establish an area within the scope of the new system and make it the "guinea pig." Install the new system in this area, and let it go by itself for a short time. It can turn up many surprises, things overlooked, miscalculated, or under-estimated. The pilot installation also indicates how long the larger installation will take, permitting revision of the installation schedule.

Selecting the spot for a pilot installation is tricky, and almost entirely dependent on the type of system being installed. Keep in mind:

a. Location

   Ideally, it should be a spot slightly removed from the main stream of the new system, but still a spot which will actively test all portions of the system and with sufficient volume to strain the capacities of the system.

b. Timing

   Start the test in advance of the regular installation by enough margin to institute whatever changes are indicated and test them as well. Then the larger installation can be approached with evidence to prove the workability of the new system.

c. Feedback

The pilot or test installation should be fully controlled to get the maximum feedback. "Fully controlled" means to a greater extent than planned on the full installation, because even the slightest problem needs detection before it mushrooms.

### Execution

The installation task is now ready to start. But who should bear the responsibilities of the actual installation? This new system is the brain-child of the systems analyst. Shouldn't he be the one to carry the ball? On the other hand, line managers are the ones who agreed to adopt the system because of all its inherent benefits. Why shouldn't they handle the installation task? Actually, during the planning and scheduling phase, the analyst (with the aid of line management) should have clarified who is responsible for each activity during installation. The proper assignment of tasks and responsibilities will normally vary according to the nature and scope of the system, but there are some general principles that can be applied to the execution of installation plans.

*Role of Line Supervisor*—The departmental supervisor normally is the key man in all installations. The people doing the work report directly to him, obtain their instructions from him, and frequently perform in direct ratio with his enthusiasm and drive. If the new system contains some of his own ideas, he will gladly accept the responsibility for making it work. But if the new system was jammed down his throat, his feelings are going to be something less than enthusiastic! In either event, however, he still has the responsibility and authority to make the system work.

*Role of Systems Man*—Even though the supervisor has the responsibility, the systems man can't just sit quietly by and watch him struggle! The systems man has to be readily available for advice, counsel, and to serve as an extra pair of hands whenever needed. He should actively participate in training the employees and offer whatever assistance he can. Generally, he is used as a teacher, repairman, manual laborer, and is "the cause of all this trouble" simultaneously. During the installation, he'll catch all the blame for fall-downs. (But sometimes he's also handed a portion of the credit when the thing actually works.)

*Role of Participants*—Workers may have no responsibilities for the installation as a whole, but each is the king-pin in the small area around his desk or work area. They can make or break the system, and the direction they go depends chiefly on the training and amount of information they receive. Here's where the systems man can do the most good by working directly with the participants to make sure they understand the system as a whole, the reasons for the change, and the details of their particular function.

*Duration of Installation Period*—The installation period should continue until the systems man is fully satisfied that all the problems are solved, that everyone is fully trained, and that the whole thing will run without him. He should be sure the interflows of the new system with all adjacent systems, are operational and thoroughly understood. Effective pre-planning will, of course, shorten the installation period.

*Remedial Action*—The control devices built into the installation plan will point out whether the concepts or results of the system are deviating from established standards. When this happens, swift and accurate changes must be made. Isolate the problem area, develop an alternate approach, and put it into operation as soon as possible so that the installation may proceed.

### Cost Conditions

The new system is designed to save money and time; but, during installation, costs may exceed those of the old system. This is not uncommon. The seasoned analyst anticipates this condition and informs management of this fact during the selling phase of the study. Installation costs are usually high because of:

*Duplicate Effort*—In a parallel plan, two systems are operating at approximately double the cost of the old one. In any installation, there also will be a certain amount of backtracking, or re-doing of work, despite the best laid plans. Naturally, it will prolong the time required to do the job and inflate the costs.

*More Strenuous Controls*—These mean additional clerical effort to maintain the controls, as well as slowing down the data flow by interjecting more stops and pauses. Controls may also demand that additional information or records need to be maintained—adding to the time and cost involved.

*Employees' Learning Curves*—It takes longer to do a new job than it does to do the same job after months or years of practice. The amount of additional time varies by the indivdual, the degree of change, and the complexity of the job; but don't overlook this very real factor in estimating installation costs.

*One-Time Costs*—Installations normally call for a concentrated effort of both employees and equipment, and in addition often demand special forms and machines, or people. This is mentioned in a previous chapter, but is worthy of repitition.

### Human Factors in Installation

The expression, "I never know what is going on around here" is a manifestation of the problem which often ruins a systems installation. It reflects the fact the systems analyst has not properly communicated the nature, purpose and methods of the new system to the employees. As mentioned earlier, the need to communicate and the need to recognize the importance of the system participants must be met by the analyst. His first act must be a conscious recognition of human nature and values. An employee needs to satisfy more than his economic needs in the work place. He brings with him to the job his entire person, not merely a pair of hands or a brain. The human needs of the person must be satisfied by work; they have a natural desire to "feel that they count"; a desire for recognition.

When planning and executing a system transition, remember that you are dealing with people. Their desire to be "in on things" and to feel a part of something new is a very real force. The analyst should allow them an opportunity to participate in the planning of the new system or conversion. The satisfaction of social goals in the work place is an inherent part of any job, and although the corporate goals are most important in systems planning and installation, the personal goals of the participants cannot be overlooked. The dignity of the worker is a strong force and must be recognized by the analyst.

### FOLLOW-UP

The prime purpose of a follow-up step can be twofold: to consolidate all gains made by the new system, and to eliminate any losses. After installing the new system and getting everyone thoroughly familiar with new duties and responsibilities, the systems man normally leaves the scene. But oftentimes, without his support and assistance, the new system starts backsliding, the benefits to be realized are compromised, and almost everyone involved starts adding his own interpretations or innovations to the program.

This latter point is excellent within itself, but too many people doing this will nullify any potential gain and create only confusion.

There is a good deal of confusion about what to look for during a follow-up visit, and, indeed, about how and when to do it. There is no set formula that can be applied—so much depends on the complexity of the new system, how long ago it was installed, etc. If, however, the follow-up is approached with a little planning, it can become more than just a social visit.

During the installation, a number of checkpoints were established so that progress could be immediately determined—if these points still exist, they are the logical first step in the follow-up visit. If these checkpoints were eliminated after a satisfactory installation, the follow-up visit becomes more difficult and nebulous, but there are still a few barometers available:

### Are Costs in Line?

Clerical and operating costs, if available, are the best clue you can find. If costs have gone up since installation, someone is doing too much, or there's a bottleneck, or excessive errors are being made, etc. If costs have gone down, trouble may be brewing because some procedural steps are being skipped.

### Are Facilities Adequate?

Check the office equipment being used in the new system—is it handling the actual volume of transactions? Is it holding up mechanically? Is it being operated properly? Is the filing system working? Has everyone enough room? Are the new forms doing the job, or do they need revising?

### Employee Performance and Acceptance

Is everyone happy with the change? Do workers recognize improvements (if any)? Interview all concerned, not just the supervisor, but remember to do it with the supervisor's knowledge and blessing.

### Evidence of Falling Back to Old System

This is a critical area, because this sort of backsliding can devour everything gained in making the change. When the systems man leaves the installation, there is no one immediately available to answer questions or solve problems (this, of course, isn't necessarily the case if there is a good, strong line supervisor involved). This leaves the employee to his own devices, and the only device readily available is "the way we used to do it." At the first signs of anything like this going on, it's time to get back into action and start re-installing the system from the ground up.

### When to Follow-Up

This depends almost wholly upon the complexity of the new system.

Assuming it is fairly involved and of sufficient scope, there are two alternatives:

a. *Continuous Theory*

Following-up every time the area is visited or any of the participants are encountered. (Use discretion on this).

b. *Spot Check Theory*

This entails establishing a call series at specific times for the express purpose of follow-up.

1. First Check

After installation, it's advisable to look into matters (possibly one or two key points) within 30 days. If it's a very simple change and everyone was enthusiastic about it to start with, the date of the first check can be stretched out as far as six months out. Never longer than that.

2. Frequency Thereafter

Subsequent follow-up is strictly dependent upon the complexity of the system, its acceptance, and the systems man's feelings toward it. If he feels happy with it, confident in its acceptance and techniques, then he can stretch out the follow-up visits or drop them completely after the first one. If, however, everything wasn't smooth on the first check, he'd better look into it again within the month, etc., etc.

*Who Should Follow-Up?*—The first follow-up should be done by the systems man who did the installation. No one else knows it as well. Other people are following-up, in the course of their jobs, so contact with them is frequently helpful. For example, the company internal auditor gets into every department at least once a year, and checks several things. He's interested in verifying compliance to published company policies and procedures. Obviously, this can be a big help. The line supervisor is contacting all the participants in the new system almost every day. Talk to him and get his "feel" for the system. If he's a strong supervisor, the whole matter of following-up can be dropped in his lap with the understanding that he'll call for help when something crops up.

The systems man has the prime responsibility, however. Personal contact with those involved is almost the only way to do the job. Certainly, it's the preferable way. The first stop back should entail interviewing the actual system participants. If they have a complaint, either real or imagined, the system is probably in trouble. Don't overlook the factor, either, that a brief check with each employee helps cement relations by showing personal interest in his job and in him. Who knows? The next system may be in that same area!

While talking to the participants, peruse the routine control logs or status sheets maintained as a part of the system. Unhealthy spots usually stand out. Finally, talk to those people in adjacent departments who receive information or documents from the new system. They are usually quite happy to point out all the shortcomings, both real and unreal. The biggest problem is to sort out these complaints from those truly detrimental and those that are just "sour grapes."

### Why Follow-Up?

In addition to consolidating gains and correcting misunderstandings, there are several other reasons for conducting a follow-up campaign. Here is a summary:

1. *Feedback on System*
2. *Possible Revisions*
3. *Prevent and Discourage Backsliding by Participants*
4. *New Ideas and Improvements*
5. *Justification Reports to Management*

# COST ANALYSIS

## OF A

# TECHNICAL INFORMATION UNIT

*Kjeld Klintøe*

SOURCE: From Kjeld Klintøe, "Cost Analysis of a Technical Information Unit," *Aslib Proceedings* 23 (July, 1972) pp. 362-371. Reprinted by permission of the publisher. Copyright © 1972 by Aslib.

A COST ANALYSIS programme is defined here as a project aimed at quantifying the balance of a working system by measuring the input in terms of resources allocated to the various parts of the programme and relating this to the output measured in terms of what has been accomplished. Cost analysis is primarily used when it is not possible to evaluate the benefits resulting from the programme.

Only occasionally is it possible to prove that an information service has created or contributed to the creation of real benefits, and then only with the help of users who are willing to assist in identifying and quantifying such benefits. Such a case is reported below.

At a meeting of industrial managers the accounts and cost distribution of a shipyard in Denmark were discussed. A question was raised about the total cost of internal transportation and materials handling. These costs were not known because the shipyard did not know how to identify or measure that kind of cost. A technical information officer present at the meeting was able to introduce the idea of frequency analysis techniques and arranged for two consultants to visit the manager of the shipyard.

The consultants carried out an analysis and proposed the establishment of a central service unit for transport and materials handling. This analysis was repeated one year after the establishment of the system in order to measure the extent of the improvement. The fee charged by the technical information service for the ignition process was £15. The investment in analysis, equipment and establishment of the transport system was £20,000 but the benefits gained by the shipyard were a saving running to more than £100,000 per year.

We have had a number of similar cases in our organization, not all reporting such spectacular benefits, but numerous enough to justify our belief in the economic advantages which industry can derive from a national technical information service.

Even though we may believe in the value of our work we might still be uncertain about how much to invest in an information service, especially when such units become more sophisticated in developing their techniques, creating their own science and demanding such complex and costly equipment as computers. A non-profit-making institute or unit has, just like a business enterprise, a need to carry out regularly an analysis of its activities; to analyse its market conditions, its commodities and services, as well as the development of its organization. It is a very useful exercise for the staff and it makes the discussion

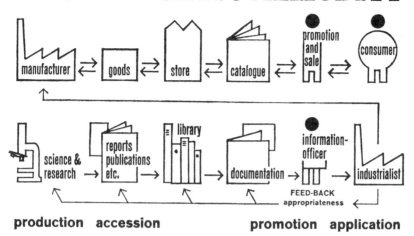

# KNOWLEDGE is a COMMODITY

FIG. I

with the board of directors and the sponsors more meaningful and professional.

I am invited, therefore, to speak about the cost analysis scheme of my own organization, Dansk Teknisk Oplysningstjeneste (Danish Technical Information Service). I thought it appropriate, even necessary, to include some general introductory remarks because it is my experience that a cost analysis scheme has no meaning unless we recognize that its main purpose is to evaluate our fundamental ideas of the business we are in and the policies which are behind our working programme.

DTO is a national service established in 1955, on the initiative of the Ministry for Commerce and Industry, to be part of the productivity programme for industry in Denmark. We are now operating as a private non-profit-making institute affiliated to the Danish Council of Scientific and Industrial Research and we rely on the Research Council, i.e. public money, for 85 per cent of our income. We are operating according to the third generation of our statutes, which means we have had two opportunities of making a fundamental business analysis, each time readjusting the programme conditions to the business environment in which we are working. What I am going to talk about is the current situation.

Our field of business is that of promoting the utilization by industry of the commodity named knowledge (Fig. 1). This means that we have to adopt the general principles of marketing technique, including:

1. Examination of the structure of the prospective market.
2. Diversification of demand.
3. Development and stimulation of demand.
4. Active presentation of assortments of 'commodities'.
5. Training in searching for and utilization of 'commodities' available.
6. Liaison between 'consumers', 'stores' and 'manufacturers'.
7. Reporting on the appropriateness of accessibility of 'commodities'.

We can analyse the market to show the geographical distribution of industry, or the distribution of manufacturing enterprises in terms of the number of people employed (Table 1), but none of these figures have any real bearing on our commodity.

TABLE 1: *Distribution of manufacturing enterprises by numbers employed*

| No. of enterprises | No. of persons employed |
|---|---|
| 5500 | 10–100 |
| 800 | 101–500 |
| 100 | more than 500 |

What we really need to know is the distribution in industry of 'receiving sets' for our commodity, knowledge, as measured, for example, by the number of engineers in industry (Table 2).

TABLE 2: *Distribution of engineers among industrial enterprises*

| No. of enterprises metropolitan area | provincial area | No. of engineers employed |
|---|---|---|
| 1450 | 1705 | 1–3 |
| 290 | 240 | 4–10 |
| 90 | 70 | 11–20 |
| 25 | 25 | 21–50 |
| 15 | 5 | more than 50 |

In these circumstances how can an information service promote better and more effective utilization of knowledge when the staff in the companies is so small that engineers cannot afford the time required for analysing and adapting, or even reading the material we can disseminate to them?

In Denmark there are a number of centres of specialized knowledge, e.g. the Technological Institutes in Copenhagen and Aarhus, R & D institutes which are part of the Academy for Technical Sciences and laboratories located at the technical universities, etc., who also employ technical staff (Table 3). How could we mobilize the potentials of these centres? The answer was to establish a personal contact and dialogue between the industrialist with a problem and the specialist with the most relevant and appropriate knowledge and experience.

TABLE 3: *Distribution of engineers among the specialized centres*

| No. of centres metropolitan | provincial | No. of engineers employed |
|---|---|---|
| 208 | 224 | 1–3 |
| 76 | 54 | 4–10 |
| 27 | 19 | 11–20 |
| 20 | 8 | 21–50 |
| 8 | 8 | more than 50 |

The programme developed covered six major aspects, viz:

1. Field liaison service.
2. Active information and loan service.

3. Conference, course and training service.
4. Question and answer service.
5. Advisory service on setting up industrial information services.
6. Specialized information service

with the emphasis on DTO taking the initiative at all times.

The first step was to establish the field liaison service with the following aims:

1. Visiting firms uninvited.
2. Stimulating demand for knowledge by interviews.
3. Promoting sources of specialized knowledge available,

and obtaining feedback on the distribution of fields of interest and receptivity and on the appropriateness of the sources and services available. Then, having adapted ourselves to the client's environment and having obtained an understanding of his needs, the next step was to provide an intelligence service on behalf of the clients, evaluating and extracting from various sources of information those items which could stimulate his demand:

1. Searching information and evaluation according to the dynamic 'profile' of the individual firm.
2. Selective dissemination,

the feedback in this case being requests for information, documents and service.

This was accompanied by a programme of conferences and courses designed to motivate a group of industrialists into demanding more information, documentation, library service, consultation, research, etc.

All these services were provided without charge as a matter of policy.

If, as a result of this activity, we were successful in stimulating the interest of industry, we expected a demand for our services in identifying sources of information, providing special services, specialist assistance, special information, etc. This would be charged for, because at this stage we would be working as a supplement to, or extension of, the staff of the industrial organization.

To maintain and develop our experience and capability, supplementary programmes are necessary. We need an intimate knowledge of the domestic network of specialized information centres: we need to acquaint ourselves with the individual centres and their staffs in order to know which organizations and individuals are best qualified to assist industry. This programme of national collaboration also covers the service and assistance given to these centres to improve their own promotion to industry.

There is also an international collaboration programme which is both bilateral and multilateral. We are given the task of representing Denmark within the field of technical information in OECD, FID, UNIDO, etc. At present we operate the secretariat for the FID/II Study Committee on Information for Industry. We also act as the Danish Referral Centre within science and technology, as well as being the adviser to the Ministry of Foreign Affairs with regard to the group of scientific counsellors and attachés in Bonn, London, Paris, Washington and New York. We are the Danish representative on the Steering Committee of the Scandinavian Documentation Center in the USA.

I must apologize for this rather lengthy account of our policies and programmes but it is necessary to explain this background in some detail in order to show why we consider it necessary to undertake a cost analysis of the programme and how we can learn from the feedback obtained from this analysis.

The analysis of the field liaison service (Table 4) indicates the total number of visits paid during one fiscal year and the distribution among categories of units visited, plus an analysis of the geographical distribution of visits, and shows how we are covering the country in relation to the distribution of industry, and also the average cost of a visit.

TABLE 4

**Costs**

**1. Field Liaison Service**

| | |
|---|---|
| Salaries, engineers (3476³/₄ h) | Dkr. 161,365 |
| Salaries, office staff (639¹/₂ h) | -    11,172 |
| Consultants | -     2,363 |
| Transportation and car-fare | -    17,880 |
| Hotel and board allowances | -     7,365 |
| Miscellaneous (promotion materials) | -     1,491 |
| | Dkr. 201,627 |

Cost per visit: approx. Dkr. 443
Cost per enterprise: approx. Dkr. 500

**Accomplishments**

455 visits to 403 diff. enterprises, i.e.
  1 organisation
  68 centres of specialized knowledge
  332 manufacturing companies
  2 public services

Geographical distribution of visits:

| District | Number |
|---|---|
| North Jutland | 12 |
| Middle Jutland | 50 |
| South Jutland | 37 |
| Funen | 39 |
| Zealand (less Greater Copenhg.) | 53 |
| Greater Copenhagen | 148 |

Free of charge.

The selective dissemination service is based upon an intimate knowledge of the industrial enterprises, obtained through our visits and by taking part in management training courses. The input to the system includes the scanning of a number of foreign sources of information, i.e. journals, reports, indexes and special materials bought or given to us by colleagues abroad (Table 5).

Whereas cuttings (abstracts, digests, summaries) and articles speak for themselves, it may be useful to say a little more about books and reports. We have a small library, one part of which contains reference books, directories, annual reports of research establishments (Danish and foreign); the other contains books and reports on individual topics within technology, research and development, industrial experience, economics, etc. The latter collection is highly selective and books are purchased only if they are relevant to a particular industry, have a bearing on the state of the art and its development, are written in language which will motivate industrialists to read them, have an attractive layout, and are of interest to at least seven companies or institutions. These books are circulated on our initiative to selected companies just to persuade them to buy the book for their own collection. Borrowers are expected to return the books within eight days.

As all users are registered by number in the files of DTO it is very easy to count the number of items of information of various kinds sent to the individual users within the different kinds of organization, and so obtain an analysis of the

TABLE 5

| Costs | Accomplishments |
|---|---|

**2. Active Selective Information Service**

| Costs | |
|---|---|
| Salaries, engineers (2407³/₄ h) | Dkr. 116,747 |
| Salaries, office staff (7542³/₄ h) | - 132,385 |
| Consultants | - 16,003 |
| Information materials | - 29,779 |
| Postage | - 34,953 |
| | Dkr. 329,867 |

**Accomplishments**

**Cuttings:**
6606 to 987 diff. receivers, i.e.
     27 organisations
     203 centres of spec. knl.
     504 mfg. companies
     253 individuals

**Articles:**
4603 to 1099 diff. receivers, i.e.
     31 organisations
     200 centres of spec. knl.
     558 mfg. companies
     310 individuals

**Lendings of books, reports etc.:**
2459 to 724 diff. receivers, i.e.
     22 organisations
     123 centres of spec. knl.
     461 mfg. companies
     118 individuals

**Total:**
13668 to 1510 diff. receivers, i.e.
     46 organisations
     283 centres of spec. knl.
     738 mfg. companies
     approx. 440 individuals

Cost per unit: approx. Dkr. 24
Cost per receiver: approx. Dkr. 218

Free of charge.
Reaction: DK/P and DK/L tasks.

total dissemination. The response from the users takes the form of orders for specific documents (Table 6). During the year 31.5 per cent have ordered reports and other publications for which they are charged the cost, plus a small fee of approximately 75p per order. The fee is kept as low as possible to ensure that they respond to DTO, even though we know that the normal policy of many companies is to purchase locally. When delivery time is important we know we will always receive the order because we can provide documents by airmail in one-fifth of the time taken by local libraries and bookshops. However, our

TABLE 6

| Costs | Accomplishments |
|---|---|

**a) Requests for documentation (DK/P and DK/L tasks)**

| Costs | |
|---|---|
| Salaries, engineers (21¹/₂ h) | Dkr. 844 |
| Salaries, office staff (2120 h) | - 37,229 |
| Purchase of information materials and postage expenses | - 34,287 |
| Miscellaneous | - 1 |
| | Dkr. 72,361 |

Cost per request: Dkr. 52

**Accomplishments**

The reactions to 6606 cuttings + 2459 lendings have been:
537 diff. clients (31.5 %) have given 1401 orders
comprising 2261 diff. reports, documents, publications etc.

| | |
|---|---|
| Fees invoiced | Dkr. 23,775 |
| Refunding of outlays | Dkr. 40,721 |
| Total | Dkr. 64,496 |

Turnover per order: Dkr. 46

policy is not to compete with the book trade in areas other than relevance of material and speed of provision.

The initiating work of DTO aims at stimulating industrialists to ask questions: to request information for solving problems, for development of products methods, processes, etc., or to be used for long-range planning (Table 7). About two thousand requests are answered by telephone, just from the files of DTO, but 213 tasks were of a more serious nature demanding some measurable time from one or other of DTO's engineers. The requests are analysed by type of inquirer and by the kind of task, just as the topic is registered so that it can be retrieved if necessary. The inquirers are charged a fee but, as you can see, one which is not, at present, high enough to balance the cost. We would like to achieve a balance because we want industry to appreciate the value of the answers, since such an appreciation will influence them in making more effective use of the material provided.

Each request is handled confidentially, mainly by consulting Danish and foreign information analysts or specialists and only to a small extent by the normal documentation service.

TABLE 7

| Costs | Accomplishments |
|---|---|
| **b) Requests for procurement of information (DK/S tasks)** | |
| Salaries, engineers (2594 h)           Dkr. 118,333 | 213 tasks from 116 diff. inquirers, i.e. |
| Salaries, office staff (914³/₄ h)        -     15,994 | 3 organisations |
| Consultants                              -      5,117 | 8 centres of spec. knl. |
| Information materials                     -      5,127 | 104 mfg. companies |
| Transportation and car-fare              -      4,875 | 1 public service |
| Hotel and board allowances               -      1,295 | |
|                                         ─────────── | 213 tasks have been: |
|                                         Dkr. 150,741 | 199 – procurement of information |
| | 14 – planning and management |
| | of courses and conferences |
| Cost per request: Dkr. 708 | Turnover per order: Dkr. 606 |
| **c) Requests by telephone** | |
| Salaries, engineers (660 h)            Dkr.  31,602 | Approx. 2000 requests. |
| Salaries, office staff (86³/₄ h)         -      1,524 | |
|                                         ─────────── | |
|                                         Dkr.  33,126 | |
| Cost per request: approx. Dkr. 17 | Free of charge. |
| **d) Requests from foreign centres or from own organisation (DK/U and DK/AD tasks)** | |
| Salaries, engineers (339¹/₂ h)         Dkr.  15,826 | 74 tasks, i.e. |
| Salaries, office staff (101³/₄ h)        -      1,834 | 58 foreign |
| Information materials                     -      1,386 | 16 internal |
| Transportation and car-fare              -        353 | |
| Hotel and board allowances               -        100 | |
|                                         ─────────── | |
|                                         Dkr.  19,499 | |
| Cost per task: approx. Dkr. 264 | Free of charge. |

Table 8 analyses the programme of courses and conferences involving groups of users or prospective users and resulting in the establishment and operation of specialized information services geared to the needs of a branch of industry.

TABLE 8

**Costs**

**Courses and Conferences**

| | | |
|---|---|---|
| Salaries, engineers (100¼ h) | Dkr. | 5,830 |
| Salaries, office staff (279¼ h) | - | 4,805 |
| Consultants | - | 3,300 |
| Transportation and car-fare | - | 1,447 |
| Hotel and board allowances | - | 11,039 |
| Information materials, including visual aids, etc. | - | 5,521 |
| | Dkr. | 31,942 |

Conference cost per day: Dkr. 7,986

**Accomplishments**

**Conferences:**

| | Fee | Outlays |
|---|---|---|
| Group of architects | 5,500 | 15,399 |
| Ass. of factory inspectors | 4,000 | 1,225 |
| | 9,500 | 16,624 |

Conference days: 4

Conference fee per day: Dkr. 6,531

**Specialized Information Services**

| | | |
|---|---|---|
| Salaries, engineers (773½ h) | Dkr. | 33,900 |
| Salaries, office staff (387¾ h) | - | 6,672 |
| Information materials | - | 1,583 |
| Transportation and car-fare | - | 3,596 |
| Hotel and board allowances | - | 130 |
| Postage etc. | - | 160 |
| Office materials | - | 136 |
| Outlays (membership fees) | - | 3,677 |
| | Dkr. | 49,854 |

**Clients**

| | | |
|---|---|---|
| Pump-manufacturing Industry | Dkr. | 40.000 |
| Radiator-manufacturing Industry | - | 21,677 |
| | Dkr. | 61,677 |

Tasks given to the DTO by the Research Council consist of collaboration with domestic centres oriented to serve industry and also the furthering of international collaboration (Table 9). Maintaining and improving the qualifications

TABLE 9

**Costs**

**6. Collaboration with Domestic Centres**

| | | |
|---|---|---|
| Salaries, engineers (2048¼ h) | Dkr. | 109,450 |
| Salaries, office staff (511 h) | - | 9,045 |
| Consultants | - | 788 |
| Transportation and car-fare | - | 2,708 |
| Hotel and board allowances | - | 734 |
| Membership fees | - | 1,388 |
| Information materials | - | 492 |
| Miscellaneous | - | 25 |
| | Dkr. | 124,630 |

**Accomplishments**

Collaboration with departments of Universities, the Academy of Technical Sciences, Engineers' Associations, organisations of trade, public authorities, etc.

**7. International Collaboration**

| | | |
|---|---|---|
| a) Support of SCANDOC (USA) | Dkr. | 45,903 |
| b) Membership of foreign learned societies | Dkr. | 8,933 |
| c) Participation in international organisation: | | |
| Salaries, engineers (619¾ h) | - | 35,786 |
| Salaries, office staff (279¼ h) | - | 4,972 |
| Transportation and car-fare | - | 7,720 |
| Hotel and board allowances | - | 4,335 |
| Miscellaneous | - | 460 |
| | Dkr. | 62,206 |

Nordforsk, SCANDOC, OECD, ETC, FID, FID/II, ASLIB, AID, National Industrial Conference Board, American Management Association, British Institute of Management, International University Contact, etc.

TABLE 10

**Costs**                                      **Accomplishments**

**8. Study Tours and Post-Graduate Training**
Salaries, engineers (2377³/₄ + 333³/₄ h)         **Study Tours**
                              Dkr. 110,191        for 3 engineers.
Salaries, office staff (196¹/₂ h)    -    2,591
Conference and course fees           -   14,416   **Post-Graduate Training**
Transportation and car-fare          -    3,722   of 6 engineers.
Hotel and board allowances           -    3,223
                              Dkr. 134,143

Study Tours:          Dkr.  20,585

Post-Graduate Training: Dkr. 113,558

**9. Planning and Administration of
   Own Organisation**
Salaries, engineers (2421³/₄ h)  Dkr. 131,275    Regular meetings and communication of
Salaries, office staff (6534³/₄ h)  -  113,958   staff and board of directors, internal com-
Transportation and car-fare      -     7,218     mittee meetings.
Miscellaneous                    -       229
                              Dkr. 252,680        Bookkeeping, secretariat, reception desk,
                                                  telephone exchange.

of the DTO staff account for about 7 per cent of the total budget, while about
12 per cent is allocated to planning and administration (Table 10). The general
expenses of DTO, i.e. those expenses which cannot be allocated to any specific
part of the working programme, amount to about 27 per cent (Table 11). Less
than 2 per cent of the engineers' working time and less than 1.5 per cent of the
working hours of the office staff are unspecified.

TABLE 11

**Costs**                                      **Accomplishments**

**10. General Expenses**
Salaries, engineers (638¹/₂ h)  Dkr. 29,114     **Engineers**
Salaries, office staff (362¹/₄ h)  -   6,378    Total of hours applied              21,814³/₄ h
Consultants (PR, visual aids,                      less sick-days    6.8%    1,478¹/₂  -
                          etc.)  -    4,818        less holidays     9.1%    1,544    -
Sick-days, engineers (1478¹/₂ h) -  71,724        less consultants  1.6%      354    -
Sick-days, office staff (1686 h) -  29,637        less over-time    7.3%    1,592¹/₂ -
Holidays, engineers (1330¹/₂ h)  -  63,946     Standard working hours           16,845³/₄ h
Holidays, office staff (1632 h)  -  28,500
Brochures, business forms,
               printed materials -  15,501     **Office Staff**
Handbooks, directories, etc.     -  32,921     Total of hours applied              24,294  h
Office materials                 -  42,995        less sick-days    6.9%    1,686    -
Office equipment                 -  15,755        less holidays     9.7%    1,875¹/₂ -
Telephone, cables, postage       -  22,099        less canteen      3.3%      796    -
Entertainment                    -   2,528        less substitutes  1.4%      332    -
Office premises, rent, etc.      - 151,673        less over-time    0.7%      169¹/₂ -
Canteen expenses                 -  15,298     Standard working hours           19,435  h
Substitute office assistance     -   5,304
Miscellaneous                    -   5,701
                              Dkr. 543,892

Total costs:  Dkr. 2,052,471

Let me try to anticipate some of the questions you may wish to ask about the procedures we use to collect and evaluate the data necessary for the cost analysis, beginning with the analysis of working hours and the conversion of this into monetary terms. We use a weekly time sheet on which the staff record the amount of time spent each day on various tasks relating to the working programme, with a precision limit below half an hour. The sheets are collected each Monday and a weekly summary is made to the technical and office staff, from which a quarterly summary is compiled for each member of the staff. Using a standard salary per hour the salary expenses allocated to the various parts of the working programme can be calculated.

The standard salary per hour is defined as the total expenses for each person for salary, pension, insurance, etc., divided by the number of effective working hours, i.e. not counting vacations, weekends, national holidays, etc. Overtime is registered, not paid, but counted in the analysis. On the back of the time sheet, notes are made of important contacts made during the week and important cases dealt with.

The statistics on jobs accomplished are compiled on a weekly basis and every quarter a cost analysis sheet is produced. This continuing programme is so well managed by the staff that I normally receive the quarterly cost analysis within ten days after the end of the quarter.

What benefit do we derive from this quarterly cost analysis? In the first place we have given the staff a means by which they can identify their own contributions to the performance of the unit. Further, we have made them cost minded, which is indeed very necessary in a non-profit-making organization supported by the taxpayer's money. We are able to discuss our working programme and policy with the staff on the basis of facts contributed by themselves, and discussions with the board of directors can be conducted on the level of policy issues, not on detailed expenditure. As an extra benefit, we found that the public auditor stopped criticizing our somewhat untraditional working methods because he can see exactly how the money is spent and the use to which it is put. For the same reason the Research Council can have confidence in the DTO, so that the decision to continue its support, now running at two million Danish kroner per annum, can be taken purely on policy grounds.

# RESHELVING CIRCULATED LIBRARY MATERIAL IN A UNIVERSITY LIBRARY

*Robert E. Erisman*

SOURCE: From Robert E. Erisman, *Returning the Circulated Book to the Shelf; a Study of the Shelving Operation in a University Library*. Unpublished Research Study. Boulder, Colorado, 1971. By permission of the author.

## INTRODUCTION

The purpose of the study was to utilize concepts of scientific management to survey, evaluate, and make recommendations concerning the shelving operation at the main library of the University of Colorado, Boulder. The objectives of the study were: (1) To collect and analyze data that could be used by the library management in the evaluation of the present shelving operation and in the consideration of improvements in that operation, (2) To collect data that could be used as the basis for cost studies and performance budgeting, and (3) To contribute to the development of performance standards.

### The Environment

Norlin Library houses about 800,000 volumes. During the twelve-month period ending December 31, 1970, there were 267,000 volumes charged out for "home-use" through Central Desk at Norlin. Approximately that same number of books was shelved during 1970.

As the size of a library's research collection grows, and more demands are made upon it, the maintenance of the collection in good order attains increasing importance. The responsibilities of the Circulation Department include charging out, through the Central Desk, volumes located in the central stacks and elsewhere in the building, and the discharging and reshelving of those books when they are returned.

### The Problem

A user survey at the library showed that one of the more common complaints was that of slow reshelving. Also, interviews with several members of the library staff have pointed to reshelving as a major problem area. Further, inasmuch as the return of books to the shelves in order to make them again available for use is an area of almost universal concern, it was thought that a study of shelf work could be of value to other libraries.

### Scope and Limitations

The focus of the study was on the processes required to return a book to its proper location on the shelf after its receipt by the library from the patron who had checked the book out and used it at home. With such a focus, it was necessary to cut across organizational lines to study processes that were outside the province of the shelving unit, though still a part of the Circulation Department. On the other hand, this focus made it unnecessary to examine *every* function of the shelving unit beyond a depth sufficient to determine whether it was a suitable area for investigation in accordance with the purposes of the study. It was found that five main processes were relevant: (1) Removing Books from the Book Drop (Routing), (2) the Conversion of Numerics (a part of the Conversion process used to charge out books which did not have the punched card needed for the automated system), (3) the Discharging of Returned Books, (4) the Sorting of Returned Books into Call Number Order, and (5) the Shelving of Returned Books at Their Proper Locations.

There were two major limiting factors in the study: time, and the transitional nature of operations in the Circulation Department. The brief duration of the study period (five weeks) affected the amount of data collected, and therefore the validity of the findings.

The transitional nature of the operation was manifested most noticeably in the job rotation plan that was in effect at the time of the study. Many staff members were performing duties usually assigned to others. Also, several key staff were to terminate soon, including the head of the department, his chief clerical assistant, and the Stack Supervisor.

It could be desired that a more stable system had existed at the time of the study. On the other hand, perhaps the present study offers a more realistic picture of library processes in their fluid environment than would a study of a seemingly placid operation.

## METHODS AND PROCEDURES

### Methods

Ordinary methods of data collection and analysis were employed. The information gathered in the study did not lend itself to sophisticated statistical techniques. The investigation was a simplified "systems study," with emphasis on the survey phase. With the exception of a few devised by the analyst to fit the needs of the study at the time, the data-collecting methods and instruments used were adapted as needed from those presented by Richard M. Dougherty and Fred J. Heinritz in their *Scientific Management of Library Operations* (New York, Scarecrow Press, 1966), and Edward A. Chapman (and others) in *Library Systems Analysis Guidelines* (New York, Wiley-Interscience, 1970).

Objectives and requirements were obtained through interviews with the library management and other library staff members, including the head or primary assistant of each branch or department housed in Norlin Library. Information on the characteristics of the system, such as the personnel, facilities and equipment, and records, was obtained by means of questionnaires, interviews, and direct observation. Operations were documented on flow process charts through observation, with occassional confirmation of findings by interview. An organization chart of Circulation Department personnel was drawn up.

Time studies were conducted using direct observation, with infrequent help from the subject in determining the number of books being handled each time. No diary studies were conducted. It was the policy of the analyst to gather data as unobtrusively as possible in order to minimize the influence on the work flow of the measurement process itself.

### Procedures

The study was conducted over a five-week period. The first week was a time of preliminary survey and development of a plan of attack. A literature search was conducted. In the second week, the beginning stages of investigation were performed in all areas to be studied. The analyst immersed himself in the system and became familiar with the system and the people operating it. Further, the analyst became familiar *to* the people operating the system so that he could later observe their work without significantly affecting their behavior because of the presence of a "stranger."

The third week saw the continuation of activities previously initiated plus the beginning of the charting of operations. The fourth week was primarily one of conducting the first set of time studies in the various areas. For the most part these turned out to be *practice* studies—both for the subjects and for the analyst. The investigation of personnel and facilities was completed. In the fifth week the second set of time studies was conducted and all other study activities were concluded.

The time studies were conducted as follows:

1. *ROUTING.* (Removing books from book drop) The analyst started the stopwatch as the subject touched the first book to pick it up. The analyst counted the books as they were placed on the Routing Shelves. As the last book was placed on the shelf the analyst stopped the watch, recorded the number of books, and recorded the time elapsed.

2. *CONVERSION OF NUMERICS.* The stopwatch was started as the first book was touched to be picked up for conversion. Upon completion of the process for that book, the subject returned the book to the shelf of the book cart. As the subject touched the next book to pick it up, the analyst took note of the elapsed time for the first book, reset the stopwatch to zero (still running), and recorded the time. This sequence was repeated for at least ten observations at any one sitting—sometimes twenty—so that "strings of observations" could be gathered in order to view the process with wide perspective, rather than to isolate individual actions in the Conversion process. For similar reasons, groups of books were timed from touching the first book on the cart to placing the last book back on the shelf of the book cart. Approxi-

mately one shelf was timed on each occassion. All times include any special problems or other interruption that the subject would normally be faced with while converting.

It was observed by the analyst that the subject sometimes performed an extra task in the Conversion process. Before starting the regular Conversion process, she would sometimes arrange the shelf of books in call number order—a kind of pre-sorting process. The analyst recorded some times for this.

3. *DISCHARGING*. The stopwatch was reset to zero (still running) each time the data collector made a sound as a result of the insertion of the book card. The time was recorded immediately after each reset. As in the Conversion process, *groups* of books were also timed. Also, the Arranging of books in call number order took place upon occasion and was timed.

At first, the analyst timed from touching the first book to touching the next book but found this visual signal to be less precise than the audible signal finally used.

4. *SORTING*. The stopwatch was started as the subject touched the first book— whether to arrange an entire shelf in order, to pick the book up to carry it to the sorting shelves, or to pick it up as one of a group of books (handful) he would then take to the sorting shelves. The subject was asked to call out the number of books picked up each time, and the analyst recorded the number. When the subject returned to the cart to touch the first book of the next group, the time elapsed was noted, the stopwatch was reset to zero (still running), and the time was recorded for the number of books that had been announced previously. Near the end of the study the analyst was able to position himself to count the books without having the subject call out the number of books.

5. *SHELVING*. The stopwatch was started as the shelver touched the first book of the group (handful) he was to shelve. The analyst noted and recorded the number of books picked up each time. When the shelver returned to touch the first book of the next group on the cart, the analyst noted the time, reset the stopwatch (still running), recorded the time elapsed, and looked up to count the number of books in the next group.

## RESULTS

### Characteristics of the System

1. *OBJECTIVES AND REQUIREMENTS*. Interviews with the library management indicated that the objectives of the library are to serve the research and teaching needs of the university and, in a more specific sense, provide the right book to the right person at the right time. Objectives for the shelving operation are mainly refinements of those for the Circulation Department as a whole. They include the maintenance of the collection in good order to maximize accessibility, and the prompt return of the book to the shelf after its receipt from the patron.

Performance requirements for the shelving operation include a 24-hour turn-around time for the shelving of books. Reports are required for statistics and for evaluation of the operation. The average rate of shelving is expected to be about 100 books per hour.

2. *PERSONNEL*. The staffing of the Circulation Department includes one professional, nineteen clerical staff (18.5 FTE), and about 600 hours per week of student assistant time. It was determined that only certain members of the Circulation staff were involved in any way with the shelving operation. All of the clerical staff completed Job Description Questionnaires but interviews were not conducted for the library guards or the Reserve staff.

3. *FACILITIES, EQUIPMENT, AND CONDITIONS*. A floor plan of the Circulation work areas was drawn to show the spatial relationships involved. Photographs were taken of the various work areas of the shelving operation to show the facilities and equipment and give some idea of the conditions. The Circulation Department is located on the main floor of the library, just inside the entrance. The book drop is approximately thirty feet from the door. The Sorting area is in one corner of Tier 3 of the stacks, directly behind and adjacent to the other work areas of Circulation.

The equipment used in the shelving operation, in addition to ordinary library furnishings, consists of (1) an open bin file for punched book cards awaiting conversion, (2) seventeen "three shelf display" book trucks, and (3) the discharging data collector, a C-DEK 3213 Data Entry Station, manufactured by Colorado Instruments. The data collector is connected to a 3864 Central Controller (Colorado Instruments).

Conditions are difficult to describe objectively. Responses by the staff to the question on the Job Description Questionnaire indicated a concern for dust, dirt, noise and temperature problems. Distances between work stations were measured but nothing significant was discovered. In fact, the analyst was struck by what seemed to be crowded conditions; but the staff did not express serious concern about such a problem.

4. *FORMS, RECORDS AND REPORTS.* The analyst performed an elementary forms analysis on all forms used in the shelving unit but found only one form used directly in that part of the shelving operation surveyed in this report. Other records and reports were compiled on a weekly or monthly basis. Most of them drew statistics from the individual Shelv-Stat forms.

Only the Shelv-Stat form was used directly in the operation under study. The shelver would record for each cart load of books: (1) Time Left the Sorting Room, (2) Number of Items Shelved, and (3) Time Returned to the Sorting Room. The Shelv-Stat form was also used to record individual statistics in the Clearing operation.

## Processes

Time studies were conducted for each of the five main processes, using as subjects those persons primarily responsible for their elements of the shelving operation. Additionally, some studies were conducted of other subjects in order to provide some basis of comparison. A few related activities, such as Clear Shelving, were studied briefly to add perspective for evaluating the results of the primary time studies.

A flow process chart was written for each process, describing the activities involved in the process. Time study observation sheets were used to record the raw data. From these sheets the data was organized into a time study analysis table for each process.

Finally, all time study results were gathered into a Summary Analysis of Time Studies.

## ANALYSIS OF RESULTS

The results have been presented so as to allow a maximum of self-interpretation. The following comments, however, are appropriate and should help to place the results in better perspective.

### Characteristics of the System

*OBJECTIVES AND REQUIREMENTS* of the system were readily produced by management and by one class of "users" of the system, the various department heads in Norlin Library. However, these did not seem to be understood below the level of the head of the Circulation Department.

Interviews with the *PERSONNEL* revealed that many of the Circulation staff did not know why they were doing what they were doing other than because that is what they were told to do. One-half of the staff did not know their job titles.

The organization of the department leaves something to be desired. The organization chart shows a rather unwieldy arrangement of responsibilities. The Circulation Librarian (department head) has a span of control of eight persons, six of whom report to him on a direct, individual basis. The other two staff members directly answerable to the librarian are the Reserve Supervisor and the Circulation Supervisor. The former has direct control over two full-time staff and 140 student assistant hours per week.

The Circulation Supervisor has direct authority over ten staff members and almost 500 hours of student assistant time per week. This includes primary responsibility for the operation of the circulation desk, shelving, and Faculty Delivery Service, and general authority over all other persons in the department except for the Reserve Staff and the Information Supervisor.

There seemed to be a generalized morale problem but the analyst was not able to determine its cause or its specific consequence. The most frequent complaint was of a kind of arbitrariness imposed by the

## SUMMARY ANALYSIS OF TIME STUDIES

| Process | Name | Units per Hour | Number of Observations | Time in Seconds | Number of Books | Seconds per Book |
|---|---|---|---|---|---|---|
| Removing Books | Joy | 900 | 5 | 368 | 87 | 4 |
| Removing Books | Marion | 450 | 4 | 775 | 93 | 8 |
| Removing Books | Dorothy | 720 | 4 | 421 | 77 | 5 |
| Removing Books | John | 600 | 4 | 607 | 108 | 6 |
| Removing Books | Juanita | 900 | 2 | 159 | 37 | 4 |
| Removing Books | Tom | 600 | 2 | 676 | 107 | 6 |
| Removing Books | Greg | 900 | 1 | 206 | 47 | 4 |
| Removing Books | Bill | 720 | 1 | 111 | 24 | 5 |
| | TOTAL | 600 | 23 | 3323 | 580 | 6 |
| Conversion | Marion | 138 | 90 | 2360 | 90 | 26 |
| Conversion (1 shelf) | Marion | 138 | 8 | 6065 | 233 | 26 |
| Discharging | Dorothy | 252 | 80 | 1101 | 80 | 14 |
| Discharging | Marion | 222 | 60 | 951 | 60 | 16 |
| | TOTAL | 240 | 140 | 2052 | 140 | 15 |
| Discharging (1 shelf) | Dorothy | 240 | 7 | 2977 | 193 | 15 |
| Arranging (1 shelf) | Marion | 720 | 7 | 1000 | 195 | 5 |
| Arranging (1 shelf) | Dorothy | 720 | 6 | 825 | 161 | 5 |
| Sorting | Chris | 198 | 81 | 6936 | 391 | 18 |
| Sorting | Bill I | 210 | 30 | 3135 | 183 | 17 |
| Sorting | Bill II | 300 | 35 | 2875 | 240 | 12 |
| Sorting | Dave | 186 | 20 | 1717 | 90 | 19 |
| | TOTAL | 222 | 166 | 14663 | 904 | 16 |
| Loading Cart | Bill | 1800 | 2 | 745 | 324 | 2 |
| Shelving | Dave | 168 | 125 | 8066 | 385 | 21 |
| Clear Shelving | Charlie | 168 | 51 | 3695 | 176 | 21 |

department head and his primary clerical assistant, the Circulation Supervisor. Some staff members claimed that they did not participate in the decision-making process concerning the activities in which they were involved.

The *FACILITIES, EQUIPMENT AND CONDITIONS* were neither remarkably superior nor particularly inadequate. As in most libraries, one could wish for a more ideal situation but the present situation was accepted by most of the staff as "adequate."

The analyst was struck by seemingly poor lighting and crowded working conditions.

However, no one complained about the lighting and few mentioned anything about the lack of space. Again, as in many libraries, there were times when more book trucks would have been helpful, but there was no serious concern on the part of the staff in that area.

The automated circulation system had just been installed. It was impossible for the analyst to evaluate the equipment and its effect on the shelving operation and the people involved in this transitional phase.

There were four specific problems observed by the analyst: (1) The book drop is so constructed that books are almost certain to be damaged when they are deposited by the patron. (2) The Conversion File is on a work table that is too high for the subject to reach the cards comfortably and efficiently. A bin or tub file would be more appropriate.

(3) The entrance to the Sorting Room from the Discharging Area presents a barrier in the form of a crack in the floor separating the stacks from the main building, and a metal bar in the center of the crack. Book carts must be forced over this barrier. Often three of four attempts are made before a loaded book cart can be forced over the barrier. The analyst found only one cart with damaged wheels, but it seemed certain that more would appear in time. In any case, books sometimes fell off in the attempt and time was wasted.

(4) The elevator must be paged with a key on the first three tiers. In the past, keys had been available to all who needed them. However, at present there is only one elevator key. The Sorting Area is on tier three, so all shelvers with loaded carts to shelve must run upstairs to tier four to page the elevator. Likewise, those returning from tiers one and two must make special arrangements. Some shelvers on tier one ask the Periodicals Annex attendent for a key, page the page the elevator, turn it off, return the key, and continue to the elevator.

*FORMS, RECORDS, AND REPORTS* were for the most part incidental to the shelving operation.

## Processes

Depending upon the reader's preference, the most important figures in the Summary Analysis of Time Studies are those either showing "Seconds per Book" or those for "Units per Hour."

The comments at the end of each time study analysis table explain the idiosyncrasies of individual results. Most of the time studies were brief enough to be called "preliminary studies." The secondary studies that were completed served for the most part to confirm the results of the primary studies.

The times involved in returning a book to the shelf include 6 seconds to remove a book from the Book Drop, 26 seconds to

Convert, 15 seconds to Discharge, 5 seconds to Arrange, 16 seconds to Sort, 2 seconds to Load for shelving, an unknown average time to transport to its proper location in the stacks, and 21 seconds to Shelve the book back in its place on the shelf ready for use.

Not counting the unknown transport time and various delay times, a book requires 91 seconds of processing time in the shelving operation.

## SUMMARY AND CONCLUSIONS

It was found in the library studied that a book, in being returned to its proper location on the shelf from the book drop, requires 91 seconds of processing time, exclusive of an undetermined amount of transportation time between Sorting Room and stacks, as well as unknown delay times at various points in the operation.

Other factors that affect the shelving operation are the organization and management of human resources and some specific problems such as that of the placement of a conversion card file on an ordinary table, too high for efficient access. Specific problems are mentioned, implicitly recommending that action should be taken to correct the deficiencies.

The study is primarily a documentation of the present system and presents few conclusions. A shelving time of 21 seconds per book indicates that books *can* be shelved at a rate of over 100 per hour, and that a 24 hour turn-around time is within the realm of possibility.

Further study is necessary to increase the validity of the average times derived in the present inquiry. The samples were so small that it was decided that no claim for validity would be appropriate.

Further study is necessary to determine the delay times involved in the various processes. The processing time is a beginning, but the actual time that elapses between the dropping of the book into the slot and the placement of that book on the shelf is an important result that was unfortunately not available from data provided by the present study.

A more thorough investigation of personnel and of forms should be conducted. A reorganization of the Circulation department seems called for. A more extensive forms analysis would be useful.

The present study has defined the areas that require more detailed investigation to obtain a more complete conception of the shelving operation. Time limits imposed on the study forced a decision between depth and breadth. The latter was chosen. It seemed appropriate in a systems study, no matter how "simplified," to insure that the system under investigation be studied in every area, even though time did not permit all areas to be studied in depth.

## BIBLIOGRAPHY

Battles, Dean D. (and others). "Motion and Time Study of a Library Routine," *Library Quarterly*, 13:241-244 (July, 1943).

Benford, H. L. (and others). "Analysis of Book Reshelving," in Burkhalter, Barton R. *Case Studies in Systems Analysis in a University Library*, Metuchen, N.J., Scarecrow Press, 1968.

Brown, Charles H. *Circulation Work in College and University Libraries*, Chicago, American Library Association, 1933. 179 p.

Chapman, Edward A. *et al. Library Systems Analysis Guidelines*, New York, Wiley-Interscience, 1970. 226 p.

Dougherty, Richard M. and Fred J. Heinritz. *Scientific Management of Library Operations*, New York, Scarecrow Press, 1966. 258 p.

Ekendahl, James E. "Analysis of the Shelving Function and the Achievement of Control in Stack Maintenance in College and University Libraries," (unpublished research paper at the Graduate School of Librarianship, University of Denver, 1968. 47 p.)

Flexner, Jennie M. *Circulation Work in Public Libraries*, Chicago, American Library Association, 1927. 320 p.

Gupta, B. S. "Library Stack Room," *Indian Librarian*, 10:154-157 (March, 1956).

Jesse, William H. *Shelf Work in Libraries*, Chicago, American Library Association, 1952. 68 p.

Mehta, S. N. "Work Flow in Book Stack and Maintenance Section," *Annals of Library Science and Documentation*, 13(4):173-184 (December, 1966).

Schunk, R. J. "Stack Problems and Care," *Library Trends*, 4:283-290 (January, 1956).

# RECLASSIFICATION: FLOW CHARTING SUCCEEDS

*Leslie R. Morris*

SOURCE: From Leslie R. Morris, "Reclassification: Flow Charting Succeeds," *Catholic Library World* 43 (Feb., 1972) pp. 337-342. Reprinted by permission of the author.

In 1963 a decision was made at the State University College, Fredonia, New York, to switch from the Dewey Decimal Classification to the Library of Congress Classification. During the following 7 years the only books reclassified were continuations or standing orders.

In March 1970 a decision was made to reclassify the Dewey Decimal collection in toto. One hundred thousand volumes were in the Library of Congress collection and 50,000 were in the Dewey Decimal collection. Only one librarian, two days per week, and 3 clericals, full time, would be available to staff the reclassification project. A fast, cheap, easy reclassification system was needed.

The following paragraphs describe "how we do it at Fredonia." The system described is a compendium of the 15 or 20 different systems that we investigated. We feel we have designed an inexpensive, fast, low labor method that works. Many libraries which struggle with Dewey Decimal Classification collections can copy this process to get their Dewey Decimal Classification books merged with their Library of Congress classification books. Our reclassification project averages 1700 volumes per month.

The flow chart, at end of the article, illustrates the manner in which technical service operations can be broken down into bits and pieces. Flow charts can be used to subject all technical services operations to administrative scrutiny, and to teach tasks to unskilled personnel.

A major timesaver in a reclassification project is weeding. We invited department heads to send one of their staff to weed any section in which he felt qualified. The faculty has been fairly cooperative.

Faculty members have come to the library to look over the sections of books. Some have decided that they want to keep everything, some have decided that they want to get rid of everything but most were in some middle range, very reasonable and not too difficult to deal with. Where faculty in a particular area showed no interest, we as librarians, decided to weed those sections ourselves. The weeding was made as simple as possible for the faculty. All they had to do was pencil a W on the title page. Any book that we found with a W was withdrawn, and a book without a W was reclassified. We may well have missed books in circulation, out of place on the shelves, etc., but we felt that these were being used in some manner and therefore were worth reclassifying. All suggested withdrawals were checked in *Books for College Libraries,* and anything listed was automatically reclassified. The reclassification librarian also was empowered to check the withdrawals of any faculty member. A librarian could decide to reclassify anything that looked worthwhile. The librarians were also empowered to withdraw multiple copies of books, based on usage indicated by the date-due slip pasted in the book. The librarians found that in certain cases a book that circulated a great deal should be withdrawn in the student's best interest. (Career selection guides copyrighted in the 1920's and 40's were still circulating.) Weeding enabled us to get rid of almost 10% of the total books to be reclassified. Without the weeding procedure, we would have had to spend a longer time on reclassification.

The accompanying diagram is a flow chart of the reclassification procedure. It was done as a flow chart to enable the clerical help to work almost independently, after a brief initial training. The flow chart has proved valuable in several ways: because it is in small steps, it is easy for the clericals to understand; it is available when the librarian is not; and it enables the clericals to work independently.

A flow chart of this type breaks the work into step by step segments. There are only three types of boxes on the flow chart. The rectangle is a command. The hexagon is an explanation. The diamond is a decision. Decisions can only be yes or no. A decision box has three lines; one is the question going into the box and the other two lines are either the yes or no decision. Lines always proceed down or to the right. If a line goes up or to the left, there will be an arrow showing the direction. I cannot stress too highly the amount of success we have had with the flow chart. We have since flow charted a great number of the technical services activities, e.g., L. C. card ordering, searching, etc. We have found both students and clerical help are quite able to work from these flow charts with a minimum of supervision. Please do not be frightened by the complex look of the flow chart. It is much easier to follow than it looks.

You will notice as you proceed through the flow chart that if we do not have all the copies and/or volumes of a given set of books, we return that set of books to the shelves. We find that we have 90% of the material available immediately. Our second time through the Dewey col-

lection, we will be forced to reclassify everything. Abbreviations are sometimes necessary within the flow chart to conserve space. In box 8, we tell the employee to pull the shelf list card when in fact we mean remove the shelf list card from the shelf list file. A great many times the books go to the cataloger for some sort of correction or interpretation as designated in the flow chart, but they are immediately returned to the flow. Clericals are able to leave books for an absent cataloger by leaving a note specifying the flow chart box number applying to that particular problem If the cataloger gets a book with a number 25, he knows that he must go to box 25 to find out the problem that the clerk is having with that book.

You will notice that we have made a basic decision to Xerox the catalog cards needed if the set is missing or requires two new cards. We have agreed to type a single missing card for any set of cards. This is an arbitrary decision. No cost studies were made to justify this decision. By rule of thumb it seems to be correct, and seems to be working out in practice.

We use a label to cover the old call number on the catalog cards. This label (5/8″ by 1¼″) is a pressure sensitive label that may be purchased from a number of library supply houses. It does not come off the cards, the typing does not smear, and the labels are relatively inexpensive. Labels do create an accordian effect in the card catalog, but since we have twice as many regular catalog cards as we have cards with labels, we feel that this will not be a severe problem. The problem will, of course, diminish as more and more cards are added to the card catalog.

Three clerical assistants have been assigned to the reclassification project. Clerical A is assigned 20 to 30 hours of student help to assist in pulling cards. The team consisting of the clerical and the student assistants, pulls the books, pulls the cards, relabels the cards, etc. Clerical B has the same amount of student assistance and does an identical job. The third clerical was assigned to lettering or end processes. Since we were cataloging approximately 2,000 books a month and adding reclassification of 1500 volumes a month, the third clerical was needed to do lettering, pasting, etc. The two clericals (one regular and one extra) assigned to lettering and pasting seem to be able to handle the work load generated by our normal classification and the recataloging project. If we had a fourth clerical, she would be assigned card filing. However, by having the two lettering clericals assist the filing clerks, we have managed to get by without the fourth clerical. Essentially, of the three clericals, two do reclassification and one does lettering.

A cataloger is available three hours a day, five days a week. In that fifteen hour time period, the cataloger is able to check the cards, oversee the processes, check series in the series authority file, assign call numbers to books without L. C. classification, assist in searching NUC for the call number and reclassify and recatalog the books that are extremely badly catalogued. We have been fortunate that the cataloging in our library was of a standard that enables us to proceed with the reclassification with a minimum of recataloging. If a large recataloging procedure is necessary, no system will work.

You must separate the reclassification project from the recataloging project. A recataloging project is not a reclassification project. Recataloging requires a different type of staffing, and a great deal more of the cataloger's time. Do not let the catalogers insist that everything that goes by be perfect. A recataloging project can be done separately. Both projects will go faster if the two are separated. If you have a backlog of new materials you have

absolutely no business trying to do a recataloging project. Reclassification is bad enough.

Most expenses involved in the reclassification procedure are listed below. No figures are available for heat, light, rent, etc. Most college libraries don't account for that type of expense. The average month has 21 working days.

3 clerks x $2.50 per hour x 7.5
  hours per day x 21 days = $,1181.25
6 students x $1.50 per hour x 2
  hours per day x 21 days = $  378.00
150 Xerox x .05       = $     7.50
1500 volumes x .05 supplies = $    75.00
1 librarian x $5.00 per hour x
  7.5 hours x 9 days    = $   337.50
    Total           $1,979.25

$1,979.25 divided by 1700 volumes, average volumes reclassified per month = $1.16 per vol.

The technical services librarian must demand from his head librarian, and the head librarian must demand from his administration, that the staff will be available to work on reclassification uninterrupted. There is nothing worse than a reclassification project that goes on for years and years with little or nothing being done. Do nothing until you have staff. Once you have a staff available proceed quickly.

Even if the library is crowded, do not move the books around on the shelves. Leave the empty shelves for the faculty to see. Empty shelves are more impressive than statistics. The figures are not nearly as impressive as empty shelves. Do not let yourself be waylaid into letting your clericals drift away for other important jobs. Keep them busy on reclassification. Don't allow them to get tricked into other projects. A useful tool is a bar graph showing the number of books reclassified each month. Post the graph in a prominent place. Reclassification can work in your library.

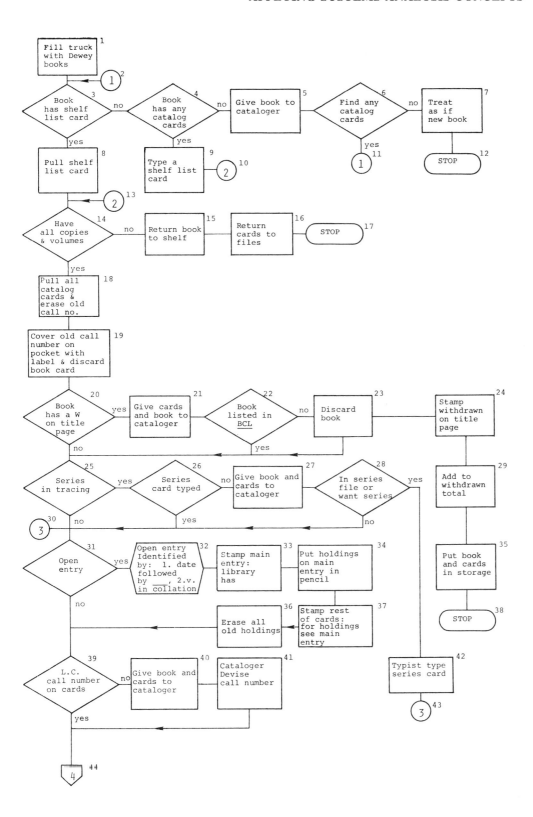

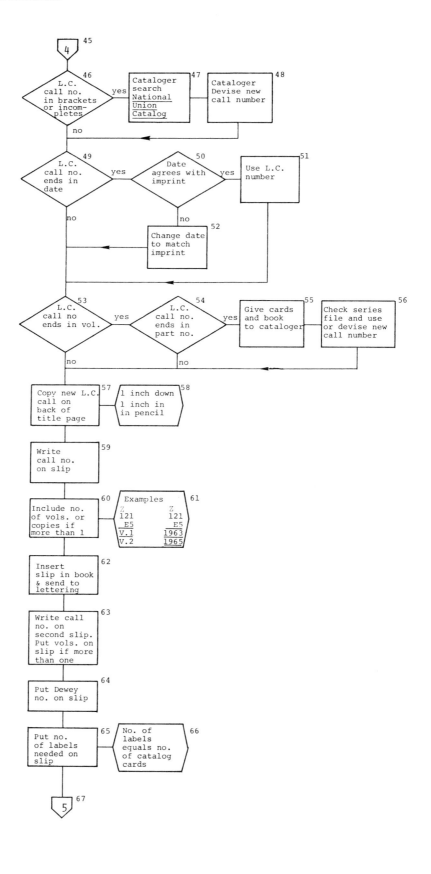

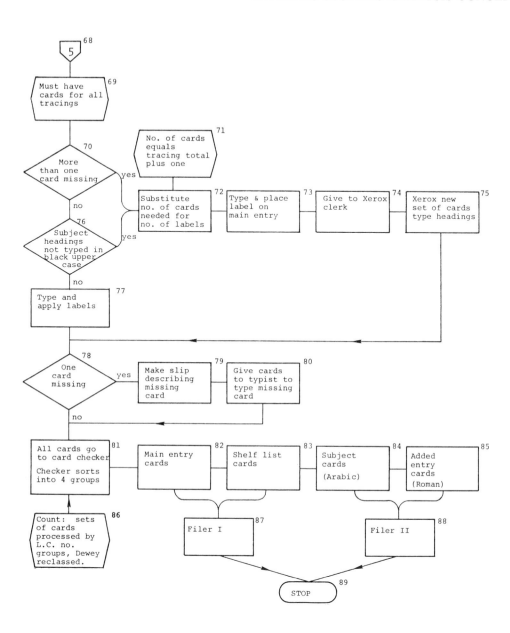

# V

# ELECTRONIC DATA PROCESSING AND SYSTEMS ANALYSIS

**Design and Implementation**
**The Use of Computer-based**
 **Data for Management**
 **Decisions**
**Computers and Libraries**

## INTRODUCTION

As indicated by its heading, this section is concerned with library systems analysis directed toward the design and installation of computer-assisted operations. Consideration is given to the potential of such systems for improving service to users and providing data of timely usefulness to managerial decision-making and planning. The accounts found here are founded on extensive experience in the study and development of computer-based systems for libraries and their actual operation and use.

The authors in this section while endorsing and encouraging the use of the computer as a tool for better library management and service, directly or by implication draw attention to the extraordinary drain on staff time required over long periods in converting to computer-based systems; and to the magnitude of developmental and capital expenses. Such expenses were not conceivable in many libraries not too long ago and doubtless are still inconceivable in many others restrained to "economy" performance budgeting consistent with reluctant administrative funding.

Total operating costs of computer-based systems similarly are off-scale mainly due to equipment rental and service charges. However, *unit* costs measured against increasing production and improving services tend to decrease to levels certainly unattainable without the machine and its associated peripheral equipment. There doubtless is a limit to the unit cost reducing capacity of computer-based systems. Such limit has yet to be reached with the capability of new generations of equipment growing and skill in its use for library operations improving.

Not an inconsiderable factor in the progress made in the use of the computer in library operations and services, has been federal incentive funding of research in computer-based library systems. Wisely these funds in the main have been allocated for projects to develop systems encouraging, indeed requiring, recognition of the goal of interinstitutional cooperation. Participation in regional, national and international library resources information and technical processing networks interlinked by the machine, is required for most libraries to reach competent levels of service to their users. It seems certain that the small to medium-sized libraries without such participation will continue incapable of meeting constantly growing user needs and demands within their own generated resources.

The small to medium-sized library is well advised not to go it alone with computer-based systems except in housekeeping functions if the local processing load warrants. It should anticipate and plan for joining the interconnecting "power grid" of resources information and materials processing promised by the research in comprehensive computer-based library systems. As is well known, this type of cooperation is a long standing and indeed required practice of other "public service" organizations for maximum service to their users upon demand. Why not libraries.

Bernstein's account supplies practical guidance based upon the experiences of the author in planning and introducing electronic data processing in a library. Bernstein lays down twelve principles "which absolutely must be applied during the introduction of a mechanised system." The principles all requiring a planning and systems analysis approach, are presented in an order suggestive of the steps to be followed in introducing and developing an integrated computer-based data processing system. The concern in the essay is with improving the efficiency, effectiveness, timeliness and economy of library housekeeping functions such as acquisitions, journal check-in, cataloging and so forth.

Lubans' paper discusses some of the possibilities of using library computer-based data in decision-making. It has long been held that computer-based systems (particularly circulation control) have the potential of providing answers to many common questions within libraries. The reported research examines the feasibility of the use of such data and through a selection of examples indicates some possible value in on-going computer-analysis of circulation data. It is shown that studies, for example, of the most frequently used books are feasible and can be beneficial to librarians and users by providing information on which planning for improved book collections can be based.

As one of the longtime leaders in exploring and developing the uses of the computer, particularly in the vital area of bibliographic control, Veaner authoritatively discusses the complexities that have been and continue to be encountered in library automation efforts. His paper consists essentially of two major parts: the process of systems analysis and design for library automation and the problems to be solved in this process; and an extensive review of the principal modes of computer processing of bibliographic data, supplemented by case studies of approaches to library automation used in four developmental projects in the United States. All in all this is a fundamental account furnishing guidelines of lasting value in library automation efforts.

Mather's paper is a comprehensively detailed case history of the design and operation of computer-based systems for data processing in the "new" Simon Fraser University Library (Vancouver) with no "manual operations" past to contend with. The systems reviewed include the usual library housekeeping functions: circulation control, acquisitions, cataloging, accounting and supply of management and planning data. Interesting and pertinent facets of Mather's account embrace the effect of automation on library users, on the library's service, librarians and space. The costs involved in automating are considered as are such problems as programming errors and human carelessness in general. Reporting on computer-based data processing systems in successful operation the author is in a position to say, ". . . Libraries and librarians will be judged not on the amount of core storage of their computers, the number of systems in operation or on the writeups they receive in journals, but rather on the service they give."

# SOME ORGANIZATIONAL PREREQUISITES
## FOR THE
# INTRODUCTION OF ELECTRONIC
# DATA PROCESSING IN LIBRARIES

*H. H. Bernstein*

SOURCE: From H. H. Bernstein, "Some Organizational Prerequisites for the Introduction of Electronic Data Processing in Libraries," *Libri* 21 (1971) pp. 15-25. Reprinted by permission of the author.

## 1. Introduction

Every, or nearly every responsible librarian, or head of a library, is forced to think about improving the efficiency of his department. The result of his reflections is often a decision to install electronic data processing. The ideal situation would be to take the well-known book "How to Mechanize a Library" and to follow the recipes given therein to achieve this mechanisation in record time. Unfortunately neither this book nor a general recipe exists for the introduction of a mechanised system into a library. At the present moment it is therefore easier either to make a list of the errors which should be avoided, or of the principles which absolutely must be apllied, during the introduction of a mechanised system. This latter possibility will be dealt with in the following paper.

I am basing my remarks upon the experience which has been gained during the development and introduction of the "Integrated Library System" [1] [2], now under development for the Library of the Ispra Nuclear Research Establishment*). I will not describe this system itself – its value will become apparent after it goes into operation –, but we have a well-founded hope

* Euratom, Joint Nuclear Research Centre, Ispra Establishment.

that the system will bring us great advantages in time, personnel and finally, in costs.

I will define some principles concerning prerequisites and means for mechanisation projects, which have resulted from our experience.

According to the empirical method, it cannot be pretended that these principles are complete. Nevertheless I am convinced – and a number of discussions with Head Librarians during past months have reinforced my convictions –, that the problems encountered when introducing a mechanised system are at least as vast as the problems of the system itself. From the point of view of the organisation expert, some of the principles may be called trivial. On the other hand, librarians are no organisation experts. The aim of this paper is to try and help other librarians to save effort and expense, which also coincides with the aims of this Seminary.

## 2.1 Expense and return

The efficiency of a library can be described only with difficulty in terms of economic facts and figures. Thus, at the beginning of many mechanisation projects there are acute emergency situations (backlog and delays) or a vague feeling that such an emergency will occur any minute. The reasons for such emergency situations need not be mentioned here. As far as the growth of literature in libraries is concerned, it is sufficient to refer to the figures by Dolby[3].

The lack of figures on costs and returns is a decisive weakness of libraries when dealing with sponsors or sponsoring bodies: in many cases it is already difficult to convince the sponsor of the necessities of a conventional library function. It is far more difficult to prove the necessity of far-reaching rationalisation measures with an economically intengible or cameralistically valid "alibi" for the project. But even if for the moment an economic proof seems to be unnecessary – one should be available for when the first expenses of the project occur.

> *Principle 1:* Never mechanise a library or an important part of a library without being able to state clearly the advantages and disadvantages of the intended solution, in comparison with the present situation, as far as time, finance and personnel are concerned.

*) The system has been developed since 1965 by Capobianchi and Petrucci on the data processing side, and by Rittberger, Möbius and others on the the Library side. Realisation of the programme began around 1967. Today the development work is nearly finished. Nevertheless the introduction of the system faces some difficulties, which depend neither upon the system itself, nor upon the Library.

This proof can often only be supplied by library statistics. This fact emphasizes the importance of library statistics[4] and we should here ask ourselves whether our statistics are able to satisfy this requirement*).

A new collection of facts and figures is often necessary. This is in itself a task which should not be underestimated, and which requires a certain amount of economical thinking on the part of the librarian.

## 2.2 Planning of the budget, personnel and timing

When the librarian has solved the above-mentioned problem, he will have obtained the agreement of his sponsor on general principles. Now the work begins. From now on, and until the final putting into operation of the new system, the library has two tasks. It must cope with ist daily routines, sometimes in an emergency situation, which may even grow worse from time to time, and it must work on its project. This is a situation which can make a librarian desperate, or even ruin him, if he does not obtain support and understanding. He must therefore observe some principles in the organisation of his work.

> *Principle 2:* Never mechanise a library or an important part of the library, without having a plan containing sub-sections for budget, personnel and timing, and which clearly explains for each person involved the work load which he will have to carry during the realisation of the project.

Sponsors still often have a tendency, either from ignorance or burocratic thinking, to regard rationalisation projects as the private hobby of the librarians concerned which can be easily satisfied by providing a new typewriter and perhaps a cabinet for punched cards. There is great astonishment, when, during the course of the project, rent for data-processing machinery has to be paid, and this without having economised on even one librarian – quite the contrary[9]. Henderson and Rosenthal have indicated with proved figures the effect of rationalisation upon the section of library work[10].

> *Principle 3:* Never mechanise a library or an important part of a library without having first obtained the complete agreement of the sponsor to the plans.

*)  The demands made by the well-known optimisation methods upon library statistics and efficiency evaluation cannot be discussed here, but they are in excess of the examples given by Kortzfleisch[5], Möbus [6], Gebhart [7], Jestes a.o. [8].

## 2.3 Organisation of the project

The planning should take into account the different stages of the project: Systems analysis, systems design, flow-charting, programming, and programme tests, test application and matching, beginning of operation (Fig. 1). These different stages demand the interlocking of a number of factors from the fields of personal politics, budget politics and budget distribution, selection and acquisition of machinery. A librarian in his daily routine, or even the head librarian of a big library will soon find it extremely difficult to supervise them all, while keeping up with normal business. It is therefore inescapable to appoint a project leader whose main task is to promote the work of the project[11].

> *Principle 4:* Never mechanise a library or an important part of a library without having appointed a project leader.

This project leader should coordinate all activities related to the project. He should be a librarian, but have knowledge of data-processing, experience in planning methods, and in the technique of respecting deadlines, and experience also in the organisation of budgets and personnel. With small projects, this project leader can execute secondary tsaks within the framework of the project. The project leader should apply modern planning methods which can be learned easily by the librarian. The best moment to establish a planning network is at the end of the flow-charting process. The estimations of programme times, times for tests, time for the contributions of the librarians, times for the preparation of test material, can all be entered as well as the deadlines for installation of equipment, and the execution of large scale experiments.

## 2.4 Activity reports and descriptions of the project

System analysis refers as well to the actual situation as to the characteristics of the new system. Systems design develops the new system. Both activities need very close cooperation between the librarians on the one hand, and systems analysts and designers on the other hand. A good systems analyst will in the course of time, become a library specialist. The librarian will obtain deeper understanding of his own methods by the logic inherent to data-processing.

Independently, whether the analyst and designer are members of the library team or not, there must be a guarantee that the principles they elaborate in collaboration will continue to serve as guidelines during the future development of the system.

> *Principle 5:* Never mechanise a library or an important part of a library, without taking accurate notes during the system analysis and during the systems design, on the analytical activities and on the concepts leading to the new system.

It is the task of the project leader to take these notes, and to insist that even the apparently unimportant details are taken care of. It is imprudent to postpone problems *because* they have not yet been considered in detail. Problems can de postponed only *if* they have been considered in detail. A few days erroneously saved on analysis expense can lead to months of additional work on development or on the introduction of the system. This cooperation requires a high standard of discipline on both sides, i.e. the analysts and the librarians.

> *Principle 6:* Never mechanise a library or an important part of a library, without concluding the systems analysis with a detailed report which the analysts and librarians are obliged to follow.

This report is, during the course of the work, the constitution of the project. Modifications will be unavoidable while the work of flow-charting or programming is in progress. The project leader has the responsibility of keeping this "constitution" up to date. The details which will have to be changed in the concept of a system during its realisation are too many and too various to be hung up somewhere as hand-written notices.

## 2.5 Utilization of routines previously developed for other libraries

The framework of the programme will be developed on the basis of systems analysis and sytems design, breaking down the system into single software packages. This work, carried out in collaboration with analysts, designers and programmers, (who are often identical) are of special interest to the library, in so far as a logic modular concept of sub-programmes will give him an advanage in time during the introduction of the system. This is the latest possible moment for deciding which of the partial solutions already developed in other libraries, and description of which have been found in the specialised literature, can be adopted, together with their corresponding programmes.

*Principle 7:* Never mechanise a library or an important part of a library without exploiting to the maximum the results of other libraries' work.

## 2.6 Collaboration between the programmer and the librarian

Although in theory the programmers should by now be in posession of all the necessary data to put up the programmes, the programming procedure is in practice a continuous process of iteration between the programmer and the librarian, and one which should be closely followed by the project leader. At the very least, cooperation is necessary on the following subjects:

Definition of input formats
Definition of output formats
Elaboration of input rules
Elaboration of cataloguing rules
Preparation of test material/programme tests
Checking of test results and proposals for modification
Preparation of new test material/programme tests
Elaboration of rules for corrections
Preparation of test material for large-scale experiments
Checking of results of large-scale experiments
Handing over and declaration "operational" of the programme.

This table shows the work load to which a librarian is exposed if he fails to introduce a scheme for the division of labour soon enough.

*Principle 8:* Never mechanise a library, or an important part of a library, without formally appointing a librarian or a group of librarians (according to the volume of the project) for the systems development.

It is a problematic request, resulting from principle 8, which has to be put to the sponsor: – will he allow extra library personnel, in addition to the programmers? However, if principles 2 and 3 have been observed, it should not be difficult to be firm with the sponsor.

It may be taken as an indicative figure that (according to the quality of the analysis), approximately three programmers will create work for one librarian.

The efforts of the project leader become decisive during the programming: on the one hand he must check up on the deadlines, and on the other hand, keep track of the work results. Programmers have the tendency to treat a programme as finished as soon as it "goes". Firstly there is no longer any intellectual stimulus to go on with the programme, secondly programme descriptions are a necessary evil which it is better to avoid. Moreover, there are often considerable differences between what a programmer and a librarian call "a programme goes". For the programmer a programme "goes" if the routines work logically and without error. For the librarian, the programme "goes" if he can really use it. There can be a noticeably gap between these two attitudes. The task of the project leader is to bridge this gap.

> *Principle 9:* Never mechanise a library or an important part of a library, without having established in the plans who is responsible for which parts of the realisation of the system.

The results of the work should be collected in one place, and this place is most usefully with the project leader. As far as the guides, descriptions, rules and contents of auxiliary files, necessary for an understanding of the system, are concerned, the data should be assembled in the form of a loose-leaf collection. They should be duplicated several times, and distributed to the people most involved in the systems development. This reference system should be updated continuously.

> *Principle 10:* Never mechanise a library or an important part of a library, without appointing a secretary to the project.

This secretary should not only collect the results of the work, and update the loose-leaf collection under the supervision of the project leader, but she will certainly be able to make herself useful with the preparation of test material, assembling bibliographical data, and writing it on punched tape typewriters.


## 2.7 Training of the staff

The amount of time needed for the programming should have been established in the time-diagram, and when the programming begins the end of the introduction phase of the system is in sight. The remainder of the time should be used to prepare for zero hour, apart from the necessary continuing collaboration between programmer and librarian.

Zero hour is the moment at which the system is put into operation. According to the volume of the project, this moment may bring profound changes to the working methods of several departments, or of the whole library. It would certainly be no exaggeration to say that a librarian is, in general, conservative. And the requirements of his profession demand stability and prudence, since certain decisions may influence the functioning of a library positively or negatively for years afterwards. A thorough preparation of the personnel in the new techniques and the new way of working is, for this reason, all the more necessary. The electronic computer should not only be a tolerated tool, but also a respected means for improving the library service. Fussler[12] indicates that 10% of the time of librarians who collaborated in systems development was spent on training alone.

The value attributed to the permanent interpenetration of librarianship and data-processing in the library is shown by Payne[13]: "The library needs staff members trained both in library operations and in computer science. It has been our experience that computer people will underestimate the library requirements by at least an order of magnitude even when the requirements are documented. The library needs a voice that can talk to the computer policy committee and look after the longrange interests of the library. The library should participate in the planning of computer facilities and see that its future plans and requirements are taken into consideration."

> *Principle 11:* Never mechanise a library or an important part of a library, without first thoroughly familiarising the library personnel with the workings of data-processing.

This preparation should be graded according to the different levels of library service:

- A general introduction to data-processing for all staff: coding on punched cards and punched tape; input equipment (perforators, typewriters); reading apparatus and computer printers; main components of computers; and performance (speed of input and output equipment).
- for the executive staff: a presentation of the new system based on the block diagram.
- for the staff of the departments involved: a thorough discussion of the performance of the system, as far as it concerns the activities of these departments.
- for the librarians involved in the systems development: participation in programming courses.

In addition, as many of the staff as possible should be sent to visit other libraries, which already employ electronic data-processing equipment. It has been proved by experience that the best course is still not half as persuasive as a library which is already using the new technique.

Although everybody would agree in principle with what has just been said, it is nevertheless difficult to put it into practice. Even if the personnel can find time for these educational events (in spite of the previously mentioned emergency situations) there still remain the periods during which a library is open to the public, which regularly bind a considerable proportion of the staff.

Consequently it is obvious that these educational events must be well prepared. A rewarding task would be to prepare a guide for this, using the experience of advanced libraries.

### 2.8 Working off backlog

Zero hour needs an additional preparation, however. In only a very few cases would one need to process only the data occurring from the moment of introduction of the new system. The emergency situation began earlier, and it will be necessary to work off the backlog.

> *Principle 12:* Never mechanise a library or an important part of a library, without having established the conditions for working off the backlog.

These conditions are of various kinds:

- the rules for backlog processing must be laid down.
- there must be staff able to apply these rules, concerning both data preparation (cataloguing or re-cataloguing), and perforation or writing on punched tape typewriters.
- the input equipment must be available.
- there should be a provision in the budget for allocating the working off of the backlog to a service firm, if this should become necessary[14].

### 3. Planning methods and efficiency

Between the old system and the implementation of the new system, lies a long and often twisted road, which passes through problems of personnel, budget, delivery delays, unpredicted difficulties with software and hardware, etc. In order to retain a minimum of supervision, the project leader must employ a method which allows him to check up at any moment upon the progress of the project and which enables him to make even substantial modifications to the plans where necessary. For this it is advisable to use a network similar to the networks used in the PERT method (Fig. 2, Fig. 3). In this method, the arrows indicate the time necessary to execute a certain work, and the knots represent the beginning or the termination of the work itself. It will not always be necessary to compute the whole network to determine the critical path. For the development of library systems an approximate calculation will often be sufficient. This diagram is an indispensable aid, especially during programming, because it is here that important non-linear processes occur.

In addition to this, it is understandable that the sponsor demands a regular statement of accounts, in order to see whether or not the funds he has made available have been usefully employed. The library of the University of Chicago has gone so far as to create the post of "cost/benefit analyser" within the library[15] whose task it is to ensure the continuing improvement of the system.

*4. Final remarks*

If one considers that in the field of industrial organisation, it is quite normal to trust oneself to specialised consulting firms for the introduction of new techniques (without anybody casting doubts upon the organisational capabilities of the managements concerned) one can roughly estimate how much work the libraries take on, when replacing old established procedures by modern methods. Almost never is the librarian sufficiently familiar with the new techniques. Almost never can he rely on proved methods for a step by step introduction of the new technique.

If the development and optimisation of the systems, which are being carried on industriously everywhere, should really pave the way for a broad application, then more attention should be paid in the future to the question of project management in libraries.

NOTES

1.  Capobianchi, S.; Petrucci, A.; Rittberger, W.: Introduction of mechanization in scientific libraries. The integrated system of the Ispra library. EUR 4250 e, Luxembourg 1969.
2.  Petrucci, A.: Integrated management system of a scientific library in an »on-line« environment. IAEA/SM-128 Symposium, Vienna, Feb. 1970.
3.  Dolby, J. L.; Forsyth, V. J.; Resnikoff, H. L.: Computerized library catalogues: their growth, cost and utility. Cambridge, Mass. 1969, p. 4.
4.  Dolby, J. L.; Forsyth, V. J.; Resnikoff, H. L.: Computerized library catalogues: their growth, cost and utility. Cambridge, Mass. 1969, p. 18.
5.  Kortzfleisch, H. v.: Gutachten über Rationalisierungsmöglichkeiten in wissenschaftlichen Bibliotheken, (1967) Bad Godesberg.
6.  Möbus, R.: Arbeitsstudium an wissenschaftlichen Bibliotheken, Zentralblatt für Bibliothekswesen 82 (1968), 322–332.

7.  Gebhardt, W.: Gedanken zum Personalbedarf einer wissenschaftlichen Bibliothek. Festschrift für Wilhelm Hoffmann in Libro Humanitas, Stuttgart (1962), 55-65.

8.  Jestes, E. C.; Laird, W. D.: A time study of general reference work in a university library.

9.  Dolby, J. L.; Forsyth, V. J.; Resnikoff, H. L.: Computerized library catalogues: their growth, cost and utility. Cambridge, Mass. 1969, p. 141.

10. Rosenthal, J. A.; Henderson, J. W.: Library catalogs: their preservation and maintenance by photographic and automated techniques. MIT-Report no. 14, Cambridge, Mass. (1968) p. 60-90.

11. Fussler, H. H.: Development of an integrated, computer-based, bibliographical data system for a large university library. Clearinghouse, Springfield, Va. (1967), PB 176 469, p. 11.

12. Fussler, H. H.: Development of an integrated, computer-based, bibliographical data system for a large university library. Clearinghouse, Springfield, Va. (1967), PB 176 469, p. 14.

13. Payne, C. T.: An integrated computer-based bibliographical data system for a large univeersity library: problems and progress at the university of Chicago.
    Paper given May 1, 1967, at the Clinic on Library Applications of Data Processing, Graduate School of Library Applications of Data Processing, Graduate School of Library Science, University of Illinois. PB 167 469, p. 12.

14. Dolby, J. L.; Forsyth, V. J.; Resnikoff, H. L.: Computerized library catalogues: their growth, cost and utility. Cambridge, Mass. 1969, p. 23.

15. Fussler, H. H.: Development of an integrated, computer-based, bibliographical data system for a large university library. Clearingshouse, Springfield, Va. (1967), PB 176 469, p. 11:

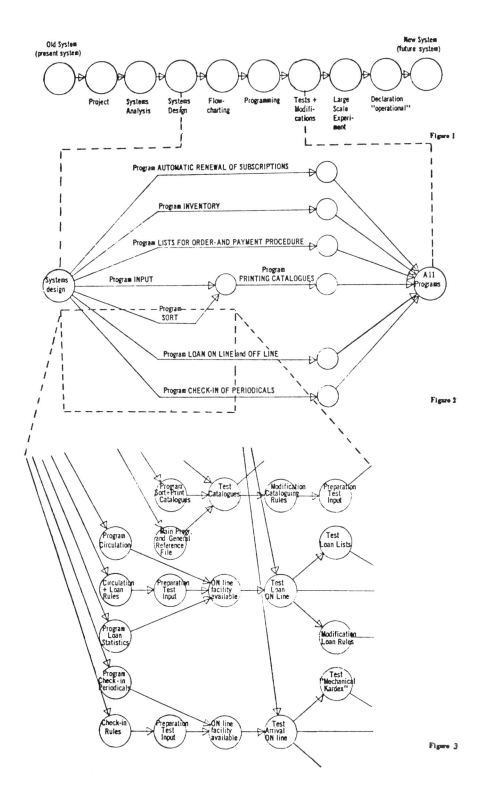

**Old System**
(present system)

**New System**
(future system)

Project  Systems Analysis  Systems Design  Flow-charting  Programming  Tests + Modifications  Large Scale Experiment  Declaration "operational"

Figure 1

Program AUTOMATIC RENEWAL OF SUBSCRIPTIONS

Program INVENTORY

Program LISTS FOR ORDER- AND PAYMENT PROCEDURE

Program INPUT

Program PRINTING CATALOGUES

Systems design

Program SORT

Program LOAN ON LINE and OFF LINE

Program CHECK-IN OF PERIODICALS

All Programs

Figure 2

Program Sort + Print Catalogues

Test Catalogues

Modification Cataloguing Rules

Preparation Test Input

Program Circulation

Main Progr. and General Reference File

Test Loan Lists

Circulation + Loan Rules

Preparation Test Input

ON line facility available

Test Loan ON Line

Program Loan Statistics

Modification Loan Rules

Program Check-in Periodicals

Test "Mechanical Kardex"

Check-in Rules

Preparation Test Input

ON line facility available

Test Arrival ON line

Figure 3

419

# Electronic Data Processing and Systems Analysis
## The Use of Computer-based Data for Management Decisions

SYSTEMS ANALYSIS,

MACHINEABLE CIRCULATION DATA

AND

LIBRARY USE RESEARCH

*John Lubans, Jr.*

SOURCE: From *A Study with Computer-based Circulation Data of the Non-use and Use of a Large Academic Library* by John Lubans, et al., U.S.O.E. Grant No. OEG 8 72 0005 (590), Boulder, Colorado, University of Colorado Libraries. Final Report. June, 1973. Reprinted by permission of the author.

Through the logic of systems analysis and the view of the circulation system and its data as part of the *total library system* it should be clear that many important studies of library use and non-use and library effectiveness could be made.

Computer-based circulation systems, it is widely believed, can be utilized to provide data-bases for such studies. The study described in this paper involved using such a data base to analyze aspects of library use and types of users. Another major objective of this research was the testing of machine-readable circulation data serving as the resource for a variety of computer-based studies. These studies were designed to supply information for decision-making in libraries in such areas as collection development and book budget allocations.

## REVIEW OF THE LITERATURE

A literature search revealed that the design of computer-based systems has not gone beyond an emulation of the manual system. Invariably a sentence or two about user and other studies now made possible (Gull, Hayes, Surace) is provided in discussions of mechanized circulation, but few have made any application of these concepts.

Historically, Becker has pointed out that mechanized circulation control began in the 1930's, when edge-notched cards were used. When punched cards were introduced (Parker) in place of the edge-notched variety, circulation systems were able to use card sorters to help maintain the circulation files.

The concept of a machine-readable book card and borrower's card was introduced in the 1940's by IBM, when it designed such an installation for the Montclair Public Library (Quigley). This system required that a punched book card and borrower's card be inserted in a "record control unit" which would, via another keypunch, reproduce the inserted information. Quigley noted in 1941 the numerous by-products with such equipment:

It is perfectly possible, for instance, . . . to learn by sorting the cards what books the doctors among the Library borrowers had read . . . how

many detective stories were borrowed during a certain time, what non-fiction had been borrowed by boys of a stated age. . . . The possibilities of obtaining information on 'who reads what' are so unlimited that Montclair librarians will probably need to guard against seeking curious bits of information or riding professional hobbies.

In 1959 IBM produced the 357 Data Collection System for circulation control. The output of this system was fed into the computer and a magnetic tape produced for computer processing. Since the early 1960's a number of other systems have been introduced, including Standard Register's *Source Record Punch*, Colorado Instrument's *C-Dek System*, and other commercial data collecting devices.

Most of the literature on circulation systems describes the installation phases of mechanized circulation systems. Economic justifications, systems analysis, and other aspects of how and why the change-over from manual to machine techniques was made are discussed. Little attention is paid to the use of this information as an aid for administrative or service decision-making.

The important study by Cammack (1967) describes data that can be collected and analyzed to assist library management. The data include charging activity patterns to assist in manpower scheduling, heaviest used portion of the collection, student and faculty usage, students and fields of study correlations, grades and library usage, etc. This study is a fine example of what can be done with circulation data.

Another report that was published in 1971 is of particular value. This is the research done at the University of British Columbia by Simmons and deals with the analysis of machine readable loan records and their application to book selection. Essentially the major accomplishment of the British Columbia study was the listing by computer of those titles in the library in need of additional copies because of their being in high demand as revealed through circulation.

DeGennaro, in his article on Harvard University's Widener Library shelf list conversion program, states that records of circulation data at Harvard have been kept since 1965 and these "constitute an invaluable and unique data base from which statistical analyses of the use of the collection have been made." The study referred to (Palmer) is an analysis by Harvard book classification number of the use of books during 1965–69. DeGennaro goes on to say that "such potentially useful management information has never before been available to library administrators."

J. McNee Elrod of the University of British Columbia says that:

> Perhaps the most important result of computerized circulation has been the ease with which months of loan records can now be analyzed . . . with the demand for each book measured . . . the library could do a much more efficient job of meeting readers' needs. The figures will indicate when a reserve book should be taken off short-term loan and when a stack book should be put on reserve or duplicated.

Writing in 1967, C. D. Gull, Professor of Library Science at Indiana University stated: "There is a very real need for information about the present capabilities and future potentialities of computerized circulation control systems . . . there is almost no information on the effective use of library materials and the need for them, on the potential application of computers for circulation statistics, on the introduction of rational management to supplant intuitive management or on the use of circulation information to improve collections and service."

Since the time of Gull's call for research in this area, more sophisticated systems have been developed (at least three on-line circulation systems have become operational) and more attention has been paid to cost/benefit analysis. But as of the date of this study, Gull's call for research has barely begun to be answered. This is unusual, because circulation data have been analyzed frequently in the past in *manual* systems (Davidson, Jain, Steig). These analyses were usually accomplished through the laborious method of tabulating the information on each book's circulation card and/or date-due slip. That these manual techniques have not been adapted to mechanization may be explained by the systems designers' disenchantment with such basic research or that they are bogged down more with just making the hardware and software work than with innovating and improving library services.

A state-of-the-art report by Cecily Surace of the Rand Corporation, published in March, 1970, commented:

> Since the principal purpose of the library is to disseminate information found in documents and one of the instruments for doing this is the circulation system, it is obvious the circulation system should not be treated as a purely mechanical operation which does not require statistical analysis and feed-back . . . It can . . . assist management in analyzing the various circulation operations, reveal weakness in the collection, and provide data on user reading habits, etc.

## THE CIRCULATION SYSTEM

The following chart, "Overview of Circulation System," describes how most computer-based circulation systems are designed to work. For every circulation transaction (or book loaned) the book's call number and the borrower's ID number are combined. With this information a variety of computer studies are possible. It is possible to find out which books circulate and by elimination, which books are not borrowed, who are the users and non-users, and what departments or areas of study make the greater or lesser use of the library. As a more specific example, if very few books about Icelandic literature are being borrowed, and if after investigation it turns out that the university is no longer interested in this subject, it can be assumed that research-level materials in the one area need no longer be purchased.

## THE STUDY

A stylized view of this study shows the comparison (C) of the circulation transaction files (A) (over 250,000 circulation transactions accumulated during February, 1971–March, 1972) and the borrower address files (B) to produce two areas: use (D) and non-use. (F) (The narrative of this section is coded by letters to the flow chart "Overview of Circulation Data Study") This is made possible by the machine listing identification numbers that have been used to charge out books and preparing a separate list of those ID numbers that have not been used to charge out materials. In addition,

## OVERVIEW OF CIRCULATION DATA STUDY

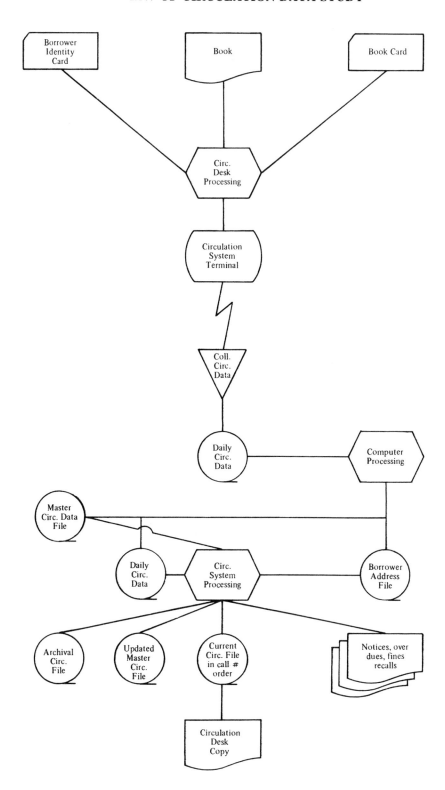

## OVERVIEW OF CIRCULATION SYSTEM

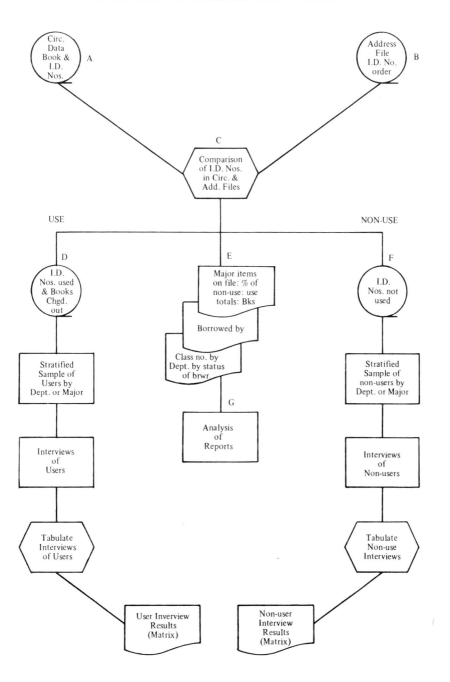

printouts (E) were made of the major items on these files, *e.g.* the totals and percentages of non-use and use, the total number and percentages of books borrowed per their classification numbers and by departments by classification numbers and the total number of percentages of books borrowed according to the academic status level of the user (freshman to graduate).

A number of interesting and administratively useful computer-based studies have been produced through this research. Among the findings of the study is that information (such as lists of books most frequently used) produced by computer manipulation can be useful in collection development and the provision of improved services to library users.

One of the tabulations shows the use of the library's collection by each level of user (from freshman to graduate). This table (I) is arranged (reading across) by classification number, the total circulation for that classi-

fication number, the fraction of the total borrowed by each class of borrower and the percentage of this fraction. The summary of this one report reflects the incremental nature of circulation frequency by level of user: Freshmen account for 10% of the circulation; Sophomores, 13%; Juniors, 16%; Seniors, 22%; and Graduate Students for 3% of the total number of books charged out by students. This finding bears out the general opinion of librarians that library use is progressive. This, of course, does not reflect in-house use of the library.

Table II reveals a section of the results of a search via computer of all books charged out 7 or more times during the year of the study. This includes renewals by the same borrower. (Another similar study was made not counting renewals) Over 1600 volumes representing a total of 13,254 uses were printed-out and are likely candidates for purchase of additional copies. This finding (as does a similar one in the Simmons study at

## TABLE I
### SUBJECT ARRAY BY CLASSIFICATION NUMBER, DIVIDED BY TYPE OF USER AS TO NUMBER AND PERCENTAGE

| Code | Class Name | Total | Freshman | Sophomore | Juniors | Seniors | Grads. | Misc. |
|------|-----------|-------|----------|-----------|---------|---------|--------|-------|
| PA | | 2404 | 504.21 | 433.18 | 267.11 | 275.11 | 902.38 | 23.01 |
| PB | | 98 | 02.02 | 14.14 | 06.06 | 20.20 | 48.49 | 08.08 |
| PC | | 266 | 09.03 | 11.04 | 17.06 | 19.07 | 208.78 | 02.01 |
| PD | | 1013 | 54.05 | 69.07 | 139.14 | 164.16 | 537.53 | 50.05 |
| PG | | 2298 | 144.06 | 192.08 | 231.10 | 508.22 | 1209.53 | 14.01 |
| PJ | | 1095 | 78.07 | 96.09 | 177.16 | 288.26 | 400.37 | 56.05 |
| PN | | 4352 | 416.10 | 560.13 | 794.18 | 1046.24 | 1468.34 | 68.02 |
| PQ | | 5531 | 535.10 | 580.11 | 657.12 | 889.16 | 2770.50 | 100.02 |
| PR | | 11996 | 1384.12 | 1931.16 | 2090.17 | 2413.20 | 3936.33 | 242.02 |
| PS | | 8544 | 1286.15 | 1652.19 | 1747.20 | 1784.21 | 1898.22 | 177.02 |
| PT | | 2619 | 297.11 | 375.14 | 315.12 | 452.17 | 1081.41 | 99.04 |
| PZ | | 565 | 98.17 | 91.16 | 145.26 | 109.19 | 104.18 | 18.03 |
| Q* | | 286 | 10.04 | 29.10 | 53.19 | 62.22 | 125.44 | 07.02 |
| QA | | 638 | 30.05 | 52.08 | 138.22 | 169.27 | 201.32 | 48.08 |
| QB | | 213 | 45.21 | 79.37 | 51.24 | 21.10 | 17.08 | 00.00 |
| QC | | 442 | 24.05 | 34.08 | 79.18 | 92.21 | 208.47 | 05.01 |
| QD | | 130 | 13.10 | 06.05 | 22.17 | 38.29 | 34.26 | 17.13 |
| QE | | 96 | 23.24 | 16.17 | 13.14 | 23.24 | 18.19 | 03.03 |
| QH | | 365 | 26.07 | 46.13 | 71.20 | 76.21 | 128.35 | 18.05 |
| QK | | 59 | 00.00 | 09.15 | 08.14 | 16.27 | 25.42 | 01.02 |
| QL | | 272 | 13.05 | 21.08 | 57.21 | 117.43 | 58.21 | 06.02 |
| QM | | 20 | 00.00 | 01.05 | 08.40 | 05.25 | 05.25 | 01.05 |
| QP | | 140 | 17.12 | 16.11 | 19.14 | 53.38 | 31.22 | 04.03 |
| QR | | 18 | 02.11 | 01.06 | 03.17 | 11.61 | 01.06 | 00.00 |
| R* | | 54 | 07.13 | 08.15 | 08.15 | 16.30 | 14.26 | 01.02 |
| RB | | 166 | 16.10 | 23.14 | 46.28 | 35.21 | 43.26 | 03.02 |
| RK | | 08 | 01.13 | 00.00 | 01.13 | 04.50 | 02.25 | 00.00 |
| RT | | 03 | 00.00 | 00.00 | 00.00 | 01.33 | 02.67 | 00.00 |

## TABLE II
## MATERIALS USED SEVEN OR MORE TIMES IN A PERIOD OF ONE YEAR

| Classification Number | ID Number | YRDY | Hour | TP | Number |
|---|---|---|---|---|---|
| PG*3328*Z6P25*C.1 | 1000427207 | 1300 | 1558 | 00 | 0006 |
| PG*3328*Z6P25*C.1 | 8000000723 | 1195 | 1830 | 00 | 0007 |
| PG*3328*Z6P25*C.1 | 8000000723 | 1195 | 1830 | 00 | 0008 |
| PG*3328*Z6W3 | 1000341711 | 1186 | 2198 | 00 | 0001 |
| PG*3328*Z6W3 | 1000371185 | 1234 | 1582 | 00 | 0002 |
| PG*3328*Z6W3 | 1000371185 | 1263 | 0932 | 00 | 0003 |
| PG*3328*Z6W3 | 1000416329 | 1116 | 2077 | 00 | 0004 |
| PG*3328*Z6W3 | 1000416355 | 1137 | 1591 | 00 | 0005 |
| PG*3328*Z6W3 | 1000427207 | 1316 | 1331 | 00 | 0006 |
| PG*3328*Z6W3 | 9000000321 | 1287 | 1503 | 01 | 0007 |
| PG*3335*N3*1961*C.2 | 1000354335 | 2027 | 1596 | 00 | 0001 |
| PG*3335*N3*1961*C.2 | 1000355024 | 1261 | 1073 | 00 | 0002 |
| PG*3335*N3*1961*C.2 | 1000355024 | 1284 | 1591 | 00 | 0003 |
| PG*3335*N3*1961*C.2 | 1000360573 | 1334 | 1321 | 00 | 0004 |
| PG*3335*N3*1961*C.2 | 1000361402 | 2006 | 1666 | 00 | 0005 |
| PG*3335*N3*1961*C.2 | 1000361402 | 2006 | 1666 | 00 | 0006 |
| PG*3335*N3*1961*C.2 | 1000382009 | 1202 | 1180 | 00 | 0007 |
| PG*3337*G6Z78 | 1000350050 | 1110 | 1736 | 00 | 0001 |
| PG*3337*G6Z78 | 1000350050 | 1175 | 1849 | 00 | 0002 |
| PG*3337*G6Z78 | 1000350050 | 1200 | 1749 | 00 | 0003 |
| PG*3337*G6Z78 | 1000350050 | 1224 | 1444 | 00 | 0004 |
| PG*3337*G6Z58 | 1000350050 | 1224 | 0933 | 00 | 0005 |
| PG*3337*G6Z78 | 1000350050 | 1239 | 0879 | 00 | 0006 |
| PG*3337*G6Z78 | 1000350050 | 1262 | 2189 | 00 | 0007 |
| PG*3361*T5Z66 | 1000355024 | 1280 | 1484 | 00 | 0001 |
| PG*3361*T5Z66 | 1000355024 | 1309 | 1492 | 00 | 0002 |
| PG*3361*T5Z66 | 1000355024 | 1337 | 1009 | 00 | 0003 |
| PG*3361*T5Z66 | 1000355024 | 1355 | 1072 | 00 | 0004 |
| PG*3361*T5Z66 | 1000355024 | 2018 | 1472 | 00 | 0005 |
| PG*3361*T5Z66 | 1000355024 | 2046 | 1244 | 00 | 0006 |
| PG*3361*T5Z66 | 8000001653 | 1231 | 1561 | 00 | 0007 |
| PG*3365*A6*1967*T.2 | 1000382196 | 1134 | 1694 | 00 | 0001 |
| PG*3365*A6*1967*T.2 | 1000382196 | 1159 | 1529 | 00 | 0002 |
| PG*3365*A6*1967*T.2 | 1000382196 | 1186 | 2034 | 00 | 0003 |
| PG*3365*A6*1967*T.2 | 1000382196 | 1214 | 1442 | 00 | 0004 |
| PG*3365*A6*1967*T.2 | 1000388721 | 2017 | 1299 | 00 | 0005 |
| PG*3365*A6*1967*T.2 | 1000388721 | 2047 | 1537 | 00 | 0006 |
| PG*3365*A6*1967*T.2 | 1000427207 | 1239 | 1387 | 00 | 0007 |
| PG*3366*A13 | 1000321236 | 1093 | 1673 | 00 | 0001 |
| PG*3366*A13 | 1000346621 | 1124 | 2353 | 00 | 0002 |
| PG*3366*A13 | 1000376723 | 1137 | 1856 | 00 | 0003 |
| PG*3366*A13 | 1000440619 | 1311 | 1803 | 00 | 0004 |
| PG*3366*A13 | 1000444144 | 2030 | 1800 | 00 | 0005 |
| PG*3366*A13 | 1000446360 | 1242 | 1596 | 90 | 0006 |
| PG*3366*A13 | 8000000357 | 1187 | 1582 | 00 | 0007 |
| PG*3366*A15D65 | 1000288394 | 1145 | 1036 | 00 | 0001 |
| PG*3366*A15D65 | 1000376723 | 1096 | 1478 | 00 | Number |
| PG*3366*A15D65 | 1000392896 | 1278 | 1315 | 00 | 0003 |
| PG*3366*A15D65 | 1000421030 | 1165 | 1537 | 00 | 0004 |
| PG*3366*A15D65 | 1000421030 | 1193 | 0843 | 00 | 0005 |
| PG*3366*A15D65 | 1000422080 | 1118 | 1801 | 00 | 0006 |
| PG*3366*A15D65 | 1000430557 | 1228 | 1417 | 00 | 0007 |
| PG*3386*M36*C.1 | 1000206000 | 1092 | 1380 | 00 | 0001 |
| PG*3386*M36*C.1 | 1000334910 | 1301 | 1297 | 00 | 0002 |
| PG*3386*M36*C.1 | 1000415507 | 1263 | 1078 | 00 | 0003 |

the University of British Columbia) indicates the most heavily used items in the library, pinpointed to call number, the identification number of the borrower and the time they were borrowed. By checking the shelf list for the necessary bibliographic information, this information can be of significant value in deciding to purchase duplicate materials based on measured use.

In an attempt to measure use by departments or study majors, Table III is provided as a sample page. This study measures the numbers of uses of books by the students associated with a particular department. The use is shown (left to right) and ranked by frequency for each classification number. This report involved the use of certain information in the master address file pertaining to each student's major. The findings show a wide range of use by students in various departments. This data could be correlated with that of the book budget allocation and should be of some value in deciding amounts to be allocated. Presumably,

## TABLE III
### SUBJECT AREAS OF FUNDING DIVIDED BY COLLECTION CLASSIFICATIONS ORGANIZED BY USER ACTIVITY ON ASSIGNED MAJORS

| | | | | | | | | | | |
|---|---|---|---|---|---|---|---|---|---|---|
| English | PR | 4487 | PS | 2010 | PN | 672 | PQ | 341 | 82 | 296 |
| English | B* | 281 | PA | 276 | PT | 183 | E* | 167 | PG | 134 |
| English | LB | 128 | PD | 122 | HQ | 121 | BR | 113 | JK | 110 |
| English | 81 | 107 | DA | 84 | P* | 81 | 79 | 76 | BL | 76 |
| English | BF | 72 | DT | 70 | DS | 61 | M* | 60 | 59 | 59 |
| English | HC | 56 | 75 | 55 | HM | 55 | ML | 53 | 42 | 53 |
| English | 60 | 52 | 77 | 51 | 80 | 50 | 57 | 49 | F* | 48 |
| English | 70 | 47 | JC | 47 | PJ | 45 | 33 | 45 | 72 | 45 |
| English | 66 | 44 | DE | 43 | GN | 42 | 64 | 42 | 68 | 40 |
| English | HX | 40 | LJ | 39 | 36 | 39 | 52 | 37 | 34 | 37 |
| English | 87 | 36 | ND | 36 | 48 | 36 | 49 | 35 | 45 | 34 |
| English | PZ | 33 | DK | 33 | 71 | 33 | 39 | 33 | 35 | 32 |
| English | 69 | 32 | 62 | 31 | 56 | 31 | 73 | 31 | 65 | 30 |
| English | 76 | 30 | 67 | 30 | 37 | 30 | 43 | 29 | 46 | 29 |
| English | 63 | 29 | 74 | 29 | 32 | 28 | 78 | 27 | 54 | 27 |
| English | CB | 26 | 44 | 25 | LA | 25 | D& | 24 | 61 | 23 |
| English | 41 | 23 | MT | 21 | H* | 20 | 38 | 20 | 50 | 20 |
| English | 83 | 20 | 94 | 19 | 29 | 18 | N* | 17 | HB | 17 |
| English | 84 | 17 | 91 | 16 | G* | 16 | 51 | 16 | 31 | 15 |
| English | 47 | 15 | Z* | 14 | DC | 14 | 88 | 13 | GV | 12 |
| English | 92 | 10 | 53 | 10 | BM | 9 | CC | 9 | 40 | 9 |
| English | 55 | 8 | 18 | 8 | 19 | 8 | QH | 8 | GT | 7 |
| English | PC | 7 | VK | 7 | 86 | 7 | 97 | 6 | 22 | 6 |
| English | NA | 6 | RB | 6 | JA | 6 | CT | 6 | Q* | 6 |
| English | QL | 5 | UA | 5 | 85 | 5 | 89 | 4 | 93 | 4 |
| English | 27 | 4 | DU | 4 | JF | 4 | QC | 4 | A* | 3 |
| English | QD | 3 | QB | 3 | K* | 3 | DD | 3 | S* | 3 |
| English | 28 | 3 | 30 | 3 | 12 | 3 | 17 | 2 | 26 | 2 |
| English | SB | 2 | T* | 2 | QM | 2 | CR | 2 | 95 | 2 |
| English | 96 | 1 | 90 | 1 | GB | 1 | HG | 1 | HJ | 1 |
| English | J* | 1 | QE | 1 | QK | 1 | QA | 1 | PB | 1 |
| English | BP | 1 | AE | 1 | SK | 1 | RK | 1 | TK | 1 |
| English | U* | 1 | TP | 1 | 23 | 1 | 24 | 1 | 11 | 1 |
| English | 20 | 1 | 07 | 1 | 10 | 1 | | 0 | | 0 |
| English | | 0 | | 0 | | 0 | | 0 | | 0 |
| English | | 0 | | 0 | | 0 | | 0 | | 0 |
| English | | 0 | | 0 | | 0 | | 0 | | 0 |
| English | | 0 | | 0 | | 0 | | 0 | | 0 |
| English | | 0 | | 0 | | 0 | | 0 | | 0 |
| English | | 0 | | 0 | | 0 | | 0 | | 0 |
| English | | 0 | | 0 | | 0 | | 0 | | 0 |

Total of uses listed
012843

POSSIBLE STUDIES WITH MACHINEABLE CIRCULATION DATA FILES

| Data Elements in Available Files | Volume of Circulation (by Subject) | Work Loads by Hour, Session | Subject Use of Books | User Affiliation (by Dept.) | Collection Evaluation by Use & Non-Use | Non-Users and Users | Fines—Overdues | Inventories | Circulation of Books by Type of Purchase | Predictions of Future Use |
|---|---|---|---|---|---|---|---|---|---|---|
| Book's call number | X | | X | X | X | X | | X | X | X |
| User's status: | | | | | | | | | | |
|   Frosh | | | | X | X | X | | | | X |
|   Soph | | | | X | X | X | | | | X |
|   Jr. | | | | X | X | X | | | | X |
|   Sr. | | | | X | X | X | | | | X |
|   5th Yr. | | | | X | X | X | | | | X |
|   Gr. | | | | X | X | X | | | | X |
|   Fac | | | | X | X | X | | | | X |
|   Staff | | | | X | X | X | | | | |
| Dept/Major of User | | | | | | | | | | |
| Registration date (e.g. QPA of users) student number | | | | X | X | X | | | | X |
| Date & Hour of transaction discharge-charge | | X | X | | | | X | | | X |
| Charge-out record for each book (cumulative) | X | | X | X | X | X | X | | | X |
| Overdue notices issued | X | | | | | | X | | | |
| Financial (fines) information | | | | | | | X | | | |
| Holds placed on a book | | | | | | | | | | X |

(all things being equal) those departments showing more activity (per capita) in borrowing books should be better supported than those showing less. At least it is a factor (not previously so readily available) to be considered in allocating book funds.

The above are some of the studies done under the grant. Beyond the activity within the grant, additional studies are quite feasible and should be investigated. A partial list of these for future investigation are charted in the following matrix of "Possible Studies with Machineable Circulation Data Files." It should be mentioned that a variety of computer-based files other than the circulation file alone may be used in circulation studies. The more files available to a library

the more types of studies are possible. For example, a machine-readable shelf-list is essential if a study is to be made of the nonuse of a certain subject area. For studies such as this the need to have access to the "universe" of books is important. The same rationale exists for a user study. To attain total figures of use and nonuse, the universe of users (potential and real) must be known. The accuracy of non-library computer files should be evaluated. A case in point would be university-wide address files. If the address file has not been kept current or been purged of obsolete records then a number of problems will be encountered in using such a file for a specialized library application.

## SELECTED BIBLIOGRAPHY ON COMPUTER-BASED CIRCULATION SYSTEMS AND LIBRARY USER STUDIES

Barkey, Patrick. "Patterns of students use of a college library." *Coll. & Res. Lib.* 26:115-118, March, 1965.

Becker, J. "Circulation and the computer." *ALA Bull.* 58:1007-1010, Dec., 1964.

Boyer, C. J. and Jack Frost. "On-line circulation control; Midwestern University library's system using an IBM 1401 computer in a time sharing mode." In *Clinic on library application of data processing, 1969.* Univ. of Illinois Proceedings. Univ. of Illinois, Grad. Schl. of Lib. Sci. 1970 pp. 135-195.

Braude, Robert M. "Automated circulation systems." *Colo. Acad. Lib.* 5:1-6 Autumn, 1969.

Brown, Norman A. and Paula M. Strain. "Use of an automated shelflist." *Sci-Tech News* 36-37, Summer, 1967.

Bush, C. G. *et al.* "Attendance and use of the science library at MIT." *American Documentation.* 7:87-109, 1956.

Cammack, Floyd and Donald Mann. "Institutional implications of an automated circulation study." *Coll. & Res. Lib.* 28:129-132, March, 1967.

Clayton, H. "An investigation of various social and economic factors influencing student use of the library." (Unpublished Ph.D. dissertation, University of Oklahoma, 1965).

Davidson, John S. "The use of books in a college library." *Coll. & Res. Lib.* 4:294-297, September, 1943.

Dawson, C. S. *et al. Increasing the effectiveness of the MIT Science Library by the use of circulation statistics.* Unpub. MIT, Cambridge, 1962.

DeGennaro, R. "Harvard University's Widener Library shelflist conversion and publication program." *Coll. & Res. Lib.* 31:318-331, September, 1970.

Elrod, J. McRee. "Letter to the editor." *Coll. & Res. Lib.* 32:145, March, 1971.

Flannery, Anne and James D. Mack. *Mechanized Circulation System.* Lehigh University Library, Bethlehem, Pa., Center for Information Sciences, Lehigh University, 1966. (Library systems analysis, report no. 4) 17 p. + appendices.

Gull, C. D. "Automated circulation systems." *in Library automation: a state of the art review. Institute on Library Automation, San Francisco, 22-24. June, 1967*; edited by Stephen R. Salmon, Chicago, American Library Assoc., 1969, pp. 138-148.

Hamilton, R. F. "The Illinois State Library on-line circulation control system." in *Proceedings of the 1968 Clinic on Library Applications of Data Processing.* University of Illinois, Grad. School of Lib. Sci. 1969.

Hayes, Robert and Joseph Becker. *Handbook of data processing for libraries.* N.Y. Wiley, 1970. (See pages 481-547 for discussion on circulation systems.)

Jain, A. K. *A statistical study of book use.* Ph.D. thesis, Purdue University, January 1968. (PB 176525) includes bibliography of use studies.

Johns Hopkins University. *Progress report on an operations research and systems engineering study of a university library.* Baltimore, Milton S. Eisenhower Library, Johns Hopkins University, 1965. 110 p. (PB 168187).

Knight, Douglas M. and E. Shepley Nourse, eds. *Libraries at large,* New York, Bowker 1969, pages 101-120. (This is the resource book based on the materials of the National Advisory Commission of Libraries).

Lane, G. "Assessing the undergraduates' use of the University library." *Coll. & Res. Lib.* 27:277-282, July, 1966.

*Library Technology Reports*. Three systems of circulation control. Library Technology Project, May, 1967, 40 p. (also see the reports of July, 1965 and March, 1966).

Lubans, Jr., John. "On non-use of an academic library: A report of findings." in *Use, Mis-use and Non-use of Academic Libraries. Proceedings of the New York Library Association—College and University Libraries Section. Spring Conference, May 1, 2, 1970*. The Association, Woodside, New York, pp. 47-70. (Includes bibliography pp. 105-126).

Lubans, Jr., John. "Student use of a technological university library." *IATUL Proceedings*. 4:7-13, July, 1969.

McCoy, Ralph E. "Computerized circulation work: a case study of the 357 data collection system (Southern Illinois University)." *Library Resources & Technical Services*. 9:59-65, Winter, 1965.

McCune, L. C. and S. R. Salmon. "Bibliography of library automation." *ALA Bull*. 61:674-675 + June, 1967.

McDowell, B. A. J. and C. M. Phillips. *Circulation control system*. Southampton: University of Southampton, Library. 1970 (SOUL/Automation Project Report No. 1) 64 pp.

MacKenzie, A. Graham and Ian M. Stuart, *ed. Planning library services, Proceedings of a research seminar held at the University of Lancaster 9-11, July, 1969*. Lancaster; University of Lancaster, 1969. (University of Lancaster Library Occasional Papers, No. 3). various pagings.

Palmer, Foster M. *Widener Library circulation statistics 1965-1969 book use and stack space*. Harvard Univ., (unpublished) March, 1970, 18 pp.

Parker, Ralph H. "The punched card method in circulation work." (University of Texas Library) *Library Journal*. 61:903-905, December, 1936.

Quigley, Margery C. "Library facts from IBM cards." *Lib. J.* 66:1065-1067, December, 1941.

Salmon, Stephen R. "Automation of library procedures at Washington University." *Missouri Library Association. Quarterly*. 27:11-14 March, 1966.

Simmons, Peter. *Collection development and the computer*. Vancouver, Canada, School of Librarianship, the University of British Columbia, 1971.

Steig, Lewis. "Circulation records and the study of college-library use." *Library Quarterly* 12:94-108, January, 1942.

Surace, Cecily J. *Library circulation systems—an overview*. Rand Corp., Santa Monica, Calif., March, 1970. 25 pp. (ED 039001).

"Total systems concepts in the design of a computer-based circulation system." in *Library systems analysis guidelines* by E. Chapman, P. St. Pierre and John Lubans, Jr., N.Y., Wiley, 1970, pp. 197-207.

Trueswell, Richard W. "A quantitative measure of user circulation requirements and its possible effect on stack thinning and multiple copy determination." *American Documentation*. 16:20-25, Jan., 1965.

Woods, William Edward. *Factors influencing student library use: an analysis of studies*. (1930-1964) M.A. thesis, University of Chicago, School of Library Service, 1965.

A PPROACHES

TO

L IBRARY A UTOMATION

*Allen B. Veaner*

SOURCE: From Allen B. Veaner. "Approaches to Library Automation," *Law Library Journal* 64 (May, 1971) pp. 146–153. By permission of the publisher and the author. Copyright © 1971 by the *Law Library Journal*.

## I. Introduction

There was a time not long ago, about the time third generation computing equipment was introduced, when it was widely believed that library automation was one of the less demanding intellectual tasks for computerization. With their promise of rapid access to files through devices like cathode ray tube terminals, third generation computers created a tremendous amount of excitement and expectation among not only librarians but also academic administrators. Thus, by about 1964 it was confidently predicted that most of the bibliographic and related activities in major libraries would soon be performed by computer. To be sure, there were a few people in 1964 who knew better and were courageous enough to say so, but most potential users in libraries were readily swept away by the waves of enthusiasm and expectation generated by this new hardware and the promises of manufacturer-supplied software, particularly the concept of time-sharing.

That we nave not achieved in 1971 what we anticipated in 1964 cannot altogether be attributed to the computer industry or our professional colleagues in computer programming. Librarians have long believed that they thoroughly understood bibliographic processes and could describe them in sufficient detail to facilitate new system designs. Experience of the past 7 years demonstrates conclusively that this premise was a false one; we are just now beginning to appreciate not only the complexity of our traditional systems of bibliographical control but also the variety and exquisite subtlety of the innumerable intellectual decisions associated with each facet and step of each bibliographic procedure. This hypothesis has been well documented by Shoffner.[1]

Library automation activity of the past 7 years has been dominated largely by research and learning, as distinct from production. Prior work in automation has either been conducted in parallel with manual systems, has supported manual systems, or been in the nature of experimentation. There have been no cases where computerized production systems have superseded manual ones and realized budget savings; there have been a few instances of improved reliability, but not *both* economy and reliability. Between the development of a conceptual plan for library automation and the achievement of dependable, operational systems, there is an arduous and difficult path, but the rewards of travel on that path, if successful, can be tremendously satisfying to the individual, the institution, and the profession. The challenge might be compared with the invitation issued in 1900 by the famed polar explorer, Ernest Shackleton:

Men wanted for hazardous journey. Small wages, bitter cold, long months of complete darkness, constant danger, safe return doubtful. Honor and recognition in case of success.

Except for the "small wages," one might well be describing an enterprise in library automation.

## II. The System Development Process

But system design for library automation, system design for any enterprise, differs from polar exploration in one important aspect: one begins with the goal, the end point. One pretends that a fully realized system is at his disposal and, working backwards, the user specifies his system requirements. An important mission of the team of system designers and programmers is to convert requirements into specifications, and to realize the complete, functional system.

We are describing here the system development process, which is really a strategy for the achievement of any complex goal. The system development process evolved from the need to coordinate the many interrelated activities in large-scale development projects, particularly where such activities have a degree of time dependency with each other and with the main goal, and where they might be carried out by widely separated development groups. Undoubtedly, the largest such activity to date is our country's aerospace program. Its complexity may be judged from the fact that the annual aerospace budget is 2,000 to 10,000 times the amounts commonly available for *all* library automation development efforts. This comparison will also help us appreciate the complexity of the library automation effort itself.

The system development process consists of six distinct but overlapping phases:[2]

The first is preliminary analysis. In this phase, current operations are reviewed, a conceptual model of the new system is constructed, and a series of alternatives proposed for the realization of the new system. If there is any intellectualizing (in the derogatory sense of the term) in a project, it may be done only at the very beginning of this phase. Everything after that is hard work. In preliminary analysis, one may hypothesize a variety of technical solutions to the general objectives of management. Through a process of negotiation, the users and the designers select the most technically feasible and the most economically realistic alternatives; their recommendations are presented to management for approval. This will not be a brief report; it may easily take 75 to 100 pages to detail the scope of a project.

The second phase is detailed analysis. In this phase, current operations are thoroughly analyzed in order that the users and designers may understand them in complete detail. Where operations tend to be functionally divided, as is the case in library technical processing, this approach also assures a complete understanding of all interdependent and interacting processes in the library, as well as congruence of vocabulary among designers and users. The output of this phase is a substantial report, probably in several volumes, documenting the current system.

After the second phase is well underway, enough has been learned from the first two to begin general design. In this phase, the user specifies his requirements for the new system. He states his need in terms of files, printed outputs, inputs, services, turnaround time, and reliability. He may also propose performance specifications. General design is far more de-

2 For a complete and detailed explanation, *see:* Stanford University. SPIRES/BALLOTS Project. *System Scope for Library Automation and Generalized Information Storage and Retrieval at Stanford University.* Stanford, Cal.: Stanford University, 1970. (Available from ERIC Document Reproduction Service as ED 038 153, or, while the supply lasts, from Documentation Office, Cypress Hall Annex, Stanford, Cal. 94305. $7.50.)

tailed than one might expect from its name. In it are specified all processes, procedures, and steps necessary to perform each predefined operation in a system. All new processes and procedures are flow charted, all forms are designed, all personnel and training requirements are defined, files are defined down to the last data element, and each data element is unambiguously specified. The general design document is a tool for the programmers who will develop the detailed design. Theoretically, a programmer could take the general design document, work independently, and if nobody made any mistakes, the end product would be just as the user had specified in the first phase. There is no known occurrence of such an event.

Phase four, detailed system design, is concerned with the development of software to support the general design. Its output will be computer programs that perform the jobs specified in the previous phase. Substantial amounts of computer time will be needed during this period for testing and debugging programs. Once again, if all our programmers were perfect, no testing time would be needed. Generally speaking, a program must be logically perfect to perform the task expected of it; programs may be made more efficient (use less computer resources) by fine tuning, but a program that misformats a call number or cannot properly sort bibliographic entries is useless no matter how efficiently it may run. At this stage, program testing is typically done with dummy data.

The fifth phase in the system development process is implementation. In this phase the system is installed and operated in a manner that most clearly simulates the intended real life environment. Hardware is brought in, files are converted, people are trained, a great deal of practicing is done in parallel with the manual system, and programs are refined. This stage is much like the shakedown cruise of a new vessel or the test flights of a prototype aircraft.

The final phase of the system development process is production. The new system replaces the old, the development team is broken up or goes on to another assignment, and a small crew of maintenance programmers remain.

Even the smallest and simplest design effort is made up of these six elements, which in essence define what you are going to do and how you are going to do it, and then do it. Let us examine in some detail a few of the problems that arise in the preliminary analysis phase.

## III. Problems Encountered during System Development

Because the computer is an expensive tool, few computer projects should be started without a thorough knowledge of one's current costs. The library profession has not yet developed work definitions and cost accounting methods that make data from one institution readily comparable with those from another. Sometimes locally developed cost data are defective in one or more important parameters, such as accounting for overhead, nonproductive time, and indirect costs such as space and utilities. Indirect costs are exceptionally important in computer applications, because of constant pressures from users to expand services and facilities. In a manual library system, it may be uncomfortable but possible to crowd several more people into an area to increase productivity. People really do not take up much space. But to add a new piece of computer gear may require an addition to a building or supplementary facilities for electricity, air conditioning, communication lines, and staff.

During the earliest stages of conceptual planning, it would be a good idea to examine the impact of standardization upon one's internal procedures. All automatic systems, whether mechanical or electronic, depend upon standards because the behavior of mechanized systems is deterministic. There is no human being to interact with, to explain the exceptions to. All specifications, instructions, and procedures must be built into the system, in the form of software or hardware. One inevitable result is that, compared to manual systems, some computer systems will appear to have little flexibility. Whether provided manually or through the computer, flexibility costs money. Although it is often said that librarians pay only lip service to national bibliographical standards, if one examines the kinds of changes made locally to centrally distributed bibliographic data, one is struck by the remarkable acceptance of the intellectual product. Discounting for a moment necessary local changes or additions, such as local call number and shelf location, one tends to find an undue preoccupation with the cosmetic aspects of card reproduction and the desire to retain record formats and designs that may no longer be optimum for the user, *i.e.,* the public. (We often forget that our bibliographic tools are created primarily for the *users,* not the technical processing staff.)

Now, manual systems for manipulating and reformatting bibliographic data are very labor intensive; typically, the work is performed by that section of the staff where labor turnover is highest and by a staff whose work must be closely supervised. If each library or each branch in a large system requires a unique format for its bibliographic files and products, the chances are great that the cost of computerization will be substantially higher than manual costs. Only by reducing internal processing of bibliographic records to a series of standardized, repetitive tasks can we take advantage of the speed of the computer. In the final analysis, if flexibility and customizing of the record are required, one cannot beat a manual system, at least in the current state of the computer art. The basic question is: Can management afford to pay for highly flexible manual systems in an era of rapidly rising labor costs and tremendous expansion of publication and service demands by our clientele?

## IV. Basic Computer Methodologies

It will be useful at this point to review available computer methodologies for bibliographic applications.

The first thing to appreciate about bibliographic records is their very wide range of activity. Probably most of the records in a large research library never get accessed more than once in a few decades. Inactivity of a record, however, can be very misleading. *Even inactive records must be kept up to date.* As librarians, you are keenly aware of the enormous file-organization problems we face. Libraries in the fields of government and law face especially complex problems with corporate entries and jurisdictions. The situation is extremely frustrating with publications originating from areas of great political instability. Besides, your files may range over many centuries of published materials, and cover hundreds of thousands, or millions, of persons and entities. To maintain any kind of order and consistency, all types of entries in your files must conform to established formats and styles. This file maintenance work must be performed even if no one uses your files. Library files may therefore be said to be extremely update sensitive.

Of course many library files are subject to intensive use. Familiar examples are your circulation files and the files of outstanding orders or books in process. In a large library system, these files might generate hundreds or thousands of transactions per hour. Let us see how certain transactions are handled in computer systems.

The oldest and most familiar transaction-oriented system is batch processing. In batch processing, an entire file of information is printed out on a regular schedule and distributed as a report, each new report superseding an old one. In a computerized batch system, all records must be processed to update part of the file. Creation of the update transactions themselves is typically also a batch process. The basic idea is to cluster into manageable batches each kind of work associated with posting a file: creating new records, updating old ones, deleting obsolete records, processing the changes, and printing out the results. Printing a daily list of all books in circulation in the library is a representative batch processing application. The end product is always out of date by the processing cycle, usually 24 hours or more.

The second major kind of computer methodology is known as the on-line system. Here the user conducts a transaction with only a record of interest at a given moment. Records not affected are not accessed, and in theory, information is processed as it is input. There are many varieties of on-line systems for managing files; few have interchangeable hardware or software. In general, on-line systems are vastly more expensive and complex then batch systems, because the file or data base must be accessible very rapidly. The devices and communications systems that provide rapid access to the records are many times more costly than the relatively simple reel of tape or deck of punched cards used to store files in batch systems.

Almost all on-line transactions are initiated by the user, working at a communication device, commonly called a terminal. A terminal may be a computer-controlled electric typewriter or a silent, fast writing cathode ray tube. While many transactions are activated by the user, it is important to observe that many are initiated by computer programs and are not under direct control of the user. On-line systems perform the same basic functions as batch systems: creation or deletion of records, update and change, and query. The main differences are in cost, and in speed and style of user interaction. For instance, the query function in a batch system is supported by the user reading a printed report, such as a circulation list or a book catalog. In an on-line system, a user may query a record that has been updated or changed

a few minutes earlier. Instead of reading a printed document, the user views a "soft" copy on the screen of a cathode ray tube terminal (CRT).

In reviewing the distinction between these two major modes of processing data, it is important to recognize that, for file management, the library typically employs an on-line *manual* system, which, with certain exceptions, is extremely cost effective and rapid in response. This is the case particularly with small to medium-size public libraries and special or technical libraries that do not perform archival functions. That is, such libraries do not grow indefinitely nor do they maintain files without limit to size. The small library probably does not get along too well with a batch system; its response to update is much too slow, even if the turnaround time is 24 hours. Yet this is all most small libraries can afford. Economics permitting, it may be possible for the small library to participate in a larger on-line system. However, attached to this seemingly simple concept are innumerable details of problem definition, system development, economics, telecommunication, reliability of service, and others. Although there currently is much experimentation with networking, mainly in private industry and in government-sponsored research, it will be some time before a library can conduct its file management business through remote terminals as easily as we communicate verbally by direct distance dialing.

## V. Four Approaches to Library Automation

I should now like to review a series of approaches taken by several development projects in the United States. This part of my review is based upon a series of presentations given at the Collaborative Library Systems Development Conference, held in New York in November 1970. This meeting was convened to review the present state of the art of library automation in academic libraries, as well as to detail and differentiate a variety of approaches, especially those at Chicago, Columbia, and Stanford. The full proceedings of that conference will soon be published by the MIT Press under the general editorship of Paul Fasana, assistant to the director of the Columbia University Libraries.[3]

[3] Fasana, Paul J., ed. *Collaborative Library Systems Development.* Papers Presented at the Second CLSD Conference, Nov. 9-10, 1970, New York City. Cambridge, Mass.: MIT Press, 1971 (in press).

My intention here is to summarize four different approaches, along with their advantages and disadvantages.

For the substance of this part of my paper, I am indebted to the speakers at that meeting. They are David C. Weber, director of university libraries at Stanford University; John McDonald, director of libraries at the University of Connecticut; David Kaser, director of libraries at Cornell University; and Carlos Cuadra, manager, Library and Documentation Systems Department, System Development Corporation. They represented, respectively, individual effort at one institution with external support, a regional network contracting technical development to a commercial firm, a consortium representing consolidated efforts of five institutions, and, finally, commercial services offered by the profit-seeking firm.

### A. In-house Development

According to Weber, the most striking aspect of individual effort in developing a complex on-line system is the tremendous investment required in staff time and the time scale of the development project.[4] The staffing effort at Stanford has ranged from 10 to 25 persons, in accordance with need during a given phase of a system development process. Obviously, the staff must be larger after the midpoint, because programming responsibilities are heavy. The staff may be smaller in the beginning and at the end. It is estimated that approximately 25 man years of local effort have been contributed by the professional staff of the library. Local contributions have required heavy commitments from the best professional people in the library; it is not a job that can be assigned to junior professionals. The director of libraries and major department heads have sacrificed many thousands of man hours for development work.

Material contributions are also needed in the form of adequate quarters for the development staff. This is difficult to achieve in older library buildings where space is not very flexible. Eventually, proximity to the computation center becomes essential for the large project. To support its development staff, the Stanford Library has budgeted for something over 2,000 square feet of office and conference space near the computation center. This has been provided by means of portable buildings.

Weber cites the numerous advantages to individual effort. The first obvious advantage is one's ability to direct resources towards preferred priorities. Second, the users can be intimately involved with the design of the system they ultimately will use. Third, communication between the designers and the users will be facilitated; many meetings will be held and problems thrashed out long before there is any dangerous divergency of understanding. A fourth advantage is that progress is not curtailed by a weak member in a group. Finally, the development effort may be less costly, but only if the project is very effectively managed, with carefully established milestones and rigorous scheduling and monitoring.

I would like to add that scheduling in the library environment is not easy to achieve. Librarians are unaccustomed to the notion of meeting deadlines and sticking to schedules in a large development project with many interdependent parts. This is not meant in a derogatory sense; it is simply a new experience for librarians.

Weber also recounts some of the other disadvantages to individual effort. Among them are constraints imposed by the local computing environment. Not many libraries can justify a computer of their own, except for mini-computers and the low-cost range of third-generation equipment or second-generation equipment. So one must work within a facility that must service many users. A library employing a batch processing system will require many hours of printing time; this means tying up the facility's printing equipment for a long period each day. This will obviously have a detrimental effect upon other users of the facility, and the facility management may defer a large printing job. This is an example of the type of problem libraries face in getting the desired machine priority for doing their jobs.

A second disadvantage is the uncertainty of external funding. An external agency does not usually wish to fund a project specific to one institution; some degree of extendability or transferability must be demonstrable. I cannot discuss the transferability problem at this time; for a brief outline of it, I refer readers to my paper, "Major Decision Points in Library Auto-

---

4 Weber, David C. "Independent Development." *Ibid.*, at 98 *ff*.

mation."[5] A position paper on transferability is badly needed but has not yet been written.

A third hazard is the risk of mismanagement. In fact, this is probably the greatest hazard in library automation projects of any size, owing chiefly to librarians' inexperience in mechanization. The danger is perhaps much greater for solo efforts than for cooperative enterprises where checks and balances as well as a critical and questioning clientele provide early warning signals of management troubles. First-rate management talent is both scarce and expensive.

### B. Interinstitutional Cooperation

Let us turn next to an example of interinstitutional collaboration. Here I will be paraphrasing the remarks of Mr. John McDonald.[6] Mr. McDonald's own library, the University of Connecticut, had experienced extremely rapid growth, having increased its acquisition rate by a factor of eight in 10 years. He did not feel it wise to impose a local automation effort upon the already-existing problems of rapid growth. Connecticut and a number of New England academic libraries formed NELINET, New England Library Network, to tackle automation cooperatively. McDonald describes three kinds of problems arising from this approach to automation: relations between the libraries themselves and the contractor, internal relations among members of the academic constituencies, and external relationships with the regional library community at large.

In the category of external relationships, there are bound to be differences in the degree of institutional commitment, no matter how well the collaborating libraries get along with each other. McDonald also mentions differences in will or ability, a degree of impatience with the rate of development, and different ideas of service to the consuming library. Counterposed to these admonitions is the advantage that having technical work done by an outside contractor avoids "big brotherism" if one of the libraries is by far the largest of the group. Balancing this advantage is the risk that, because outsiders are not in actual day-to-day library

activities, they may lose sight of specific local requirements. A final problem with an outside contractor is the question, Who owns what products? What part of the software is proprietary and belongs to the contractor? What part belongs to the collaborative? If public funds have been used for development, these questions can become difficult to resolve.

In reviewing the second category, internal relationships, McDonald points out an irksome problem: difference of opinion and commitment between library management and the technical services staff. I am certain I need not elaborate on this problem before this audience.

Confirming Weber's case for in-house development efforts, McDonald concedes that it is harder to get good local staff commitment when much of the intellectual work is done by outsiders.

What are the good points to this approach? One is the promotion of standardization and consistency with national bibliographic standards. McDonald believes that librarians have overvalued internal consistency and that one of the salutary effects of automation may be "to lessen our desire to maintain what might be called a foolish consistency at the expense of national standards."

A final good effect is that a collaborative project provides a ready answer to that favorite question from faculty and administration: "When are you going to automate your library?"

In category three, relationships with the larger library community, there is the ever-present hazard to any successful effort: envy and criticism. Complementing other cooperative activities and avoidance of duplicate effort are described as the remedies. However, I should like to point out that, because we are still in a research and development phase, it is almost impossible to avoid work that is both duplicative and competitive. We now know that library automation is sufficiently complex that it cannot be simply broken up into a series of predetermined tasks, which are then assigned to the best teams. We are still in the era of defining and understanding the tasks themselves.

### C. The Consortium Approach

David Kaser, director of libraries at Cornell, reviews the experiences of a consortium of five

5 Veaner, Allen B. "Major Decision Points in Library Automation," 31 *College & Research Libraries* 299-312 (1970). (This is an abridged version; the full text, with more citations, appears in *Minutes* of the 75th meeting of the Association of Research Libraries, pp. 2-33.)

6 McDonald, John. "NELINET: One Approach to Library Automation," *op. cit.*, Fasana, at 102 *ff.*

university libraries, working together in upstate New York.[7] The group is known as FAUL— Five Associated University Libraries.[8] A number of proposals were formulated and discussed, and some were attempted. Included were the creation of a machine-readable union catalog for the five libraries, a MARC service center, the concept of creating a service bureau dedicated to performing computerized library functions, and a joint on-line serials control system.

It is apparent that a great deal of learning occurred in the FAUL approach, since a number of the functions proved too expensive or impractical in the present state of the computer art. The union catalog was too expensive to operate; the MARC processing center functioned successfully but was discontinued when none of the members used or planned to use the data within a reasonable time frame. A healthy degree of self-criticism led to a decision not to implement the service bureau proposal, apparently because the plan was regarded as overly ambitious. Work is currently underway on the joint approach to a serials system.

Although Mr. Kaser is optimistic of success, he is reserved enough to realize that this endeavor might not succeed. He concludes that "cooperation when it costs nothing is easy, but when it is expensive, as this is, even among men with the best will in the world, cooperation requires of all of its participants the wisdom of Solomon, the patience of Job, and the prophetic powers of blind Tiresias thrown into the bargain. Unfortunately, there is no evidence to indicate that the directors of FAUL are any more endowed with these multiple traits than are normal men elsewhere."

### D. Commercial Services

What can commercial services offer? Carlos Cuadra of System Development Corporation suggests three areas: consulting, continuing service, and sale of software for local use.[9] What are the respective advantages and disadvantages within the three areas?

[7] Kaser, David. "FAUL: A Consortium Approach to Library Automation," op. cit., Fasana, at 106 ff.

[8] FAUL members include: Cornell University, State University of New York at Binghamton, State University of New York at Buffalo, Syracuse University, University of Rochester.

[9] Cuadra, Carlos. "The Commercial Service Approach to Library Automation," op. cit., Fasana, at 110 ff.

In consulting, the library gets more experienced (and we hope better qualified) help than they themselves can afford to hire; time will be saved; money will be saved; finally, performance can be guaranteed. This last point is certainly very important. If a commercial firm cannot deliver, the library has no financial obligation. At least that is the way it should be if the contract has been properly written.

Here are some of the disadvantages. The library might not find the best firm to do the job, and even if it did, the people best qualified to do the job might be busy on other work and not be available. Secondly, a contractor's work may be less creative in the design area, because as Cuadra points out, the contractor does not have to live with the system. A third disadvantage is reduction of the ability to develop one's in-house staff, and a fourth is the risk of poor documentation. The documentation problem is a pretty serious one, and Cuadra rightly points out that sensible users sometimes reject free programs if they are poorly documented. It is simply too much work to learn the program and fix the bugs; it is simpler to write a new program. Then there is the question that McDonald raised: Who owns what?

In reviewing commercial services, Cuadra suggests that this approach might be called "avoidance of automation." A prime advantage is the avoidance of huge capital investment. A disadvantage is the markup, needed to assure a reasonable profit to the commercial firm; Cuadra suggests this may range from 60 to 200 percent. Then there is the danger of becoming excessively dependent upon a specific commerical service. What does one do if that service goes out of business or shifts to another line of work?

It is true as Cuadra suggests that use of a commercial service protects the user against "surprises by the computer center." The most annoying surprise is one he cites: changing the computer without much notice. In some environments, such changes have been known to occur with no notice. But I would counter with a question I hinted at a few moments earlier: What protection does the user have against surprises from the commercial service?

Turning to the third facility of the commercial service, sale of off-the-shelf software, Cuadra points out the advantages to the "have not" libraries, those that not only cannot develop systems with their own resources but that probably also cannot gain external support for development work. External software may be relatively cheap to buy—an obvious advantage—but it can be very expensive to operate, because the user must pay for all computer resources, including time, storage of data, communications, and terminals. There will also be an additional cost in training operators, as well as the costs of internal dislocation caused by organizational changes imposed by a new system. In other words, every new system imposes startup costs.

But there are advantages to this off-the-shelf software: It is cheap, available, and maintainable at reasonable costs. On the negative side, the user cannot know the details of how the programs operate, cannot fix bugs found after the warranty has expired, and may find that some of the programs are not congruent with the library's actual needs. So in a certain sense, buying off-the-shelf software is pretty much like buying a private automobile. It may be relatively cheap to buy this complex piece of machinery, but if it has lots of power-operated accessories and other extras, the owner is probably not going to be able to maintain it himself, and there will be many cheaper ways of getting from one point to another.

## VI. Conclusion

Which of these four distinct approaches to library automation is the best? Ten years from now, we may be able to answer this question. I would expect that by that time the American library community will have focussed its goals and objectives much more clearly than is presently the case. The striving for perfection in bibliography and the felt need for customizing bibliographic records may be blurring the view. Of particular concern is the phenomenon of technical services as a self-consistent, self-justifying, self-perpetuating activity in our libraries, and the relative lack of success in promoting national bibliographic standards. The financial crisis inherent in this situation is already upon us. We know that the cost of computing and computer hardware is continuously going down on a unit-cost basis and that labor costs are rising rapidly. Beginning clerical wages in libraries have just about doubled in the past 10 years, whereas cost of performing a given computer function is approximately cut in half every 3 or 4 years. Although there may be a technical limit to cost reduction in computer

power, it certainly seems obvious that it won't be long before we will simply be financially unable to operate our libraries without computer assistance.

To obtain that assistance, we need to continue current research and development efforts, selecting a few promising projects for implementation. Libraries and librarians must acknowledge that this current R & D phase means incurred costs without immediate benefits. We must not repeat or continue the error of underestimating the intellectual difficulties associated with achieving bibliographical control and accessing bibliographical records by machines. We must adjust our expectations to reality, to that which is technically feasible, electronically dependable, and, above all, maintainable. If it is man-machine interaction and communication on a large scale that we are talking about, then we still have a great deal to learn. Our economic constraints do not give us much time to do the learning. I recommend that we try to learn fast.

## GENERAL REFERENCES

1. Cuadra, Carlos. "Libraries and Technological Forces Affecting Them." Presented at Conference on The Library in Society—Towards the Year 2000, School of Library Science, University of Southern California, Los Angeles, Apr. 25-26, 1968.
2. Fasana, Paul J. *The Collaborative Library Systems Development Project: A Mechanism for Inter-University Cooperation.* New York: Systems Office, Columbia University Libraries, 1970 (Technical Memorandum No. 4).
3. "Information Retrieval and Library Automation." A chapter on this topic has appeared in each volume of the *Annual Review of Information Science & Technology,* edited by Carlos Cuadra and published by *Encyclopaedia Britannica.* Chapters have appeared by Black and Farley, Kilgour, Parker, and a chapter for the forthcoming volume (vol. 6) is in preparation by H.D. Avram.
4. Knight, Douglas M., and E. Shepley Nourse, eds. *Libraries at Large: Tradition, Innovation, and the National Interest.* The Resource Book Based on the Materials of the National Advisory Commission on Libraries. New York: Bowker, 1969.

# DATA PROCESSING
## IN AN ACADEMIC LIBRARY:
## SOME CONCLUSIONS
## AND OBSERVATIONS

*Dan Mather*

SOURCE: From Dan Mather, "Data Processing in an Academic Library: Some Considerations and Observations," *PNLA Quarterly* 32 (Summer 1968) pp. 4–21. Reprinted by permission of the author.

443

The majority of articles on library applications of data processing describe systems which either exist only on paper or which have been thoroughly debugged. The growing pains, set-backs, and disasters are seldom emphasized. Anyone who has worked with data processing to any great extent knows that the system worked out on paper differs from the one which finally goes into operation, that target dates are not always met, and that not all objectives are attained. Without belittling in any way the contributions of those who took part, it will be useful to discuss in rather frank terms the history of the two systems currently in operation in the library at Simon Fraser University, and to look at some of the effects of automation on faculty, students, and librarians. While every institution has a unique set of problems, resources, and service requirements, it should be worthwhile to share some of our experiences.

### Development of the University and the Library

In order to understand why things happened the way they did, it is necessary to see how the University developed and the bases on which it was established. It is the newest of the provincially-supported universities in British Columbia, the others being the University of British Columbia and the University of Victoria. The creation of a third university was one of many recommendations made by John B. Macdonald (then president of U. B. C.) in a report on higher education in British Columbia.[1] In March 1963, the B. C. Legislative Assembly authorized the establishment of the new university. Although the Macdonald Report had emphasized that the new institution be oriented toward undergraduate work, the lack of graduate programs proved to be a barrier in attracting senior faculty. University aims therefore were modified to include certain areas of graduate study. Enrollment was projected to grow from 1,700 in 1965 (there were actually 2,500 students registered for the first classes) to 18,000 by 1984. The campus was designed to be built in stages for accommodation of the anticipated student population growth. The library building was planned to expand to a capacity of approximately 1,000,000 volumes. Within a period of two and one-half years, administration and faculty were hired and the initial phase of construction was completed. Classes opened in September 1965. It was truly an "instant university".

The library came into being with the appointment of Donald A. Baird as University Librarian in May 1964. The library's first quarters were in a warehouse in the industrial section of Vancouver, next-door to a tannery. An instant university requires an instant library. There was no time for a gradual gearing up of capacity for acquisitions, cataloging, and circulation. Projected allocation of funds to the library indicated that an annual accession rate in the range of 50,000 volumes was likely to be reached by 1966. The dimensions of the work load clearly suggested that automation of technical services was, if not the obvious solution, at the very least an approach to be seriously investigated. Accordingly, a system outline was developed, based on Baird's ideas and on his observations of activities in other academic libraries which had already gone into data processing (Florida Atlantic and the University of Missouri).

Although the computing facility (an IBM 1440 with two disk drives) was planned to serve faculty and administration, the way in which the university developed resulted in a reversal of the usual trend in computer usage. In short, major administrative functions such as accounting, registration, purchasing, supplies, building maintenance and the like had to be in full operation before the university opened. Faculty, on the other hand, were completely involved with course planning and had little time for research. Indeed, many faculty members arrived on campus only a few weeks or days before classes began. Under the circumstances, it was natural that computing effort was solely devoted to non-academic requirements. Since the availability of adequate computer time was a certainty, library systems planning was begun in earnest. However, it was apparent that the time of the three-man computing staff would be completely taken up with the needs of accounting and registration and thus could not handle library system design and programming. IBM was consequently persuaded to take it on.

In common with many librarians, Baird believed that information generated during the initial processing should be stored for re-use. It was decided that

Simon Fraser University

automation of the technical services should begin with acquisitions to be followed by cataloging and serials; it was hoped that input for the circulation system could be created as a by-product of cataloging. However, the fast-approaching opening date dictated that circulation receive first priority. It was therefore necessary to plan a circulation system which did not rely on by-products. Since there was no systems librarian, design of the system took place during a series of conferences between Baird and representatives from IBM and the university computing center. Decisions were made on equipment, format for input, and output requirements. The same approach was used in the planning of acquisitions (which began as soon as circulation design was finished), although it became necessary to set up a manual system to acquire material for opening. Taking all things into account: the lack of data processing background on the part of the librarians, the lack of familiarity with library procedures on the part of the people from

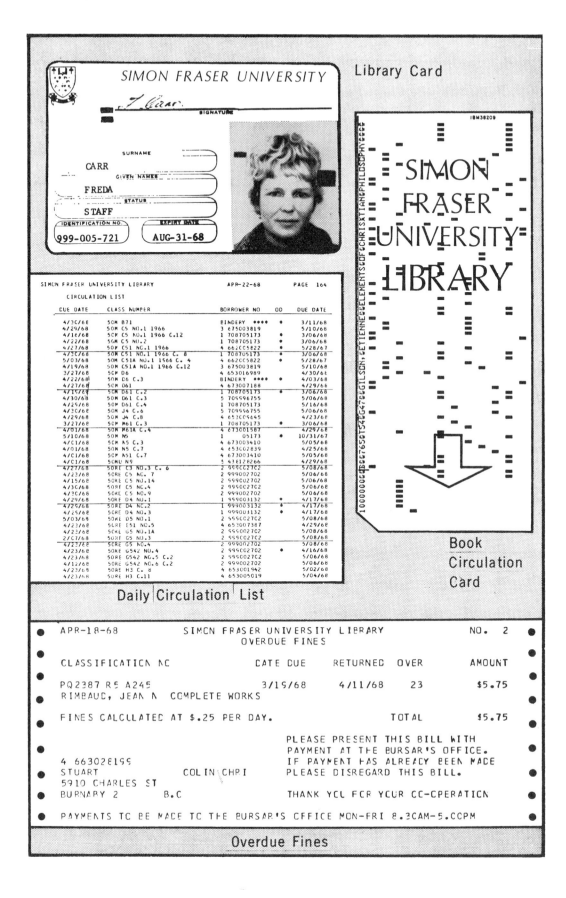

Library Card

Book Circulation Card

Daily Circulation List

Overdue Fines

IBM and most particularly, the limited amount of time available, the programming and systems work done by IBM for circulation, cataloging, and acquisitions were satisfactory. It would have been naive to expect sophisticated systems under these circumstances.

## DESCRIPTION OF SYSTEMS

### Circulation

The equipment used to control circulation is the IBM 1030 system. It consists of two kinds of units: a badge reader and a remote card punch. Information fed into the badge reader is recorded in punched card form on the remote punch. To circulate, renew, or discharge a book, the following items are put into the badge reader: a laminated badge which contains the borrower number and loan period (a special badge is used for discharges); a punched book circulation card (taken from the pocket of the book) containing call number and short author/title. The information in the badge and the book card is read by the 1031 badge reader and transmitted to the 1034 remote card punch where a circulation record is created. The circulation record contains date and type of transaction, loan period, and borrower number, in addition to short author/title and call number. Circulation records are sent in daily batches to the university computing center (located in the library building) where they are transferred to magnetic tape and due dates are calculated. The day's transactions are passed against a master tape record comprising all items in circulation. New charges are added to the master record, returns are deleted and new due dates are established for renewals. A list of all books in circulation is then printed. It contains the call number, borrower number, loan period, date due, and an indication of overdue notices sent; the list is arranged in call number order. The master tape record is scanned by the computer to identify items which have become overdue and notices are printed. These are sent to the borrower by the library. When an overdue book is returned, the computer calculates the amount of the fine and prints a bill. The amount due is then fed into the university accounting system where it becomes a routine charge against the student. Bills for fines are forwarded to the central accounting office which is responsible for collection of all monies due to the university.

Listings of reserve books (of which this library has a very large number, with major changes three times a year) are produced as a by-product of the circulation system. Books placed on reserve are "charged out" in a manner similar to that used for regular loan items. A badge containing the name and number of the course is substituted for the regular library card. Consequently all reserve books appear on the daily circulation list. Reserve listings arranged by author and by course are produced weekly. Books on reserve are circulated manually.

Copies of the daily circulation list are located near the catalog, enabling students to see, before going to the stacks, if a book is checked out, and if so, when it is due for return. The borrower can also be identified by looking up the borrower number in a master list. Because the number of copies available of a given title is not indicated in the catalog, a search of the circulation list before going to the stacks can be misleading. Answers to many routine questions are provided by the information in the circulation and reserve lists, freeing staff for more important work, such as making change for the Xerox and giving directions to the washroom.

The effect of the automated system on faculty has not been altogether positive. Renewals cannot be made by phone or memo since both the library card and the book circulation card must be put into the 1031 badge reader in order to record the transaction. As might be expected, there is some unhappiness with the system on this account, but the greater majority of faculty understand why renewals cannot be handled by phone. Perhaps this limitation has produced a positive effect in that faculty, or at least their secretaries, visit the library a little more often than would be the case if phone renewals were possible.

Many snags were encountered before the system operated satisfactorily. As with a phone book, a circulation list issued once a day (or once an hour) is out of date before it is printed. In a library such as ours was at the time it opened (with relatively few books per student) it was imperative that circulation control be tight. Inevitably, interruptions in the printing of the circulation list were experienced. Sometimes as many as three days went by without a new list. Some of the delays were caused by down-time (euphemism for "the machine doesn't work") while others resulted from circulation being bumped off the computer by high-priority jobs (writing pay checks, for example). Defects in the programming were encountered, such as a mysterious dropping of charges from the master tape. In such instances, transactions would simply disappear. Whenever one of the "lost" items was renewed or returned, the machine became confused. While the computer was able to get out of its dilemma by printing an error message, the circulation staff had no such easy means of resolving the problem. Once dropped from the master tape record, there was no way to identify such "lost" books, let alone the one who had originally checked them out. Also, there was a large number of improperly recorded transactions, usually the result of key punch errors either in the book circulation card or in the badge. The computer would reject these invalid transactions and they would not, therefore, appear on the circulation list. Manual records had to be maintained for such books.

During the initial period of operation it became evident that certain essential features were lacking from the system. When these were identified, the library was understandably anxious to incorporate them "immediately". An example of the kind of problem, and the need for a quick solution, was the discovery of a large group of book circulation cards which could not be handled properly by the system since they were incorrectly formatted. Alteration of programs was necessary in order for these items to be processed; the alternative would have been to re-punch several thousand cards, once the books had been identified. No allowance had been made in the programming for the discharge of a book which was returned on the same day it was checked out; the computer churned out overdue notices as soon as the loan period had expired since it did not "know" that the book had been returned. Problems also resulted from a decision to simplify machine sorting of records by using only the last five digits of the borrower number. Unexpected increases in student population had the inevitable consequence of two or more students winding up with the same digits in the last five positions of the student number. Students received overdue notices for books they had never checked out. When social insurance numbers were substituted, the same five-digit convention was continued. Again the inevitable happened. It was finally necessary to expand the record to include the complete social insurance number. Also there was a programming inadequacy in handling renewals. Often the computer did not recognize renewal transactions. Thus one book would be carried on the circulation list as being charged out twice to the same person and an overdue notice would be generated when the due date for the initial charge was reached. Not unexpectedly, students resented receiving an overdue notice for a book which they had renewed.

In order to correct errors and to build in new features, programs were patched. Patching is always hazardous and is especially so if it is done under pressure. In several instances the demand for changes precluded a careful assessment of the consequences of program alterations. Side-effects of patching were sometimes immediately apparent, although not infrequently they became evident only when it was too late to take corrective action. A series of badly-needed diagnostic and control reports was implemented which provided librarians and programmers with a means of identifying the major problem areas. The reports permitted the programmers to clean up the existing system, since a complete re-write was not possible.

While programming errors were the source of several of our headaches, some of them were attributable to IBM. The number of 1030 installations in the Vancouver area increased sharply starting in 1965. The two-terminal Simon Fraser configuration was a minor element in the growth pattern, compared to the multi-terminal University of British Columbia system, installed during the same period. Industrial use of the equipment also turned upward, resulting in heavy pressure on IBM's service department. It was soon apparent that there was an insufficient number of service men trained to handle 1030 maintenance. Understandably, this created serious difficulties for the libraries, as well as for IBM. Preventive maintenance for the Simon Fraser units could not be set up because the few qualified service people were busy answering calls during the day. Although desired by the library, the company was reluctant to provide preventive maintenance outside regular working hours. Trouble-free service is essential, particularly in a public service function such as circulation. Of course, machines are no closer to perfection than are their creators, but a dissatisfied customer (whether he is a student or a librarian) will not be consoled by philosophy.

Serious problems result if the 1030 system is down for more than a few minutes. Since there is no way of predicting when the system will return to normal operation, interim manual records of charges and renewals must be made, to be fed into the machines after service has been restored. Provision for this circumstance has been made by IBM in the form of the "data cartridge" (a plastic device with a series of movable bars) enabling the borrower number and loan period to be read by the 1031 badge reader. When the equipment is again operating properly, the data cartridge and book circulation cards (which has been held in a file) are put into the 1031 to create the circulation record. Book circulation cards for manual charges are then transferred to another file, to be put back into the books when they are returned or presented for renewal. The bars on the data cartridge must be reset for each transaction, reflecting changes in borrower number and loan period. This is a livable arrangement as long as the number of manual charges is small, but when a breakdown occurs during a busy period or if it lasts for an hour or so, a backlog of cumbersome data cartridge work accumulates. Other side-effects of breakdowns are lineups at the circulation desk (a manual charge requires about 3.5 times as long as a regular charge), errors in the manual records and mistakes in converting from the manual records to the data cartridge. In the event that down time continues into the next day, renewals and returns are a day late. The computer, not "knowing" this, blindly cranks out invalid overdue notices, creating headaches for those who have to cope with the justifiably unhappy customers.

The most baffling problems are those which are intermittent. They are difficult to diagnose, short of dismantling the equipment and testing each component. Since the system is in operation 18 hours a day during the week, there are no large blocks of time available for extensive testing and servicing. But as the operators have become more familiar with the peculiarities of the machine, (those things which are not covered in the instruction manuals), it has been possible to cope fairly successfully with intermittent malfunctions. A recent example of this kind of problem involved a malfunction which, according to the manual, could not happen under any circumstances. Although the library staff maintained that the problem existed, the repair men were not convinced, until the supposedly impossible malfunction took place while a service man happened to be on the scene. There was considerable satisfaction in telling the industrial giant "I told you so".

In time the major service problems were licked and a preventive maintenance schedule was established. IBM worked closely with the people in circulation, giving them the know-how to handle many routine problems. The library, for its part, restricted operation of the equipment to those people who were familiar with it or who had been briefed, by means of careful instructions, on its idiosyncracies. Communication with the service men was improved in that complaints tended to be more specific, thereby improving the odds that the trouble could be quickly identified and rectified. This was, to put it mildly, a trying period for the circulation staff. It required a great deal of patience, understanding, tact and not a little courage to deal with unhappy students and faculty protesting unjust fines, overdue notices for books they had never checked out, and the like. For the most part, the users were understanding but became somewhat bitter toward "the machine".

## Cataloging

Although cataloging has not been automated, the book circulation cards have enabled us to produce an extremely valuable by-product. Master listings by main entry and call number are regularly produced. These are distributed to public service areas and are used as finding lists. They sometimes save a trip to the catalog but in no sense do they substitute for it since they contain only the call number and a truncated author and title. Special lists are sent to faculty based on broad LC classifications: for example, the Philosophy Department receives B, BC-BD, BH-BX, Psychology receives BF, DF, RC, and so forth. Not all departments make use of the lists, but whenever the delivery schedule is interrupted and a few faculty remind us of the fact, we know they are useful. Recently, one of the lists was sent to England so that a newly-hired department head could get an idea of the library holdings in his subject (he also received a list of books which had been ordered by members of his department but which had not yet been cataloged). Similar lists have been sent to faculty overseas as an aid in selecting books from antiquarian dealers. Reference books and curriculum materials are also printed out, as required.

There is increasing use of the lists by departments to screen new requests before transmittal to the library; I would like to be able to say that it is conscientiously done by all departments. I would also like to be able to say that these lists have substantially increased faculty awareness of the make-up of the collection.

Cataloging statistics are gathered by means of a punched card system. Instead of the traditional method where each

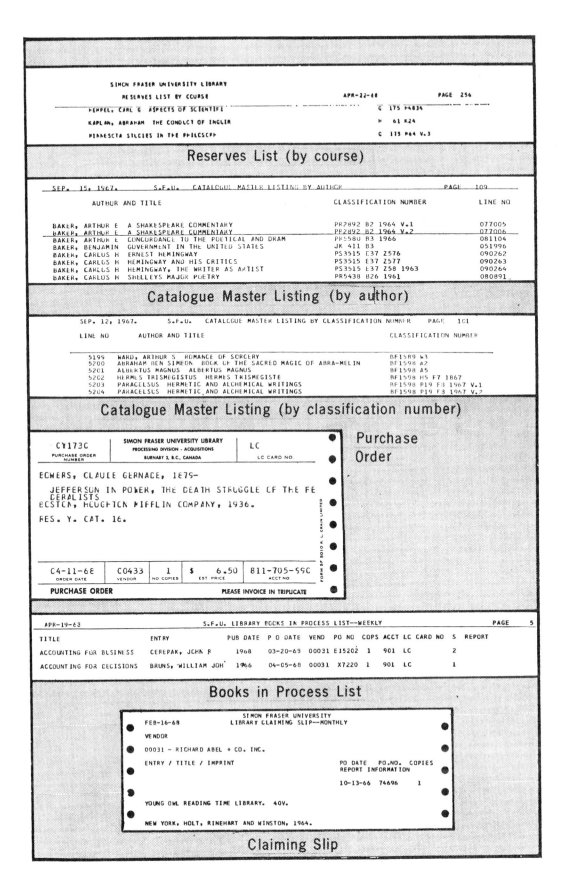

SIMON FRASER UNIVERSITY LIBRARY
RESERVES LIST BY COURSE                    APR-22-68          PAGE 256

PEMPEL, CARL G  ASPECTS OF SCIENTIFI                    G  175 P4834

KAPLAN, ABRAHAM  THE CONDUCT OF INQUIR                  H  61 K24

MINNESOTA STUDIES IN THE PHILOSOPH                     Q  175 M64 V.3

## Reserves List (by course)

SEP. 15, 1967.    S.F.U.  CATALOGUE MASTER LISTING BY AUTHOR                    PAGE  109

        AUTHOR AND TITLE                      CLASSIFICATION NUMBER          LINE NO

BAKER, ARTHUR E  A SHAKESPEARE COMMENTARY     PR2892 B2 1964 V.1          077005
BAKER, ARTHUR E  A SHAKESPEARE COMMENTARY     PR2892 B2 1964 V.2          077006
BAKER, ARTHUR E  CONCORDANCE TO THE POETICAL AND DRAM  PR5580 B3 1966     081104
BAKER, BENJAMIN  GOVERNMENT IN THE UNITED STATES  JK 411 B3              051996
BAKER, CARLOS H  ERNEST HEMINGWAY             PS3515 E37 Z576            090262
BAKER, CARLOS H  HEMINGWAY AND HIS CRITICS    PS3515 E37 Z577            090263
BAKER, CARLOS H  HEMINGWAY, THE WRITER AS ARTIST  PS3515 E37 Z58 1963   090264
BAKER, CARLOS H  SHELLEYS MAJOR POETRY        PR5438 B26 1961            080891

## Catalogue Master Listing (by author)

SEP. 12, 1967.      S.F.U.   CATALOGUE MASTER LISTING BY CLASSIFICATION NUMBER     PAGE   101

    LINE NO      AUTHOR AND TITLE                        CLASSIFICATION NUMBER

    5199    WARD, ARTHUR S  ROMANCE OF SORCERY            BF1589 W3
    5200    ABRAHAM BEN SIMEON  BOOK OF THE SACRED MAGIC OF ABRA-MELIN   BF1598 A2
    5201    ALBERTUS MAGNUS  ALBERTUS MAGNUS              BF1598 A5
    5202    HERMES TRISMEGISTUS  HERMES TRISMEGISTE       BF1598 H5 F7 1867
    5203    PARACELSUS  HERMETIC AND ALCHEMICAL WRITINGS  BF1598 P19 F3 1967 V.1
    5204    PARACELSUS  HERMETIC AND ALCHEMICAL WRITINGS  BF1598 P19 F3 1967 V.2

## Catalogue Master Listing (by classification number)

| CY1730 PURCHASE ORDER NUMBER | SIMON FRASER UNIVERSITY LIBRARY PROCESSING DIVISION - ACQUISITIONS BURNABY 2, B.C., CANADA | LC LC CARD NO. |
|---|---|---|

## Purchase Order

BOWERS, CLAUDE GERNADE, 1879-

JEFFERSON IN POWER, THE DEATH STRUGGLE OF THE FE
DERALISTS
BOSTON, HOUGHTON MIFFLIN COMPANY, 1936.

RES. Y. CAT. 16.

| 04-11-68 ORDER DATE | C0433 VENDOR | 1 NO COPIES | $ 6.50 EST PRICE | 811-705-990 ACCT NO |
|---|---|---|---|---|

**PURCHASE ORDER**          PLEASE INVOICE IN TRIPLICATE

FORM SF 3010 R. L. CRAIN LIMITED

APR-19-68                S.F.U. LIBRARY BOOKS IN PROCESS LIST—WEEKLY                        PAGE    5

| TITLE | ENTRY | PUB DATE | P O DATE | VEND | PO NO | COPS | ACCT | LC | CARD NO | S | REPORT |
|---|---|---|---|---|---|---|---|---|---|---|---|
| ACCOUNTING FOR BUSINESS | CEREPAK, JOHN R | 1968 | 03-20-68 | 00031 | E15202 | 1 | 901 | LC | | 2 | |
| ACCOUNTING FOR DECISIONS | BRUNS, WILLIAM JOH | 1966 | 04-05-68 | 00031 | X7220 | 1 | 901 | LC | | 1 | |

## Books in Process List

                    SIMON FRASER UNIVERSITY
FEB-16-68           LIBRARY CLAIMING SLIP—MONTHLY

VENDOR

00031 - RICHARD ABEL + CO. INC.

ENTRY / TITLE / IMPRINT                  PO DATE   PO.NO.  COPIES
                                         REPORT INFORMATION

                                         10-13-66  74696     1

YOUNG OWL READING TIME LIBRARY.  40V.

NEW YORK, HOLT, RINEHART AND WINSTON, 1964.

## Claiming Slip

cataloger keeps a weekly or monthly tally of his production, he checks the appropriate blocks on a simple form (one form for each item processed), the forms are key punched and then printed out monthly, or as often as needed.

*Acquisitions*

Acquisitions was automated April 1, 1966. All information required to produce purchase orders and accounting information is keypunched to form an order deck. The data in the deck is manipulated by the computer to produce a purchase order for each item, input to the university accounting system, control of orders in process, and certain by-products. Before keypunching, the bibliographic information is edited for purposes of machine filing; this takes the form of eliminating initial articles, honorific titles, etc. Information keypunched consists of the following: purchase order numbers, purchase order dates, account numbers, estimated time of arrival, estimated price, LC card number, number of copies, vendor number, complete main entry in LC style, short title, complete imprint, edited entry, edited title, comments, ("rush airmail", "your catalog 45 item 3"). This data is used to produce the following:

1. 3 x 5 purchase orders containing entry, title, imprint, vendor number, estimated price, date of order, number of copies, comments.

2. Expense distribution cards which, when fed into the university accounting system, commit the amount of the purchase order.

3. Books in process list which contains: short author; short title; account number; vendor number; date of order; number of copies; status of the order (on order, received, cataloged); reports from vendors of delays (NYP, OS, etc.).

Order decks are sent to the university computing center daily. Books in process lists which include all books on order, received, and recently catalogued are produced weekly; a cumulative supplement to the weekly list is printed daily. Purchase orders arranged in vendor sequence are received daily from the computing center. Address labels for each vendor represented in the day's run are also printed. Using input from the acquisitions system, daily reports are made by the computer showing the number of purchase orders printed and a summary of commitments by account number.

The initial phase of the system employed 026 keypunches for input (all in upper case) and disks for storage. Order decks were verified before weekly batches were transmitted to the computing center. A bottle-neck rapidly developed in key punching where two groups were making heavy use of the facilities: cataloging, for production of book circulation cards (without which a book could not go to the stacks) and acquisitions. There was continuous demand for the library's lone verifier, not just from cataloging and acquisitions, but also from the registrar, the bursar, and the computing center. The machine responded to this pressure by breaking down with increasing frequency; it was a rare instance when it performed satisfactorily for two days in succession. Whether it was a lemon or not will probably never be known, although the operators still insist that it was. Whatever the reason, it was quickly identified as a very critical and unreliable link in the system. As a result of its failure to function satisfactorily, a backlog of unverified and uncorrected order decks was created. Acquisitions began receiving letters from antiquarian dealers who had reserved material from catalogs, but for which they had not received purchase orders. The log-jam of orders was finally broken by the establishment of a partial night shift.

This sudden surge of new orders had unforeseen consequences for the computing center. The acquisitions system utilized disk packs for storage, as noted previously. Layout of the disks had been determined by calculating the maximum number of orders which would be in process at any one time with an allowance for sufficient sorting space. After an item had been catalogued, it was then to be dropped from the disk file, thus making room for more new orders. The calculations had been based on the short "turn-around" time possible under ideal circumstances with a regular flow of orders through receiving, payment, and cataloging. The ideal conditions assumed by the system design never existed.

The first hitch was the unanticipated number of new orders which hit the system as soon as the verification problem had been solved. Coupled with this was a book budget much larger than had been anticipated during the design of the system. The increased number of order records on the disk seriously hampered the machine-sorting necessary for production of the books in process list. The computing center adopted stop-gap measures to accommodate the additional orders in process, but the period of time between receipt of a request from faculty and the printing of a purchase order lengthened. The weekly batches became bigger and bigger. As they grew in size, duplication within the batches assumed serious proportions, since there was no provision for machine checking of an incoming order against the other orders in the batch. There was a sudden increase in rush orders, largely for reserve, but many from frustrated faculty who had learned that the best way to get needed materials into the library was to mark them "rush" (it had become necessary to set up a special procedure for rush books which did not depend on the automated system for purchase order printing).

Then, to crown this phase of the system, there were two unfortunate accidents. The first involved a thoughtless smoker and some live cigarette ashes on a disk. A number of records were lost but were recreated from the punched card decks. The second was much more serious. Careless labeling of the acquisitions disk pack (containing the only machine record of orders in process) resulted in its being used for sorting. By the time the error was discovered, several thousand order records had been erased. Re-creation of these records required many man hours to remove rubber bands, paper clips, LC cards, notes, request cards and other extraneous material from the order decks retained in acquisitions, hours of sorting time, machine time, and further hours to match the decks with the notes, request cards, etc. During this period all acquisitions activities were halted.

The second phase of the acquisitions system was characterized by the rewriting of the system and the installation of IBM 1050 units, replacing standard keypunch machines. An IBM 1050 system is essentially a standard key punch machine driven by the equivalent of a Selectric typewriter. It has three major advantages: typewriter copy is produced enabling visual verification; upper and lower case characters are produced just as on a standard typewriter, it has on-line capabilities, (with the proper interface, information keyed on the 1050 can be fed directly into computer storage, eliminating punched cards). Magnetic tape replaced disks as the storage medium, thus giving the system the capacity it required. Further changes were incorporated to solve problems which had turned up during the initial period, such as machine-checking for duplication (as a supplement to manual checking), computer-produced special instructions to vendors (these had previously been typed on the purchase orders by acquisitions staff), and daily instead of weekly batches. With these major changes, the system came of age.

In contrast to the first system which was an inflexible, inefficient and unreliable means of acquiring books, and one which was often more of a nuisance than a help, the new design was flexible, responsive and had sufficient capacity to handle a relatively unlimited volume. The success of the "second generation" system is attributable to the experience gained from the old system by both librarians and programmers, and to the development of a better appreciation of one another's work. Some of the features of the re-written system, such as automatic claiming, lists of orders in process for each teaching department, and machine checking for duplicates, are outlined below.

The first claims lists of overdue orders (by vendor, in title sequence) were produced, but were soon discovered to be of little use. It became obvious that few, if any, dealers keep their files in title sequence. The major suppliers of in-print material usually have their files organized by publisher, so our claims lists were useless to them. Since we had not had the foresight to flag the publisher in the machine record, it was impossible to produce a list in publisher sequence. Suppliers who had small numbers of outstanding orders were able to make good use of the lists, however. The lengthy printouts of claims were abandoned and an individual slip for each title was substituted, containing entry, title, vendor number, date of order, record of vendor reports, and a message to the vendor identifying the slip as a claim. In this form the vendor could arrange them in any order he wished.

Each teaching department receives a monthly list of orders in process which have been charged to its account. The lists are arranged by main entry. Copies are sent to the departmental library representative and to the appropriate reference librarian. It can be argued that such lists do not have as much value to faculty as would an order slip filed in the card catalog, but they seem to provide a satisfactory substitute. Availability of these lists counteracts the tendency of faculty to maintain files of books requested. I wish that I could say that these files do not exist at Simon Fraser. Faculty are able to see the progress of their orders through the technical services and to keep informed of their colleagues' ordering. Inclusion of reports of delays on the lists reduces the number of inquiries directed to acquisitions. Used with the master listings of books catalogued (referred to above), it is hoped that there will be a significant reduction in the number of duplicate requests submitted by faculty. Reference librarians are able to keep informed of faculty ordering patterns through use of the lists.

The reader will note references to manual and machine checks for duplication. It had been expected that all duplication checks could be handled by the computer. IBM initially suggested that the punched order decks could be passed against the Catalog Master Listing (in entry sequence) in order to eliminate the printing of purchase orders for duplicate items. This proved impractical because of the briefness of the author/title information contained in the book circulation card, of which the Catalog Master Listing is a by-product. Even if the information on the book circulation card had been of sufficient length to enable this approach to be used, added copies and new editions would have presented some knotty problems. Another approach was tried. It was decided to build up a tape file of the acquisitions record, which contains more elaborate bibliographical information than does the book cirulation card. As a book was cataloged, it was transferred from the tape record of orders in process to a new tape file. New orders were then passed against this tape and duplicates eliminated. Initially this proved to be of some use in reducing duplicates but it was found that much of the entry information on this file was inconsistent and unreliable due to some slip-shod bibliographical searching and keypunching. Since titles were more likely to be correct than entries, it was decided to put the files in title sequence and to check new orders against this, first by title, then by entry within title. This variation was more successful initially, but the tape file soon became so long that an excessive amount of machine time was required to screen new orders against it. The order decks rejected as duplicates by the computer represented wasted keypunching effort. We realized that it would cost more money to use the computer to eliminate duplicates than to eliminate the bulk of them manually before they were put into the system. Accordingly, the main books in process list was put into title sequence. This enabled student assistants to identify a request for a book already on order without having to be knowledgeable about corporate entries or library filing rules.

To eliminate duplication within the daily batch of orders itself, a program was written by which the computer checks each item against all others in the batch. This is done first by title. If duplicate titles are found, the entries are compared. Where both are equal, one order is rejected. A similar program is used to screen all items in the batch against the books in process tape file (from which the books in process list is prepared). Provision has been made to accommodate added copies. However, a year after these programs were written, the percentage of duplicate orders which got through the system and into cataloging was still high. Investigation showed that there was a "hole" in both programs, the result of a minor error in logic, which allowed many duplicates to slip through. This hole has now been plugged.

Reliance on machine-produced lists has made people more conscious of the need for good input. The old saw, "Garbage in, garbage out" is nowhere more clearly demonstrated than in the lists; they have been a factor in improving the legibility and biblographic quality of requests submitted. Although an on-order list arranged by title was viewed with considerable skepticism by both faculty and librarians, they have adjusted to it.

IMPLICATIONS OF INTEGRATION

*Accounting*

Integration of library ordering with the university accounting system has produced joys and sorrows. On the plus side, direct input to the financial records is produced as a by-product of book ordering. In other words, there is no need for the library to maintain its own set of accounting records since the information required for committing (account number, amount, type of currency, vendor, purchase order number, purchase order date) is contained in the order deck; the computer extracts what it needs and feeds it into the accounting system. The semi-monthly commitment reports list all outstanding orders by department, all current expenditures and cancellations by department, and provides a summary of all expenditures for the fiscal year.

On the minus side, the library has no financial back-up records in case the university accounting system breaks down. As might be expected in a new university, the automated accounting system did run into some minor difficulties which, although they were of little consequence in the total accounting picture, had a considerable impact on library operations. To put it briefly, the library operated for several months without knowing whether its book funds were heavily over-committed or dangerously under-committed. The uncertainty dictated that ordering be curtailed until our commitments were definitely known. When the matter was straightened out, it was found that book ordering would have to be considerably accelerated if library funds were to be expended before the close of the fiscal year. In order to produce the required number of orders with the staff available, it was necessary to "streamline" many ordering procedures. The results of these short-cuts were slipshod bibliographical searching and sloppy keypunching, mentioned above. As might be expected, the chickens all came home to roost, mainly in the form of duplicates. Garbage in, garbage out.

Problems were created for the library when the purchase order numbering system was established. The Purchasing Agent did not realize that a library spending close to $500,000 a year for printed matter needs a lot more than 30,000 order numbers (since unique purchase order numbers are a necessary part of the accounting system, it was not possible to re-use numbers). A series of make-shift numbers was set up using letter prefixes. Programs had to be patched to enable the accounting system to accept and manipulate alphabetic characters in numerical fields. Then it was discovered that a peculiarity of the coding structure of the 1050 system ruled out the use of certain of the prefixes which had been established; this required further patching.

A machine-readable vendor file was created for use in all university ordering. Blocks of numbers were assigned for various parts of the alphabet so that the file would be in both numerical and alphabetical order. An alphabetic distribution table based on student names was used to organize the file. The shortcomings of the distribution pattern were soon apparent. For example, the block of numbers assigned for CAN proved to be much too small for the vendors whose names begin CANADIAN or CANADA. In addition, no one outside the library realized how many vendors are used by a university library. In a short time the alphabetical order of the file

deteriorated, making look-ups difficult. The order of the file was the responsibility of another department and the person who made the decisions was obviously unaware of the niceties of library filing rules. Maybe this was a good thing, but B. H. Blackwell wound up at the beginning of the B's, Marcel Blancheteau (whose invoices carry the words "Aux Amateurs des Livres" in a prominent location) was put at the end of the A's, the German dealer Otto Roppel was filed under O, and Les Livres Etrangers was filed under LES. Acquisitions people soon learned to live with this as they became familiar with the numbers assigned to frequently used vendors. Inevitably a complete revision of the file became necessary. A more suitable alphabetical distribution table was used and new numbers were assigned to the majority of vendors. The computing center had to set up a cross-reference tape so that commitments made under the old numbering system could be properly processed through the accounting system with the new, valid numbers. Recently a second revision of the entire file was made.

*Registration*

The decision to have a combined university identification and library card had unforeseen implications for the library. The cards are prepared and issued as part of the registration process. Only registered students (and permanent university staff) are eligible to receive the cards. Only people with these cards can check out books. The library's desire to provide check-out privileges for visiting scholars and other persons not officially connected with the university was thwarted but an arrangement was later worked out to accommodate this type of loan. Another problem has been encountered in the loan privileges of teaching assistants. Under university regulations, TA's fall into two categories:

those who are also graduate students; those who are not graduate students. If a TA is taking graduate courses, he is considered as a graduate student. If he is not taking graduate courses, he is classified as a member of faculty.

Another negative feature of the combined card is the infrequency of issue. Cards are issued annually, expiring August 31. If, for example, a student enrolls for the Fall Semester, and does not attend the Spring and Summer Semesters, he can merrily check out books for eight months without having to worry about paying fines since he is no longer a part of the university community. If new cards were issued or validated each semester there would be more control.

Substantial reliance on the university computer makes the library subject to the side-effects of hardware and software changes. In a little over two years there have been major alterations in computer configuration (from 1440 to 1401 to 360/40), one major change in storage devices (disk to tape) and one change in programming language (Autocoder to PL 1). Concurrently acquisitions input shifted from standard keypunches to the 1050 system. Improvements in speed, flexibility, and capacity have off-set the initial delays and snags encountered with each change.

## CONCLUSIONS

Use of the computer at Simon Fraser has led to some general conclusions as to its effect on the operation of an academic library. However, a *caveat* is necessary: Simon Fraser is atypical in that it is new and therefore has no "manual past" with which its "automated present" can be compared. The following comments on space, costs, capacity, and effect on librarians will be meaningful only if it is remembered that they are based on a special situation. The computer's impact on library service is implicit in the remarks which follow.

## Space

Inadequate space is a fact of life in many libraries, even in a new institution such as ours. Although maximum utilization of floor space was not a major factor in deciding to automate, if indeed it was a factor at all, it is now recognized as a most welcome by-product. That the computer was making significant savings in space became apparent when a study was made to determine the effect of converting the acquisitions system to manual operation. It was a foregone conclusion that much more floor space would be required for a conventional order file since many of the operations performed by the computer on the machine record (automatic claiming, purging, filing, statistics) would have to be performed by people working directly at the file. The study showed, among other things, that the man hours required to keep such a file up-to-date could not be satisfactorily accommodated in a regular working day.[2] Although it might have been theoretically possible to do this by assigning certain sections of the alphabet in which orders could be filed, other sections in which claims could be made, further sections in which duplicate checking could be done, etc., it was realized that such restrictions would add considerably to the time required to process orders. It was also judged that the quality of the work performed under these conditions would not be adequate to maintain a satisfactory operation. The only way in which reasonable standards of quality and quantity could be maintained would have been to establish a second shift of two or three people, thereby creating problems of supervision.

It is in circulation, however, where the saving in floor space is most apparent since the loan activity is located in an area where space is at a high premium. Literally, there is no room either for the files required by a conventional system, or for the people to tend them. The daily circulation list is located on a counter and thus occupies no floor space at all (the counter would be required no matter what system was used to circulate books).

## Costs

Although the exact costs of the automated systems in use at Simon Fraser have not been calculated, it can be assumed that manual systems of acquisition and circulation would be more economical, on a direct cost basis. This assumption would be valid even if the manual systems were to be as sophisticated as those which are computer-assisted. It doesn't take any special insight to conjecture that a comparison of automated and manual acquisitions systems producing, let us say, 20,000 orders per year, would show the manual system operating with the lowest unit costs. If, however, the number of orders was significantly increased, the cost per unit of the manual system would probably increase geometrically while the unit cost of the automated system would actually decrease until full capacity was reached. For example, it is estimated that there will be no appreciable increase in machine time requirements for acquisitions or circulation until present volume has grown by approximately 100%.

An area which might show the greatest contrasts in costs is the maintenance of the orders-in-process file (filing new orders, recording dealer reports, claiming, purging). Increased acquisitions result in larger files and more people requiring access to them. It is equally true that the unit filing time increases as the file expands; the increased cost of keeping a growing file up-dated assumes a geometric character. The accuracy of a file, particularly an order file, tends to decrease as the file expands, excluding the sequence errors caused by typing mistakes. While the quality of a manual or a machine file is governed by the accuracy of the input, the chances of misfiling in a computer-maintained file are almost zero, no matter how large the file is, or how fast it grows. In other words, the accuracy of a machine file is directly proportional to the accuracy of the input; this is not true of the manual file with its smudged carbon copies, staples, paper clips, pencilled notes and "pockets" of misfiled cards. Another advantage of the machine file is that records do not have to be removed for purposes of adding dealer report information, typing of claiming lists or for any other reason. In its printed form, the machine file is portable, easy to scan, and occupies little space.

Anyone who has tried his hand at library cost analysis knows how difficult it is to pinpoint the cost of a specific task, but it is far more difficult to put a price tag on the speed and accuracy with which a librarian can supply information to the user. For example how much is it worth for the circulation staff to be able to answer all questions relating to the location of material, i.e., what is on loan, what is at the bindery, what is on reserve, which books have been reported missing, which books have been lost and are being replaced, etc.? Further, how much is it worth to the library to enable the user to provide his own answers to such questions by consulting a print-out? What is the value of automatic claiming of overdue orders, of an accurate orders-in-process list, of orders-in-process lists available to faculty? No one would argue that these services are not desirable or necessary, but until library cost/benefit analysis has advanced further, the jurisdiction will have to be that they are worth whatever they cost.

## Management Information

Although the computer's effect on costs, space utilization and system capacity are important, there is another area (aside from retrieval which is outside the scope of this paper) which is of great interest to the library administrator. This is management information, i.e., that data which shows what is going on in the library. There is increasing comment, from both librarians and educators, that academic libraries are not keeping pace with the demands of the academic community. Few librarians would disagree with the premise on which the Monteith experiment was based: "Traditional college instruction fails to exploit fully the library resources available for it and that the average college student's experiences with the library constitute a limited and fairly insignificant part of his education".[3] This premise is echoed in the librarian's criticism of faculty ("Our faculty are just not library oriented") or students ("Our students don't know how to use a library"). To blame either faculty or students is unjustified. The mere fact that the Monteith Experiment was made should make us feel uneasy and suggest that we need to examine closely the fundamental concepts on which library service is based.

For example, the management information that can be obtained from an automated circulation system can provide some of the facts needed to identify patterns of book use, i.e., who is using what, how long are books in various categories being used, when are they being used, and what is not being used. Data processing equipment enables the librarian to collect and analyze massive amounts of such data at levels of speed, accuracy, and economy not otherwise possible. It is possible to correlate student records with book usage and thus provide information on the relationships, if any, between library use and such factors as the student's grades, his instructors, his high school record, etc.[4] Looking into the future, the librarian may be able to find answers to such questions as "What did you, the library patron, look for that you could not find?" and there

is the further possibility that a comprehensive on-line system could assist in finding out *how* people look for information.

The academic library has recently been referred to as "The Bottomless Pit."[5] There is little doubt that the university administration, hard-pressed for funds from all sides, has shown little inclination to make a basic change in the percentage of the university's resources customarily allocated to the library.[6] It is naive to assume that the librarian can improve his position by continued reliance on such pious generalities as "The library is the heart of the university." In my opinion, the librarian who is well-armed with management information, preferably in the form of print-outs since these are more likely to impress the administration, will have a better chance of getting his budget approved than will the librarian whose request is supported by truisms and lip service from faculty. The potential of the computer as a management aid is largely unexplored by librarians.

*Effect on Librarians*

Although it must be remembered that the library at Simon Fraser is not typical, our experience indicates that the computer's impact on people is following a pattern similar to that found in other areas, i.e., business and government, where the computer is used for handling large quantities of information. The ultimate effect of the computer on employment is subject to controversy. It is clear, however, that jobs have been both eliminated and created within organizations using data processing. In our case, several typist positions were eliminated when acquisitions went on the computer but the typists themselves, were given key punching training. They now find their work more satisfying because it offers greater challenge as well as higher pay. The position of library bookkeeper, which would have been required under ordinary circumstances, was never established because all accounting is done by the computer.

There is no evidence to suggest that the computer will replace any librarians at Simon Fraser since the work it does is not "professional" in any sense of the word. For example, it doesn't take a librarian to perform such routine tasks as determining when an order is overdue from a vendor, claiming an overdue order, converting U.S. dollars to Canadian dollars, keeping up-to-date records of commitments and expenditures, maintaining an orders-in-process file, determining when the loan period of a book has expired, sending out overdue notices or calculating fines. Our experience has shown that the computer does not take any meaningful work away from the librarian but rather that it makes his job more interesting and challenging. This happens in two ways. First, it relieves him of some of his routine responsibilities and thereby gives him more time to devote to work of real consequence. Second, it demands of him that he think clearly and logically about each of the steps required to produce a given end-product:

> A man cannot instruct the computer to perform usefully until he has arduously thought through what he is up to in the first place, and where he wants to go from there . . . The re-thinking process gets more difficult as the computer gets better.[7]

An example of this effect on the librarian's thinking took place in the early stages of planning the acquisitions system. The systems analysts from IBM asked embarassing questions about the form and content of some of the outputs of the system, as the librarians had visualized it. The questions were hard to answer because the person who undertook the library's part of the design of the system (the writer) was thinking in terms of a traditional acquisitions system with the usual files, arranged according to standard library filing rules. He found himself advocating and defending certain procedures simply because he was used to doing them this way. He was not taking a fresh look at the whole process but was, instead, clinging to the traditional *means* of running an acquisitions system. A good system can result only if atention is focused first on the *ends*, then on the means. The computer demands of the librarian that he clear his mind of some of the things he has been taught and of many of the procedures to which he is accustomed. He must carefully think through what he is doing and why he is doing it before he can tell the computer what to do. If the librarian does not do the re-thinking, someone else will. After all, it is the librarian who best knows what the library and its users need. There is a real danger, however, that by accepting things as they are, he will lose out to someone from outside the present library "establishment," such as the documentalist or the systems analyst.

To the librarian, the most interesting example of the kind of job the computer can create is that of "systems librarian" which, to the writer's mind, presents opportunities, challenges and satisfactions unmatched in any other library work. Except perhaps for administration, there are no positions more demanding. Not only must the systems librarian be well-grounded in data processing, he must also have an understanding of all phases of library operations—from acquisitions through reference, from pamphlets through microforms, from serials through monographs. The degree to which he is effective is directly proportional to the depth of his knowledge of computers and libraries. Library systems work has a fascination of its own sufficient to satisfy many people but, more importantly, it is work of real consequence: the results of good systems work can be seen in better service to the library patron, our ultimate aim.

The librarian who has gotten his feet wet in data processing is understandably enthusiastic because he grasps the potential of the computer as a means of handling the vast amounts of data required in the operation of a library. Once he has worked closely with it, he will not be easily satisfied with anything less (at least two of our staff have declined higher-paying jobs in libraries that did not have access to a computer). However, if the enthusiasm is not tempered by good judgment and a sense of proportion, data processing will be seen as the solution to all library problems and further, as an end in itself. The danger then arises that a disproportionate share of the library's resources will be used in automating for the sake of automating. Consequently the overall service the library can offer its patrons may suffer. After all, libraries and librarians will be judged not on the amount of core storage of their computers, the number of systems in operation or on the writeups their systems receive in journals, but rather on the service they give. There is no doubt that the computer will, for the best reasons, assume an increasingly important role as a means of assisting librarians in meeting demands for service, but a certain amount of caution is necessary to prevent us from losing sight of our objectives, and the cost of achieving them through data processing. Computers are not cheap.

To summarize, the use of the computer in circulation, acquisitions, and cataloging has been a unique experience, producing some mildly surprising effects on librarians and library operations. "Experience" is the key word. There now exists at Simon Fraser a considerable body of experience in library applications of data processing and we are thus better prepared to improve existing systems and to move into new areas.

Footnotes:

[1] John B. Macdonald, *Higher Education in British Columbia and a Plan for the Future* (Vancouver: The University of British Columbia, 1962), p. 64-65.

[2] T. C. Dobb, "General Statements re Operation of Manual System" (unpublished report, Simon Fraser University Library, Dec. 27, 1967).

[3] Patricia B. Knapp, *The Monteith College Library Experiment*. N.Y., Scarecrow, 1967. p. 11.

[4] *E.g.*, Floyd Cammack and Donald Mann, "Institutional Implications of an Automated Circulation Study," *College and Research Libraries*, 28 (March 1967), 129-132.

[5] Robert F. Munn, "The Bottomless Pit, or the Academic Library as viewed from the Administration Building," *College and Research Libraries*, 29 (January, 1968), 51-54.

[6] American Library Association. Library Administration Division. *Library Statistics of Colleges and Universities, 1965-66 Institutional Data* (Chicago, 1967), p. 8-9.

[7] Gilbert Burck, *The Computer Age* (N.Y.: Harper & Row, 1965), p. 3.

# VI

# SELECTED BIBLIOGRAPHY OF REFERENCES IN LIBRARY SYSTEMS ANALYSIS

## Introduction

This bibliography is a selective listing by author of representative articles for most aspects of systems analysis. These additional references have been provided for readers wishing to expand their interest in and knowledge of systems analysis. At the same time the reader is referred to the additional readings suggested at the end of many of the selections in the text.

This listing gives full bibliographic citations for the papers in the *Reader* and triples their number with selected additions. A wide variety of sources since 1954 have been used in the preparation of this bibliography; among which, are recently published bibliographies on library systems analysis, course-required reading lists in systems analysis for schools of library science, and the standard indexes: *Library Literature* and *Library and Information Science Abstracts*.

Adelson, Marvin, The System Approach: A Perspec ive, *Wilson Lib. Bul.*, 42:711-715 (1968).

Aslib Research Department, The Analysis of Lik ary Processes, *J. of Doc.*, 26:30-45 (1970).

Ayres, F. H., Some Basic Laws of Library Automation, *Program*, 4:68-69 (1970).

Bare, C. E., Conducting User Requirement Studies in Special Libraries, *Sp. Lib.*, 57:103-106 (1966).

Becker, J., System Analysis, Prelude to Library Data Processing, *ALA Bul.*, 59:293-296 (1965).

Becker, Joseph and R. M. Hayes, *Handbook of Data Processing for Libraries*. New York: Wiley, 1970.

Bellomy, F. L., The Systems Approach Solves Library Problems, *ALA Bul.*, 62:1121-1125 (1968).

Bernstein, H. H., Some Organizational Pre-requisites for the Introduction of Electronic Data Processing in Libraries, *Libri*, 21:15-25 (1971).

Bittel, Lester R., *Management by Exception: Systematizing and Simplifying the Managerial Job*. New York: McGraw Hill, 1964.

Bolles, Shirley W., The Use of Flow Charts in the Analysis of Library Operations, *Sp. Lib.*, 58:95-98 (1967).

Brownlow, Jane L., Cost Analysis for Libraries, *DC Lib.*, 31:54-60 (1960).

Burkhalter, Barton R., ed., *Case Studies in Systems Analysis in a University Library*. Metuchen, N.J.: Scarecrow Press, 1968.

Burns, Robert W., A Generalized Methodology for Library Systems Analysis, *Coll. & Res. Lib.*, 32:295-303 (1971).

Cammack, Floyd M. and Donald Mann, Institutional Implications of an Automated Circulation Study, *Coll. & Res. Lib.*, 28:129-132 (1967).

Carroll, P., We Need Work Measures, *Sp. Lib.*, 50:384-387 (1959).

Chapman, Edward A., Paul St. Pierre and John Lubans, Jr., *Library Systems Analysis Guidelines*. New York: Wiley, 1970.

Churchman, C. West, Operations Research Prospects for Libraries: The Realities and Ideals, *Lib. Q.*, 42:6-14 (1972).

Churchman, C. West, *The Systems Approach*. New York: Delacorte Press, 1968.

Covill, George W., Librarian + Systems Analyst = Teamwork?, *Sp. Lib.*, 58:99-101 (1967).

DeGennaro, Richard, The Development and Administration of Automated Systems in Academic Libraries, *J. Lib. Automation*, 1:75-91 (1968).

Dougherty, Richard M. and Lawrence E. Leonard, comp., *Management and Costs of Technical Processes: A Bibliographical Review, 1876-1969*. Metuchen, N.J.: Scarecrow Press, Inc., 1970.

Dougherty, Richard M. and Fred J. Heinritz, *Scientific Management of Library Operations*. Metuchen, N.J.: Scarecrow Press, 1966.

Drott, M. Carl, Random Sampling: A Tool for Library Research, *Coll. & Res. Lib.*, 30:119-125 (1969).

Erisman, Robert E., *Returning the Circulated Book to the Shelf; A Study of the Shelving Operation in a University Library*, Denver, University of Denver, Graduate School of Librarianship, Research paper. May, 1971.

Fasana, Paul J., Determining the Cost of Library Automation, *ALA Bul.*, 61:656-661 (1967).

Fasana, Paul and James E. Fall, Processing Costs for Science Monographs in the Columbia University Libraries, *Lib. Res. & Tech. Serv.*, 12:97-114 (1967).

Flood, Merrill M., Systems Approach to Library Planning, *Lib. Q.*, 34:326-338 (1964).

Ford, Geoffrey, Data Collection and Feedback *in Planning Library Services; Proceedings of a Research Seminar . . .* 9-11, July, 1969. Edited by A. Graham Mackenzie and Ian M. Stuart (University of Lancaster Library Occasional Papers No. 3) University of Lancaster Library, Lancaster, Great Britain, 1969.

Grad, Burton, Structure and Concept of Decision Tables *in Proceedings of the Decision Tables Symposium* Sponsored by CODASYL Systems Group Joint Users Group of ACM, Sept. 20-21, 1962. N.Y. Association for Computing Machinery, 1962? pp. 19-28.

Greenwood, James W., Jr., *EDP: The Feasibility Study—Analysis and Improvement of Data Processing* (Systems Education Monograph No. 4), Systems and Procedures Association, Washington, D.C., 1962.

Gregory, Robert H. and Richard L. Van Horn, *Automatic Data Processing Systems: Principles and Procedures,* 2nd ed. Belmont, Cal.: Wadsworth, 1963.

Gribbin, J. H., Computerphobia and Other Problems: User Reactions to an Automated Circulation Operation in the Library of Tulane University, *Southeastern Libn.,* 20:78-82 (1970).

Gull, C. D., Logical Flow Charts and Other New Techniques for the Administration of Libraries and Information Centers, *Lib. Resources & Tech. Serv.,* 12:47-66 (1968).

Hayes, Robert M., Library Systems Analysis *in Data Processing in Public and University Libraries,* ed. John Harvey. Washington, D.C.: Spartan Books, 1966, pp. 5-20.

Heinritz, Fred, Quantitative Management in Libraries, *Coll. & Res. Lib.,* 31:232-238 (July 1970).

Herner, Saul, System Design, Evaluation, and Costing, *Sp. Lib.,* 58:576-581 (1967).

Jensen, Ford, *A Cost Analysis and Usage Study of the Reserved Materials Collection of the University of Arizona Main Library* (ERIC Doc. ED 054 822) Arizona, The Library, 1971.

Jestes, Edward C., An Example of Systems Analysis: Locating a Book in a Reference Room, *Sp. Lib.,* 59:722-728 (1968).

Kahn, Robert L. and Charles F. Cannell, *The Dynamics of Interviewing, Theory, Techniques and Cases.* New York: Wiley, 1957.

Kee, Walter A., Must Library Surveys be Classics in Statistics? *Sp. Lib.,* 51:433-436 (1960).

Keller, PB and Cost Benefit Analysis in Libraries, *Coll. & Res. Lib.,* 30:156-160 (1969).

Kilgour, Frederick G., Systems Concepts and Libraries, *Coll. & Res. Lib.,* 28:167-170 (1967).

Klintøe, Kjeld, Cost Analysis of a Technical Information Unit, *Aslib Proceedings,* 23:362-371 (1971).

Kountz, John, Library Cost Analysis: A Recipe, *Lib. J.,* 97:459-464 (1972).

Kozumplik, William A., Time and Motion Study of Library Operations, *Sp. Lib.,* 58:585-588 (1967).

Lamkin, Burton E., Decision-making Tools for Improved Library Operations, *Sp. Lib.,* 56:642-646 (1965).

Lamkin, Burton E., Systems Analysis in Top Management Communication, *Sp. Lib.,* 58:90-94 (1967).

LARC Association, *Are Computer-Oriented Librarian Really Incompetent?* Arizona: LARC (ERIC Doc. ED 056 701), 1971.

Lazorick, G. J. and T. L. Minder, Least Cost Searching Sequence, *Coll. & Res. Lib.,* 25:126-128 (1964).

Leimkuhler, Ferdinand F., Library Operations Research: A Process of Discovery and Justification, *Lib. Q.,* 42:84-96 (1972).

Leimkuhler, Ferdinand F., Systems Analysis in University Libraries, *Coll. & Res. Lib.*, 27:13-18 (1966).

Line, Maurice B., *Library Surveys*, rev. ed. Hamden, Conn.: Archon Books, 1969.

Lubans, John, Jr., William A. Harper and Robert E. Erisman, *A Study With Computer-Based Circulation Data of the Non-Use and Use of a Large Academic Library* done under USOE Grant No. OEG 8 72 0005 (590). Boulder, Colo.: University of Colorado Libraries, 1973. ERIC ED 082 756.

Lynch, M. F., The Library and the Computer *in The Library and the Machine*, ed. C. D. Batty, North Midland Branch of the Library Association, obtainable from the Public Library, Scunthorpe (Lincs.), 1966, pp. 21-36.

McDiarmid, Errett W., Scientific Method and Library Administration, *Lib. Trends*, 2:361-367 (1954).

Mackenzie, A. Graham, Systems Analysis of a University Library, *Program*, 2:7-14 (April 1968).

Markuson, Barbara Evans *et al, Guidelines for Library Automation: A Handbook for Federal and Other Libraries*. Santa Monica, Cal.: System Development Corporation, 1972.

Mason, Ellsworth, The Great Gas Bubble Prick't: Or Computers Revealed—By a Gentleman of Quality, *Coll. & Res. Lib.*, 32:183-196 (1971).

Mather, Dan, Data Processing in an Academic Library. Some Conclusions and Observations, *PNLA Quart.*, 32:4-21 (1968).

Minder, Thomas L., Library Systems Analyst: a Job Description, *Coll. & Res. Lib.*, 27:271-276 (1966).

Moore, Edythe, Systems Analysis; An Overview, *Sp. Lib.*, 58:87-90 (1967).

Morris, Leslie R., Reclassification: Flow Charting Succeeds, *Catholic Lib. World*, 43:337-342 (1972).

Morse, Philip M., *Library Effectiveness: A Systems Approach*. Cambridge, Mass.: M.I.T. Press, 1968.

Naramore, Frederick, Application of Decision Tables to Management Information Systems *in Proceedings of the Decision Tables Symposium* Sponsored by CODASYL Systems Group Joint Users Group of ACM Sept. 20-21, 1962, N.Y. Association for Computing Machinery, 1962? pp. 63-74.

Nelson Associates, Incorporated, *Methods and Procedures for Measuring Patron Use and Cost of Patron Services for the Detroit Metropolitan Library Project*. New York, Nelson Associates, Inc., 1967 (ERIC ED 032 084).

Neuschel, Richard F., *Management by System*. New York: McGraw-Hill, 1960.

O'Neill, Edward T., Sampling University Library Collections, *Coll. & Res. Lib.*, 27:450-454 (1966).

*Operations Research; Challenge to Modern Management*. Cambridge, Mass.: Harvard University, Graduate School of Business Administration, 1954.

Oppenheim, A. N., *Questionnaire Design and Attitude Measurement*. New York: Basic Books, 1966.

Orne, J., We Have Cut our Cataloging Costs, *Lib. J.*, 73:1475-78, 87 (1948).

Pings, Vern M., Development of Quantitative Assessment of Medical Libraries, *Coll. & Res. Lib.*, 29:373-380 (1968).

Poage, S. T., Work Sampling in Library Administration, *Lib. Q.*, 30:213-218 (1960).

Raffel, Jeffrey A. and Robert Shishko, *Systematic Analysis of University Libraries: An Application of Cost-Benefit Analysis to the MIT Libraries*. Cambridge, Mass.: MIT Press, 1969.

Reynolds, Rose, comp., *A Selective Bibliography on Measurement in Library and Information Services*. London: Aslib, 1970.

Rider, Fremont, Library Cost Accounting, *Lib. Q.*, 6:331-381 (1936).

Robinson, F. *et al, Systems Analysis in Libraries*. Newcastle upon Tyne, Gt. Brit.: Oriel Press, 1969.

Robinson, F., Systems Analysis in Libraries: The Role of Management *in Interface: Library Automation with Special Reference to Computing Activity*, ed. C. K. Balmforth and N. S. M. Cox. Cambridge, Mass.: MIT Press. 101-111 (1971).

Schultheiss, Louis A., Systems Analysis and Planning *in Data Processing in Public and University Libraries*, ed. John Harvey. Washington, D.C.: Spartan Books, 1966, pp. 95-102.

Shaw, Ralph R., ed., Scientific Management in Libraries (A collection of ten articles on the topic), *Lib. Trends*, 2:No. 3 (1954).

Shaw, Ralph R., Scientific Management in the Library, *Wilson Library Bull.*, 21:349-352, 357 (1947).

Simms, Daniel M., What Is a Systems Analyst?, *Sp. Lib.*, 59:718-721 (1968).

Systems and Procedures Association, *Business Systems*. Cleveland, Ohio: The Association, 1966.

Taylor, Frederick Winslow, *The Principles of Scientific Management*. New York: Harper, 1919.

Taylor, Robert A. and Caroline E. Hieber, *Library Systems Analysis* (Library Systems Analysis Report No. 3). Bethlehem, Pa.: Lehigh University, Center for the Information Sciences, 1965.

Thomas, P. A., *Task Analysis of Library Operations* (Aslib Occasional Publication No. 8). London: Aslib, 1971.

Trueswell, Richard W., A Quantitative Measure of User Circulation Requirements and Its Possible Effect of Stack Thinning and Multiple Copy Determination, *American Documentation*, 16:20-25 (1965).

U.S. Department of the Army, *Techniques of Work Simplification* (U.S. Dept. of the Army Pamphlet 1-52). Washington, D.C.: USGPO, 1967.

Urquhart, J. A. and J. L. Schofield, Measuring Readers' Failure at the Shelf, *J. of Doc.*, 27:273-286 (1971).

Veaner, Allen B., Approaches to Library Automation, *Law Library Journal*, 64:146-153 (1971).

Voos, Henry, *Standard Times for Certain Clerical Activities in Technical Processing* (Doctoral Thesis). New Brunswick, N.J.: Rutgers University, 1965. Also in *Lib. Res. & Tech. Serv.*, 10:223-227 (1966).

Wallace, W. Lyle, ed., *Work Simplification* (Systems Education Monograph No. 1). Detroit, Mich.: Systems and Procedures Association, 1962.

Warheit, I. A., When Some Library Systems Fail—Is It the System or the Librarian?, *Wilson Lib. Bull.*, 46:52-58 (1971).

Wasserman, Paul, *The Librarian and the Machine*. Detroit, Mich.: Gale, 1965.

Whitehead, Clay Thomas, *Uses and Limitations of Systems Analysis*. Santa Monica, Calif.: Rand Corporation (Rand Corp. Paper No. 3683; DDC Report No. P-3683), 1967.

Woodruff, E. L., Work Measurement Applied to Libraries, *Sp. Lib.*, 48:139-144 (1957).

# VII

# LIST OF CONTRIBUTORS

## Contributors*

Carole E. Bare . . . . . . . . formerly Assistant Professor, Department of Education, University of California, Los Angeles, Los Angeles, California

H. H. Bernstein . . . . . . . Head Librarian, Joint Nuclear Research Centre, EURATOM, Ispra (VA) Italy

Lester R. Bittel . . . . . . . Editor-in-Chief, *Factory* Magazine

Barton R. Burkhalter . . . . . President, Community Systems Foundation, Ann Arbor, Michigan

Robert W. Burns, Jr. . . . . . Librarian for Research and Development, Colorado State University, Fort Collins, Colorado

Floyd Cammack . . . . . . . formerly Associate Director of Libraries, Georgia Southern College, Statesboro, Georgia

Edward A. Chapman . . . . . Director of Libraries Emeritus, Rensselaer Polytechnic Institute, Troy, New York

_____

*most recent position available

C. R. Clough . . . . . . . . Agricultural Division, Imperial Chemical Industries Limited (ICI), London, Great Britain

Richard M. Dougherty . . . . University Librarian, University of California, Berkley, Berkley, California

Robert E. Erisman . . . . . . Circulation Librarian, University of Colorado Libraries, Boulder, Colorado

James W. Greenwood, Jr. . . Professor, Department of Data Processing, St. Peter's College, Jersey City, New Jersey

C. D. Gull . . . . . . . . . President, Cloyd Dake Gull & Associates, Inc., Kensington, Maryland

D. S. Hamilton . . . . . . . Agricultural Division, Imperial Chemical Industries Limited (ICI), London, Great Britain

Fred J. Heinritz . . . . . . Professor of Library Science, Southern Connecticut State College, New Haven, Connecticut

Frederick G. Kilgour . . . . . Director, Ohio College Library Center, Columbus, Ohio

Kjeld Klintøe . . . . . . . . Dansk Teknisk Oplysningstjeneste (Danish Technical Information Service), Copenhagen, Denmark

John Kountz . . . . . . . . . Coordinator, Library Automation, The California State University and Colleges, Los Angeles, California

Maurice B. Line . . . . . . Director General, British Library Lending Division, Boston Spa, Great Britain

John Lubans, Jr. . . . . . . Assistant Director for Public Services, University of Colorado Libraries, Boulder, Colorado

Errett W. McDiarmid . . . Director, Graduate School Fellowship Office, University of Minnesota, Minneapolis, Minnesota

A. Graham Mackenzie . . . . Librarian and Director of the Library Research Unit, University of Lancaster, Lancaster, Great Britain

Donald Mann . . . . . . . . Assistant Director, Computing and Data Processing Center, Oakland University, Rochester, Michigan

Ellsworth Mason . . . . . . Director of Libraries, University of Colorado Libraries, Boulder, Colorado

Dan Mather . . . . . . . . . Assistant Director for Technical Services & Library Systems, Western Washington State College Library, Bellingham, Washington

Thomas Minder . . . . . . . Associate Professor, Hacettere University, Ankara, Turkey and the University of Pittsburgh, Graduate School of Library and Information Sciences, Pittsburgh, Pennsylvania

Leslie R. Morris . . . . . . Associate Librarian, Technical Services, State University College at Fredonia, Fredonia, New York

Richard F. Neuschel . . . . McKinsey and Co., 245 Park Avenue, New York, New York

F. Robinson . . . . . . . . Corporate Information Unit-Development, Imperial Chemical Industries Limited (ICI), London, Great Britain

Fremont Rider . . . . . . . Librarian, Wesleyan University, Middletown, Connecticut (deceased)

Paul L. St. Pierre . . . . . . Consultant to the Library Community for Data Processing, Albany, New York

Ralph R. Shaw . . . . . . . . Professor Emeritus, Graduate School of Library Studies, Honolulu, Hawaii. (deceased)

Allen B. Veaner . . . . . . . Assistant Director for Bibliographic Operations, Stanford University Libraries, Stanford, California

Henry Voos . . . . . . . . Associate Professor of Library Service, School of Library Service, Rutgers University, the State University of New Jersey, New Brunswick, New Jersey

W. Lyle Wallace . . . . . . . Westinghouse Electric Corporation, Sunnyvale Manufacturing Division, Sunnyvale, California

R. Winter . . . . . . . . . . Agricultural Division, Imperial Chemical Industries Limited (ICI), London, Great Britain

Elaine Woodruff . . . . . . Librarian, United States Civil Service Commission, Washington, D.C.